# Innovation, Technology, and Knowledge Management

For other titles published in this series, go to
www.springer.com/series/8124

Frederick Betz

# Managing Science

## Methodology and Organization of Research

Frederick Betz
Department of Engineering and Technology
Portland State University
Portland, OR
USA
fbetz@venture2reality.com

ISBN 978-1-4614-2756-8 ISBN 978-1-4419-7488-4 (eBook)
DOI 10.1007/978-1-4419-7488-4
Springer New York Dordrecht Heidelberg London


Softcover reprint of the hardcover 1st edition 2010


Printed on acid-free paper

Springer is part of Springer Science+Business Media (www.springer.com)

*For Nancy,*
*my dear wife who has been with me on this journey through life,*
*while also being a marvelously insightful editor.*

# Series Foreword

The Springer Book Series on *Innovation, Technology, and Knowledge Management* was launched in March 2008 as a forum and intellectual, scholarly "podium" for global/local (gloCal), transdisciplinary, trans-sectoral, public–private, leading/"bleeding"-edge ideas, theories, and perspectives on these topics.

The book series is accompanied by the Springer *Journal of the Knowledge Economy*, which was launched in 2009 with the same editorial leadership.

The series showcases provocative views that diverge from the current "conventional wisdom," that are properly grounded in theory and practice, and that consider the concepts of **robust competitiveness**,[1] **sustainable entrepreneurship**,[2] and **democratic capitalism**,[3] central to its philosophy and objectives. More specifically, the aim of this series is to highlight emerging research and practice at the dynamic intersection of these fields, where individuals, organizations, industries, regions, and nations are harnessing creativity and invention to achieve and sustain growth.

Books that are part of the series explore the impact of innovation at the "macro" (economies, markets), "meso" (industries, firms), and "micro" levels

---

[1]We define *sustainable entrepreneurship* as the creation of viable, profitable, and scalable firms. Such firms engender the formation of self-replicating and mutually enhancing innovation networks and knowledge clusters (innovation ecosystems), leading toward robust competitiveness (EG Carayannis, *International Journal of Innovation and Regional Development*, 1(3), 2009, 235–254).

[2]We understand *robust competitiveness* to be a state of economic being and becoming that avails systematic and defensible "unfair advantages" to the entities that are part of the economy. Such competitiveness is built on mutually complementary and reinforcing low, medium, and high technologies and public and private sector entities (government agencies, private firms, universities, and nongovernmental organizations) (EG Carayannis, *International Journal of Innovation and Regional Development*, 1(3), 2009, 235–254).

[3]The concepts of *robust competitiveness* and *sustainable entrepreneurship* are pillars of a regime that we call "*democratic capitalism*" (as opposed to "popular or casino capitalism"), in which real opportunities for education and economic prosperity are available to all, especially – but not only – younger people. These are the direct derivative of a collection of top-down policies as well as bottom-up initiatives (including strong R&D policies and funding, but going beyond these to include the development of innovation networks and knowledge clusters across regions and sectors) (EG Carayannis and A Kaloudis, *Japan Economic Currents*, January 2009, pp. 6–10).

(teams, individuals), drawing from such related disciplines as finance, organizational psychology, R&D, science policy, information systems, and strategy, with the underlying theme that in order for innovation to be useful it must involve the sharing and application of knowledge.

Some of the key anchoring concepts of the series are outlined in the figure below and the definitions that follow (all definitions are from EG Carayannis and DFJ Campbell, *International Journal of Technology Management*, 46, 3–4, 2009).

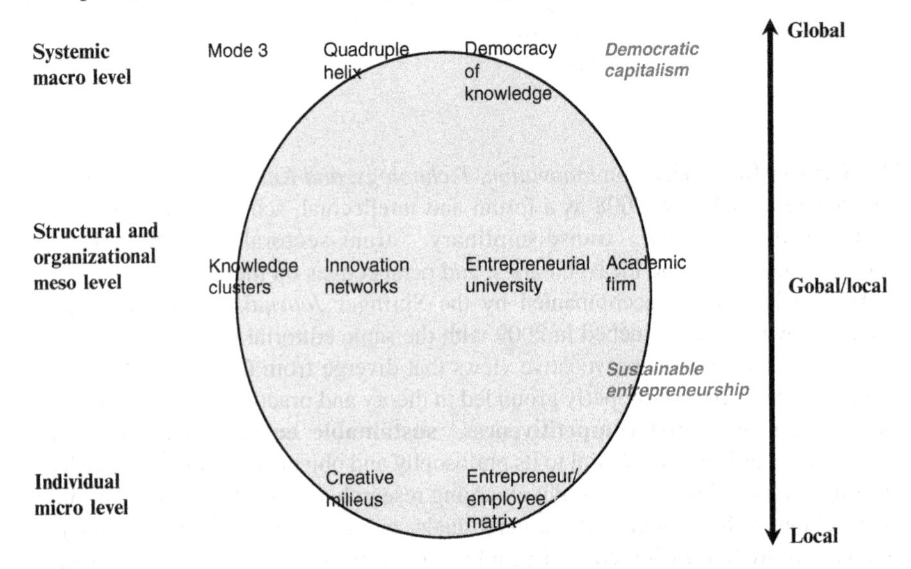


Conceptual Profile of the Series on Innovation, Technology, and Knowledge Management

- The "MODE 3" Systems Approach for knowledge creation, diffusion, and use: "Mode 3" is a multilateral, multi-nodal, multimodal, and multilevel systems approach to the conceptualization, design, and management of real and virtual, "knowledge-stock" and "knowledge-flow," modalities that catalyze, accelerate, and support the creation, diffusion, sharing, absorption, and use of co-specialized knowledge assets. "Mode 3" is based on a system-theoretic perspective of socioeconomic, political, technological, and cultural trends and conditions that shape the coevolution of knowledge with the "knowledge-based and knowledge-driven, gloCal economy and society."
- Quadruple Helix: Quadruple helix, in this context, means to add to the triple helix of government, university, and industry a "fourth helix" that we identify as the "media-based and culture-based public." This fourth helix associates with "media," "creative industries," "culture," "values," "life styles," "art," and perhaps also the notion of the "creative class."
- Innovation Networks: Innovation networks are real and virtual infrastructures and infra-technologies that serve to nurture creativity, trigger invention, and catalyze innovation in a public and/or private domain context (for instance,

government–university–industry public–private research and technology development co-opetitive partnerships).

- Knowledge Clusters: Knowledge clusters are agglomerations of co-specialized, mutually complementary and reinforcing knowledge assets in the form of "knowledge stocks" and "knowledge flows" that exhibit self-organizing, learning-driven, dynamically adaptive competences, and trends in the context of an open systems perspective.
- Twenty-first Century Innovation Ecosystem: A twenty-first century innovation ecosystem is a multilevel, multi-modal, multinodal, and multi-agent system of systems. The constituent systems consist of innovation meta-networks (networks of innovation networks and knowledge clusters) and knowledge meta-clusters (clusters of innovation networks and knowledge clusters) as building blocks and organized in a self-referential or chaotic fractal knowledge and innovation architecture (Carayannis 2001), which in turn constitute agglomerations of human, social, intellectual and financial capital stocks and flows as well as cultural and technological artifacts and modalities, continually coevolving, co-specializing, and co-opeting. These innovation networks and knowledge clusters also form, reform, and dissolve within diverse institutional, political, technological, and socioeconomic domains including government, university, industry, nongovernmental organizations and involving information and communication technologies, biotechnologies, advanced materials, nanotechnologies, and next-generation energy technologies.

*Who is this book series published for?* – The book series addresses a diversity of audiences in different settings:

1. *Academic communities:* Academic communities worldwide represent a core group of readers. This follows from the theoretical/conceptual interest of the book series to influence academic discourses in the fields of knowledge, also carried by the claim of a certain saturation of academia with the current concepts and the postulate of a window of opportunity for new or at least additional concepts. Thus, it represents a key challenge for the series to exercise a certain impact on discourses in academia. In principle, all academic communities that are interested in knowledge (knowledge and innovation) could be tackled by the book series. The interdisciplinary (transdisciplinary) nature of the book series underscores that the book series scope is not limited a priori to a specific basket of disciplines. From a radical viewpoint, one could create the hypothesis that there is no discipline, where knowledge is of no importance.
2. *Decision makers – private/academic entrepreneurs and public (governmental, subgovernmental) actors:* Two different groups of decision makers are being addressed simultaneously: (1) private entrepreneurs (firms, commercial firms, academic firms) and academic entrepreneurs (universities), interested in optimizing knowledge management and in developing heterogeneously composed knowledge-based research networks; and (2) public (governmental, subgovernmental) actors that are interested in optimizing and further developing their policies and policy strategies that target knowledge and innovation. One purpose of

public knowledge and innovation policy is to enhance the performance and competitiveness of advanced economies.

3. *Decision makers in general:* Decision makers are systematically being supplied with crucial information, on how to optimize knowledge-referring and knowledge-enhancing decision-making. The nature of this "crucial information" is conceptual as well as empirical (case study-based). Empirical information highlights practical examples and points toward practical solutions (perhaps remedies), conceptual information offers the advantage of further driving and further carrying tools of understanding. Different groups of addressed decision makers could be decision makers at private firms and multinational corporations, responsible for the knowledge portfolio of companies; knowledge and knowledge management consultants; globalization experts, focusing on the internationalization of R&D, S&T, and innovation; experts in university/business research networks; and political scientists, economists, business professionals.
4. *Interested global readership:* Finally, the Springer book series addresses a whole global readership, composed of members who are generally interested in knowledge and innovation. The global readership could partially coincide with the communities, as being described above ("academic communities," "decision makers"), but could also refer to other constituencies and groups.

Elias G. Carayannis
Series Editor

# Preface

## Philosophy and Organization of Science

Science is the societal process of generating new knowledge about nature, as scientific theories constructed upon experimental studies. Science is organized as disciplines and conducted through research projects and laboratories – the organization of science. Science is conducted with methodological techniques – the scientific method. And modern research is indeed managed, if one thinks about science in both organization and methodology. In this way, one can answer some very practical questions about research, such as:

1. How does one view science as a whole?
2. What are the intellectual frameworks of science?
3. What is scientific method?
4. What role does modeling play in scientific method?
5. How does methodology differ between social and physical sciences?
6. How does one manage a research project in a university?
7. How can one identify a research issue?
8. How can one write a persuasive research proposal?
9. How should one manage a research center in a university?
10. How can science research be effectively transferred to technology research?
11. How should a research agency plan a research-funding program?
12. How should officers in a research agency manage a research-funding program?

The title I have chosen of "managing science" is, of course, not traditional to studies on science; but I choose it to emphasize that both methodology and organization are central and interconnected in modern research. Traditionally, the studies of organization and methodology have been done separately in the areas of philosophy and of sociology and political science. And the history of science has been studied by historians. But the whole of scientific activity requires integrating these perspectives, philosophical, sociological, and historical. And this is what we will do here. We integrate these by viewing all the activities of research in a taxonomy of all research activities, based upon two dichotomies of research activity: (1) of philosophy and organization and (2) of process and content. Together, these classify research activities into method, content, strategy, and application. And we will illustrate all activities with historical cases in science. We will ground our observation the activities of science, organizational and methodological, in the history of science.

## Scope

We will explore a broad picture of methodology across all the scientific disciplines: from physical sciences to life sciences to social sciences to mathematics and computer sciences and even to management science. This scope is valuable for the research professional. Modern research (while still performed in discipline-specialty research projects) is also often performed in a research setting with other projects of other specialties, as interdisciplinary research and/or multidisciplinary research.

Multidisciplinary research settings now frequently occur in universities, in strategically focused research centers of science and engineering. Such centers support groups of research projects that interact and add up together to more than the sum of their parts: strategic and multidisciplinary research thrusts. In the modern research university (and in industrial and governmental research laboratories), researchers at the cutting edge seldom work entirely within the specialty area they were trained in as graduate students.

## Background

For scholars of science in the last decades, two books have been particularly influential: Tomas Kuhn's book (Scientific Revolutions) to sociologists of science and Karl Popper's book (*The Logic of Scientific Discovery*) to philosophers of science.

Thomas Kuhn provided the idea of an organizational distinction between normal science and scientific revolutions. He also introduced the idea of a scientific paradigm as a meta-framework for scientific theory and experiment. This book picks up on the idea of a paradigm, but I have extended the idea from physics (which Kuhn studied) across all of science. This extension is in the form of four kinds of scientific paradigms: mechanism, function, system, and logic. Paradigms are a necessary way for scientists across different disciplines to understand each other and to cooperate in multidisciplinary research.

In the positivist tradition of the Vienna Circle, Karl Popper proposed (only) an inductive approach to scientific method, in which theory is directly induced from experimental observations. Accordingly, if this were the case, then theory can never be validated but only invalidated by experiment. But this simple approach to scientific method is not substantiated by the history of science. That history abounds with the validation of many theories such as: Newtonian mechanics, Rutherford–Bohr's quantized atom, quantum theory, special relativity, Watson–Crick's double helix model of DNA, and so on. History indicates that theory construction grounded in experiment proceeds with both inductive and deductive methods in science. And we take this broader approach to scientific method, beyond Popper.

## Audiences

The book is intended for two audiences: scholars of science (about the philosophy and sociology and history of science) and practitioners of research (scientific and engineering researchers), particularly the academic researcher needing to negotiate the difficult path to publication and grants and tenure.

For scholars of science, this book integrates the three different perspectives on science: philosophical, sociological, and historical. And it extends both Kuhn's and Popper's early insights about science into a general process of science across all disciplines of science which can be tested empirically in the history of science.

For researchers, this book provides an integrated picture of science, methodologically and organizationally. Early in a research career, researchers need to have a broader methodological understanding than they ordinarily acquired within the narrow disciplinary specialty of their PhD dissertation. They also need to understand the organizational aspects of research – how research is funded, managed, and utilized.

For graduate courses in research methodology, there are many books on research methodology; but all these are wholly methodological and also extremely limited in methodology, focusing only upon measurement and the statistical techniques of measurement and inductive inference. Statistical techniques are essential topics in research methodology but conceptually limited to measurement. These include techniques for: (1) statistical inference of information on populations from data samples, (2) finding significant correlation between random variables, (3) efficient design of multivariable experiments, and (4) testing the statistical significance of a hypothesis. (In a chapter on measurement, I have discussed how these statistical techniques fit into a complete set of research techniques.)

But where do effective hypotheses come from? How can one connect hypotheses into theories? Research methodology as merely statistics does not cover the topic of theory construction (which is just as important methodologically as is measurement). This is a major difference of this book from standard treatments of methodology. The book centers upon the connection between experiment and theory, on theory construction and validation. How are experiments designed to provide insightful observations of nature? How is theory constructed upon experimental data? How can theory be validated or invalidated by experiments, and how can theory suggest new experiments.

Thus, any researcher should have a sophisticated approach to methodology including statistics techniques, modeling and theory-construction techniques, and instrument-empirical techniques. And also a researcher should have a board appreciation of methodology, cross-disciplinary research approaches. But methodological sophistication and breadth alone is not sufficient for professional survival in the modern world of research. One also needs organizational skills, particularly those of raising research funds (getting research grants) and effectively managing a research team (colleagues, graduate students, and technicians). All together, this focus on both sophisticated, broad methodology, and on effective organizational skills is what I have called the topic of "managing scientific research activity."

## Creativity and Management in Research

Such a phrase may at first appear awkward, particularly to the tradition that scientific progress is principally the creative acts of scientific geniuses and cannot be managed. Certainly, there have been and are geniuses in science. The early ones in science all performed creative acts of genius – Copernicus, Brahe, Kepler, Galileo, Descartes, Newton. From the first events of science in the 1600s, there has always been a community of scientists and scientific method and scientific paradigms. But later in the 1900s, much of scientific research began to occur in teams of scientists. Young scientists began to be mentored by older, established scientists. For example, there was the genius of Michael Faraday, after a long assistantship to Davis. And Rutherford mentored Geiger and Marsden in his famous atom experiment and later Bohr for his first quantum model of the atom. As history proceeded into the twentieth century, teams of scientists (professors and their assistants and students) performed most of that scientific research. In research teams and in funded research projects, the style of "management" had entered the activities of scientists.

For example, the famous team of Hahn and Meitner discovered the phenomena of atomic fission. The famous team of Watson and Crick (along with Wilkins and Franklin) correctly modeled DNA. Scientific research occurs not only in communities but is performed by scientific teams in research projects. Thus, the managerial idea of a research team and a research project and a project leadership do apply to scientific activities.

And now the omnipotent complement to scientific method is science funding. Scientific research is expensive. Research projects require funding by government, philanthropy, or industry. And the larger the scientific instruments have become and the more complex the projects, the costs of science have grown tremendously, so large as to have been given its own name, "big science." But even smaller scientific endeavors, individual research grants, also require funding. This is obtained procedurally by writing research proposals, submitting proposals to a research funding agency, winning a research grant, and administering the funds to assemble a research team and perform a research project. Thus organizational procedures – research teams, research projects, research funding, research leadership – are all essential to the scientific process.

We will explore three key concepts which characterize managing scientific activity:

1. *Scientific method* – as empirically grounded theory construction and verification
2. *Scientific paradigms* – as disciplinary meta-frameworks for perceiving nature
3. *National research system* – as the institutional context for funding and performing science

# Acknowledgments

I would like to acknowledge the many colleagues and friends with whom I have had the honor and pleasure of association, in my journey through the terra incognita of science. These associations began with Owen Chamberlain, my thesis advisor in physics. From him I learned the importance of research strategy, technical skills, and good physical intuition which an experimenter needs to gaze directly and deeply into the heart of nature. Also in that graduate experience, I had the stimulating comradeships of Helmut Dost, Wladislaw Troka, Klaus Schultz, Eric Arens, and Byron Dieterle, as we learned to manage research teams at the Bevatron in the new era of "big science."

Next as a postdoc, I changed fields from physics to management science. I did "research on research" at the Berkeley Space Sciences Center under the direction of C. West Churchman. Churchman was trained as a philosopher and under his tutelage I read the classics of philosophy, at the same time with Ian Mitroff, another fellow student who was moving from engineering into management. At that Center, I met and began a long research collaboration with L. Vaughn Blankenship and Carlos Kruytbosch, who taught me the value of disciplinary perspectives, seeing political interactions in all things (as a political scientist) and seeing social interactions in all things (as a sociologist).

I began teaching in business schools, learning the literatures of management and organizations and operations research. Next, I had the opportunity to work in the National Science Foundation, where I could practice all the philosophy and sociology of knowledge in the funding decisions of choosing what science and engineering research to support. At NSF, I had the pleasure of working for Marshall Lih and Lynn Preston, with whom I learned the virtues of indirect management. One does not manage professionals directly but indirectly, through budgets and peer review of quality.

I am also grateful to Nuket Yetis, President of the Science and Technology Council of Turkey, for the opportunity to lecture on science administration, pulling all this organization and philosophy together.

Finally, I would like to acknowledge the fine colleagueship of Elias Carayannis, Dundar Kocaoglu, and Tarek Khalil, with whom together we have been investigating how science interacts with technology to create innovation.

Portland, OR

Frederick Betz

# Contents

# List of Figures

# Chapter 1
# Totality of Science

## Introduction

Science is complicated because the idea includes "method" (philosophy of science), "organization" (sociology of science), "events" (history of science), and "funding" (politics of science). What we will do is integrate these different aspects of the idea of science into a complete description of scientific research – a totality of science. And we will use an odd term for this integration; we will call it "managing science."

Looking at modern research efforts in terms of funding, the idea of "managing science" is actually a practical way to think about science. Science as funded research is managed and frequently. Each year thousands of scientific research awards are made by government agencies in different nations. For example, in the USA, three Federal agencies (National Science Foundation, National Institutes of Health, and Defense Department's DARPA) each provides one-third of the funding for basic research (science) performed in US research universities. All the research projects amount to billions of dollars. For example, the National Science Foundation annually receives about 40,000 proposals and makes 11,000 awards with a research budget of about $20 billion dollars (http://www.nsf.gov).

Yet is not scientific creativity opposed to administration? Certainly, creativity is essential to science, as mediocre research produces no new knowledge, no new scientific discoveries. But in research, creativity and administration are not necessarily at odds. Firstly, creativity must be administratively captured in a research proposal. Also creativity must be administratively captured by proper evaluation of the quality of the proposed research. And creativity must be administratively captured in the successful performance and publication of the research results. We will use ideas of "management" and "creativity" together to point toward the "quality" of the research process.

We will begin our discussion of managing science by focusing upon two aspects of creativity in the management of research: (1) the interaction between method and organization in research and (2) the interaction between the macro-level of science funding and the micro level of science performance. To illustrate all this, we will review cases of actual research, occurring in the history of science. The first case is *Rutherford's Atomic Experiment in 1909* – which occurred at the micro-level of

F. Betz, *Managing Science*, Innovation, Technology, and Knowledge Management 9,
DOI 10.1007/978-1-4419-7488-4_1, 

science and primarily involved methodological issues and secondarily organizational issues. The second case is the research that gave the first direct evidence of *genetic basis of social behavior – occurring in Fire-Ant societies* – which also happened at the micro-level of science and primarily involved methodological issues. The third case is a description of the funding operations of the *US National Science Foundation in 2009* – which occurred at the macro-level of science and technology policy and involved both organizational issues (disciplinary science) and methodological issues (peer review selection procedures). What we will learn is that it is the idea of "managing science" which ties together the philosophy and sociology and history and politics of science.

## *Illustration: Rutherford's Experiment on the Structure of the Atom*

An historical example of scientific methodology can be seen in the first experiment to determine the geometric structure of the atom. This project was in the discipline of physics, but the method was general – one of experiment and new theory. The methodology is applicable to all scientific disciplines – experiment and theory. The experiment was performed in 1909 and is now often called the "Geiger–Marsden Experiment" – in honor of the post-doc Geiger, and student, Marsden, who performed it. They performed it under Rutherford's guidance (as "leadership" or "management" or "administration"). Rutherford conceived and sponsored and led the project. In the modern terms of research, now one can call Rutherford a "research-project manager" and Geiger and Marsden, the "research team."

The idea of the atom appeared far earlier in philosophy. For example in ancient Greek philosophy, the atom was proposed by a pre-Socratic philosopher, Democritus (460–370 BC). He was born in Thrace and believed all matter is made up of small, permanent units which he called "atomon," or "indivisible elements." But there is an enormous difference between a philosophical idea and a scientific idea. The difference lies in the scientific methodological idea of "experiment." When Rutherford performed his critical *experiment* on the atom, the idea of "atom" moved from the realm of philosophy into the realm of science.

Ernest Rutherford (1871–1937) was born in New Zealand, near the city of Nelson. Later he studied at Nelson College and Canterbury College. In 1883, he graduated with degrees of B.A., M.A., and B.Sc. He stayed on for 2 years to do research in electric technology. In 1895, he went to England for graduate study at the Cavendish Laboratory of the University of Cambridge. He investigated radioactivity and was able to distinguish between alpha, beta, and gamma rays in the radioactive phenomena of atoms. He introduced the terms of "alpha" and "beta" radiation. (It would later be found that "alpha" radiation was particles emitted from radioactive nuclei in decay that were equivalent to the nucleus of a Helium atom (two protons and two neutrons.) Beta radiation was electrons emitted from radioactive nuclei in decay. Gamma rays were photons emitted from radioactive nuclei in decay.)[1]

In 1898, Rutherford left England for Canada, where he obtained an appointment to the chair of physics in McGill University in Canada. There he demonstrated that radioactivity was due to the spontaneous disintegration of atoms and determined that different atoms had different times of a constant rate-of-decay, which he called the "half-life" of a radioactive atom. For this he was rewarded the Noble Prize in Physics in 1908. But just before receiving the prize, Rutherford returned to England to take a professorial chair at the University of Manchester.

Lord Rutherford (http://en.wikipedia.org, Rutherford 2007)

We need to pause in the description of Rutherford's experiment to note the scientific background to why Rutherford conceived of performing the research project in 1907. It had been 10 years earlier in 1897 that J. J. Thomson at the Cavendish Laboratory of Cambridge University had demonstrated that the electron was a subatomic particle. Thomson (1865–1940) was born in Manchester, England. He attended Cambridge University, obtaining a master's degree in 1883. He became a professor at Cambridge the following year.

J. J. Thomson (http://en.wikipedia.org, Thomson 2007)

Thomson studied the then new scientific device of the cathode tube – in which rays passed through the gas of the tube when electrical voltages were placed across each end of the tube. Thomson's first research was to discover that these rays were currents of electricity made up of a flow of particles, which he called electrons. Thomson's cathode ray tube experiment had electrons streaming from the negative cathode toward positive anode – deflected by voltage applied between midway plates and striking a fluorescent screen covering the end of the tube. For this, Thomson would be awarded the Nobel Prize in physics in 1906 – the discovery of the electron.

Next, Thomson suggested that the atom was made up of a combination of electrons and protons. And he suggested a model of their arrangement, similar to that of an English "plum pudding," with electrons embedded like plums in a positive pudding. It was this model that Rutherford intended to experimentally test. As a holder of a professorial chair in Manchester University, Rutherford was given space and a budget to run a physics research laboratory. Rutherford hired two research assistants in his laboratory and assigned them the task of performing the experiment in 1909. They were Hans Geiger and Ernest Marsden. The experiment was to look for the geometric structure of the atom. How were the positive and negative charges in the atom spatially arranged?

Hans Geiger (1882–1945) was born in Germany. He had earned his doctorate in physics in 1906 at the University Erlangen. And in 1907, he went to England to work for Ernest Rutherford. Researching radioactivity with Rutherford, Geiger invented the "Geiger radiation counter." Geiger returned to Germany, becoming head of the Physical-Technical Reichsanstalt in Berlin. Later he became a professor at the University of Keil in 1925, in 1929 at the University of Tubingen, and in 1936 at the University of Berlin. During the Second World War, Geiger participated in the German group attempting to make an atomic bomb (http://en.wikipedia.org, Hans Geiger 2007).

Ernest Marsden (1889–1970) was born in England and enrolled in the University of Manchester as an undergraduate. In Rutherford's lab, he worked under Geiger, participating in the famous experiment as an undergraduate. Later in 1914, Marsden moved to Victoria University in New Zealand. He served in World War I as a Royal Engineer. In 1924, he returned to New Zealand to found New Zealand's Department of Scientific and Industrial Research (http://en.wikipedia.org, Ernest Marsden 2007).

In the experiment, Geiger and Marsden bombarded several metal foils (including gold foil) with alpha particles. The experiment was performed in a darkened room under a low-powered microscope. Geiger and Marsden watched for tiny flashes of light – as the scattered particles struck a zinc sulfide scintillating screen (and the screen gave off light when struck by a charged particle). Most of the particles penetrated the foils, passing through with some absorbed in the foil. But once in about 8,000 times, the alpha particles bounced back from the foil toward the source – as if these particles had hit a hard object in the foil! This phenomenon was called a "back-scatter." Back-scattering in classical physics can occur when one hard object hits another hard object and scatters backwards. (For example, if two billiard balls of equal masses hit each other at an angle, then both balls scatter forward in the direction of the colliding ball; but if the hit billiard ball is larger than the hitting ball and struck directly on, then the colliding ball can rebound – scatter backwards. Back-scattering can only occur when one hard object strikes another heavier hard object and direct-on.) So in the experiment, some alpha particles were backscattered, and this could only be explained if the gold foil was composed of atoms made up mostly of space with a small hard and heavy nucleus surrounded by electrons in orbits.

In 1911, Rutherford published his analysis of the alpha scattering as the Rutherford model of the atom. His model looked like the model of the solar system, with a core atomic nucleus (analogous to the sun) orbited by particle-like electrons (analogous to planets). But later, it would be found that the atom was composed of a small atomic nucleus surrounded by a cloud of wave-like electrons in orbits (Fig. 1.1).

Yet even then Rutherford knew that an analogy of the atomic system to the solar system was impossible because of the theory of electromagnetism. Rutherford understood the model was geometrically correct but physically impossible! In classical physics, an orbiting electron would radiate away energy as electromagnetic radiation (light). Rutherford immediately understood that new physics would be necessary, a new theory. The spatial model of an atom with electrons far-out and circling a nucleus was experimentally correct, but not then theoretically possible. New theory was obviously needed. In scientific method, this demand that all scientific theory be constructed upon a base of scientific experiments of nature is called a requirement of "empirically grounded theory."

What would happen next (and a story we will later recount) is that Rutherford found a new post-doc to work on the theoretical problem of an atom whose orbiting

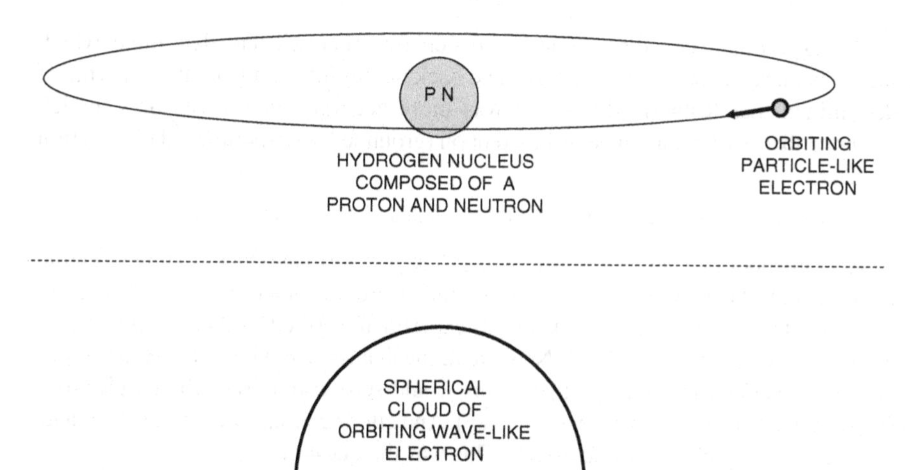


**Fig. 1.1** Classical solar analogy for a model of a hydrogen atom. Quantum mechanical model of a hydrogen atom

electrons do not radiate away energy. This would be Niels Bohr, who would quantize Rutherford's atom model and foster the development of quantum mechanical theory, wherein electrons behave as both particles and waves (as does all atomic matter). But this is the excitement of science, one discovery about nature leads to another discovery, and nature is complex and surprising.

## Methodology and Organization in Rutherford's Experiment

In this case, Rutherford (1) methodologically conceived the experiment and (2) administered/managed the project as a research-team leader. Let us first look at the methodological issues. Methodologically, we saw Rutherford using the "scientific method" as an experiment (alpha rays passing through a metal foil). We can connect this modern idea of "scientific methods" with an older philosophical idea of "epistemology." Epistemology is "knowledge of method."

> The classical idea of "epistemology" has evolved in science into the modern idea of "scientific method," wherein scientific method requires observing nature in experiments.

An experiment can discriminate between two theories and determine which theory is more accurate about nature. The back-scattering data from the experiment determined that Rutherford's model was more accurate than Thomson's model. Accurate theory of nature must be based upon (grounded) in experimental observation of nature.

Experiments determine the accuracy of theory – empirically grounded theory

Also methodologically, we saw in Rutherford's experiment, that existing scientific theory couldn't explain the experiment. If Rutherford's geometric model of the atom was correct (and it was correct in nature, as experiment verified it), then existing theory did not work at the atomic level. Newtonian mechanics and Maxwell's equations of electricity/magnetism predicted that atoms would self-destruct (as orbiting electrons should radiate electromagnetic energy away). And curving orbits of electrons do radiate photons. *But why not in atomic orbits?* This was the question.

So in scientific method, Rutherford immediately appreciated that new physical theory was required – to explain the natural phenomenon of the atom. Yet new theory cannot be simply induced directly from the experiment. The formulation of theory is more complex than simply seeing a pattern in data – induction. Theory construction also involves deduction, as theory must fit to the empirical grounds of other relevant theories. Rutherford's model of the atom still had to fit with Maxwell's theory.

Theory cannot be simply induced (generalized) from experiment – as theory construction involves both inductive and deductive methods of logical thinking.

Thirdly, methodologically we saw in Rutherford's experiment that there was a prior scientific content upon which Rutherford based his research (Thomson's plum pudding model of the atom). The modern idea of scientific content is that it accumulates over time as the current state of knowledge of nature – scientific representation. Again we can connect this to an older philosophical idea of "ontology." Ontology is the knowledge of the universe – of what "stuff" is the universe composed? This classical philosophical idea of ontology as the content of the universe has now evolved into the modern idea of scientific representation – expressing nature in the terms of a "scientific paradigm." Nature is modern science's term for the "observable universe" – everything in the world that can be observed and studied.

A "scientific paradigm" is an intellectual framework within which scientific theories of nature are constructed from the scientific method.

And fourthly, methodologically we saw in Rutherford's experiment that there was also an instrumental aspect in the experiment. Rutherford had to have instruments – technical devices and materials to do the experiment, such as: radioactive sources, metallic foils, microscope, and scintillation material. We can call this aspect of scientific research the "technology basis of science." (e.g., in Geiger's radioactivity research, Geiger invented the radiation counter, which was an essential tool in the later technology of nuclear energy.)

Scientific instrumentation often provides a basis for the invention of scientific technology and for the application of science into technology.

Next, we can turn to the administrative issues – as all modern events in science both take place in institutional contexts (*organization*) and occur as *methodologically* conceived research. *Organizationally*, we saw that Rutherford was a university professor with a funded laboratory, enabling him to hire Geiger and Marsden as research assistants and buy research supplies and equipment.

This is a modern idea of administration as an important organizational dimension of science (and does not have a philosophical antecedent tradition) – science administration.

The post-doc Geiger and the student Marsden performed the actual experiment – research team.

And in modern academic science, this is the norm: (1) professors conceive of research and obtain research-project grants to support graduate students and post-docs and (2) graduate students and post-docs perform the actual research in the project.

Rutherford held the position of professor, which provided his salary and living expenses for performing the educational duties of a professor. But he was hired not for his teaching skills but his research skills. Historically, since the German reform of universities in 1800, professors in Europe had been paid salaries for educational services but were selected for research skills. This is the pattern in modern research universities, to select upon research ability but to pay for teaching. A first organization feature in science is this which connects organization to methodology – pay for teaching (organizational) but selection on research ability (methodological).

At both macro-level of modern science and technology policy, this is the academic policy of integrating education and research functions in a research university.

Rutherford's teaching duties took care of his living, but what about funds for his research? Hiring Geiger as a post-doc and paying the undergraduate Marsden to conduct the tedious counting of scintillation spots of alpha particles took money not included in Rutherford's professorial salary. Two sources were possible. The University of Manchester at the time had likely provided a separate research budget for its famous professor. Or Rutherford may have had private sources of research money from philanthropic individuals. Manchester had attracted Rutherford after he had become scientifically famous by winning the Nobel Prize in Physics. It took a separate research budget (other than his professorial salary) for Rutherford to pay assistants, buy research equipment and supplies, in addition to having university space for a research laboratory.

The administrative aspect of Professor Rutherford's research was that he had obtained research funding to support his laboratory.

Also Rutherford had identified the scientific issue (the model of an atom) as the objective of the research. This identification of research issue derived from the current state of the scientific literature, Thompson's "plum-pudding" model of the atom. So in identifying the research issue from the current state of scientific knowledge, Rutherford was thinking strategically about research. Strategic thinking is an important function of administration.

Rutherford, as the principal scientist in leading his research team, was thinking strategically like a strategic administrator (or manager) of research.

Also Rutherford assembled a team to conduct the research tasks of the critical experiment. Assembling a team and planning a project and assigning tasks to the team are all responsibilities of administration. Rutherford was the "manager" of a research team. Rutherford (1) obtained resources, (2) strategically decided on the research direction, and (3) planned and assembled and guided a research team. In modern terms, a manager performs activities such as strategic thinking and planning and execution. A manager of a research project is called a "project manager." Of course, there was no term for "project management" back in Rutherford's day. But today, we can say Rutherford in the Geiger–Marsden experiment performed the role of research-project manager.

From this example, we can begin to infer that any significant event in the progress of science will show at least two dimensions in the activity – a methodological process of the research (e.g., Rutherford's experimental approach) and an organizational context of the research (e.g., Rutherford's university-performed science).

## Activities of Science

We can generalize what we have seen in this case about all activities of science having both methodological and organizational aspects. The philosophical dimension (philosophy of science) is concerned with how research should be performed, the *method of science*. The organizational dimension (sociology of science) is concerned with *why* research should be performed, the *administration of science*. So the philosophy and organizational dimensions of science provide one dichotomy to distinguish activities in science. The "what" research should be performed – lies in another dimension about science which is a *content* dimension – part of a state and process dimension. One can use two dichotomies (philosophy and organization and process and state) to categorize the activities of science, as shown in Fig. 1.2.

| | PROCESS OF KNOWLEDGE | STATE OF KNOWLEDGE |
|---|---|---|
| PHILOSOPHY | SCIENTIFIC METHOD (EPISTEMOLOGY) | SCIENTIFIC CONTENT (ONTOLOGY) |
| ORGANIZATION | SCIENCE ADMINISTRATION (RESEARCH) | SCIENCE APPLICATION (TECHNOLOGY) |

**Fig. 1.2** Philosophy and organization of science

The modern idea of scientific content is that it accumulates over time as a current state of knowledge of nature – scientific progress. We noted that the older philosophical idea of the content of the universe was called "ontology." For example, in the philosophical tradition, two ancient Greek philosophers identified the ontological nature of the universe differently. Heraclitus' idea of the ontology of the universe was of continual change – an ontology of Becoming. Later another Greek philosopher, Parmenides, argued that all change was merely in appearance and the ontological reality of the universe was one of "permanence" – an ontology of Being.

Now in science, these philosophical perspectives are incorporated within a scientific paradigm of mechanism. The ontological permanence in physics are unchanging laws (such as the conservation of the totality of mass and energy). The ontological change in physics is the evolution of a physical system as sequential states (such as the states of an atom as orbits of electrons, transforming from a lower energy orbit by absorbing a photon to a higher energy orbit; and subsequently emitting a photon to transform form a higher-energy orbit to an unoccupied lower-energy orbit). So in modern philosophy of science, the ancient philosophical ideas of ontology are now expressed in the idea of a scientific paradigm (which we will later discuss).

Also in traditional philosophy, the name given to the idea of a proper approach to creating knowledge was called "epistemology." This is from the Greek "episteme" (knowledge), and "logos" (theory) – theory of knowledge. Now we use the term "scientific method" as a modern approach toward knowledge – modern epistemology. Scientific methodology is conceived within research projects and performed according to techniques of scientific method. Scientific method is an empirical/theoretical process of inquiry into nature to discover and understand (1) what things exist in nature (discovery), and (2) how natural things work (understanding). Scientific method consists of research techniques for performing experiments to discover nature (empirical) and for constructing theory based on these experiments (theoretical) to understand and explain nature. Scientific method creates *empirically grounded theory*.

Viewed in totality in all its activities, science can be seen as a process and content of inquiry of research which asks and answers basic questions about nature. Inquiry into nature focuses upon the questions: what things exist in nature and how do these things interact? Science as inquiry can thus be characterized by four kinds of activities, which address the following questions:

1. Of what is the universe made? This is the basic question about nature. We can call this activity about the content of nature as *scientific content* (*ontology*). Scientific content expresses the current state of knowledge in science about the nature of the observable world.
2. How do we know this? This is the basic question about method. Science now answers that question with the methods of science in experimentally grounded theory. We can call this *scientific method* (*epistemology*). Scientific method proposes the proper philosophical approach (methodology) as a set of research tasks in the process of advancing knowledge.
3. What procedures and resources are necessary to inquire into nature, to use scientific epistemology? This is the basic question about the methodological approach and organizational resources needed to conduct scientific research. We can call

this *science administration* (*research*). Science administration organizationally funds and/or performs research in the methodological forms of scientific inquiry – as research proposals.

4. Why is science useful to society? This is the basic question about the value of science to human civilization. Science is practical to society by providing a knowledge base for technological innovation. We can call this *scientific application* (*technology*). Science administration organizationally funds and/or performs the process of advancing knowledge in terms of research tasks.

The four issues of scientific inquiry – ontology, epistemology, research, and technology – categorize the activities of science as a totality of science – scientific paradigms, scientific method, science administration, and science applications.

## *Illustration: Fire Ant Society*

This integrated approach of methodology-and-organization applies not only to the physical sciences but to all scientific disciplines. As an example in socio-biology, we look at the historical event of research into the behavior of ant societies – the first research to connect a distinct gene with social behavior – sociobiology. In both biology and the social sciences, the concept of "instinct" is that of behavior determined by genetic structure – "hard-wired" technology in biological behavior. Have biologists actually found a gene that directly determines a behavior? Yes, the first discovery of this occurred in a research project attempting to understand the social behavior of ants. Holdern described this: "Genes regulating behavior are very hard to pinpoint … But fire ant researchers at the University of Georgia say they have characterized a gene that may single-handedly determine a complex social behavior: whether a colony will have one or many queens" (Holdern 2002, p. 1434).

The discovery of how a gene can control behavior lay in understanding – how the ants sense the pheromones that identify a queen ant: "Fire ants have two basic kinds of social organization. A so-called monogyny queen establishes an independent colony after going off on her mating flight, nourishing her eggs with her own fat reserves without worker help until they hatch and become workers themselves. Polygyny queens, in contrast are not as robust and fat. They need worker aid to set up new colonies. They spread by budding from one primary nest into a high-density network of interacting colonies. Monogyny communities permit only a single queen, and those with a resident royal kill off any intruding would-be queen. Polygyny colonies can contain anywhere from 2 to 200 queens and accept new queens from nearby nests" (Holdern 2002, p. 1434).

The difference between the two kinds of queens (monogyny and polygyny) was explained as a difference in a single gene that ants possess – their Gp-9 gene. This Gp-9 gene encodes information to know how to build a pheromone-binding protein, which is crucial for recognizing fellow fire ants. In a monogyny colony, all the ants have two copies of one version of the gene called a *B* allele. But in polygyny colonies, at least 10% of the ants have a different version of the Gp-9 gene called a *b* allele. So there can be two kinds of queens in the ants, those with the gene *BB* and those with the gene *Bb*.

The behavioral result of these different versions of the gene is that ants in a polygyny community will kill off any potential *BB* queens but not Bb queens. Ants in the *Bb* portion of the colony will accept *Bb* queens, and the *Bb* queens can escape attack from the *Bb* workers. *Bb* workers cannot recognize *Bb* queens as competition to the *BB* queen. The *b* allele encodes for a faulty pheromone-binding protein that does not bind the queen's pheromone.

Fire Ant Queen (http://en.wikipedia.org, Fire ant 2007)

The importance of this research was that it connected a specific mechanism with a specific behavior: "This is the first time scientists have nailed down the identity of a single gene of major effect in complex social behavior.... Andrew Boruke of the Zoological Society of London said: The research opens for the first time the study of genes influencing social behavior across the whole span of the biological hierarchy" (Holdern 2002, p. 1434).

## Scientific Paradigms

As we see in the case of the fire-ant society, (1) social behavior was connected to sensing capability (smell) and (2) connected to genes for constructing the mechanisms for sensing. Social behavior for the ant society was *functional* to the society, enabling the society to *function* as a social group with workers serving a single queen (responsible for reproduction). This kind of explanation in biology – of connecting *biological functioning* (worker ant behavior) to physical *mechanisms* (genes) – has been called "reductionism" in biological explanation.

For example, writing about methodology in biological research, Douglas Kell and G. Rickey Welch wrote about this reductionist approach in molecular biology: "Molecular biology is widely reputed to have uncovered the secret of life, with DNA being heralded as the quintessential component of biological systems.... The biologist's problem is the problem of complexity. According to conventional wisdom, the way to study complex systems is by breaking them down into their parts" (Kell and Welch 1991, p. 15).

Reductionism assumes that living systems are mechanistically arranged hierarchically in space – so that larger organizations of life can be explained in smaller levels of organization. Kell and Welch used as an example: "Hydrolysis of energy-rich molecules causes protein filaments to slide, sliding filaments cause muscles to contract, muscle contraction causes the acquisition of food and the escape from predation, and group behavior centered around common defense and protection of food supply causes the formation of social systems. This analytical approach may be termed "hierarchical reductionism" (Kell and Welch 1991, p. 15).

Kell and Welch asserted that while this approach provides much insight, it in itself cannot provide a complete explanation in biology: "Methodologically speaking, reductionism is fruitful; it yields much utilitarian detail about observable phenomena. However, reductionism can never be complete; there is always a "residue" (in biology) …" (Kell and Welch 1991, p. 15).

What they meant by "residue" is that biological *explanation* requires the idea of *function* in addition to *mechanism*. As a biological explanation, the idea of *function* is not a mere intellectual residue from the idea of *mechanism*. This methodological idea of *function* is a *scientific paradigm* equal to and independent of the *scientific paradigm of mechanism*.

A scientific paradigm is an intellectual framework within which scientific theories and experiments are expressed.

In science, the paradigmatic concepts of *function and mechanism* are independent ideas and co-equal ideas for scientific explanation.

The explanatory concept of "function" introduces into science other ideas beyond those in the concept of *physical mechanism*. For example, an essential to the concept of behavior in organisms with neural systems is the concept of the "intention" of the organism. The term of *intention* denotes a deliberate action by an organism. Functional concepts, such as intention, are needed in biology to explain complicated behavioral patterns of sentient beings. A sentient being is one who is conscious of its state of being. Sentient consciousness can be kind of "hard wired" in the genes of behavior, as in the case of the fire ant. Or consciousness can be a mixture of hardware and software, as in the learned behavior of more intellectually complex organisms – such as the fiery-tempered humans (in comparison to the fire ants).

For any society, living behavior needs to be described by scientists not only by the *mechanism* of the living processes (e.g., pheromone-binding protein in the ant case) but also by the *function* of the process (e.g., identification of the queen ant). And in this idea of function, one can assign (from the perspective of a *BB* fire ant queen) a kind of "will" in the functional activity – the will of the *BB* queen to have no competitors. In sensing out other competing *BB* queens and killing them, the *BB* workers can be said (from the queen's perspective) to be carrying out the queen's will.

The functional idea of a "will" may sound odd applied to fire ants, but there are socio-biology parallel tales in human societies – of a similar kind exclusive leadership. For example in the history of England's monarchs in the sixteenth century, the protestant Queen Elizabeth had her contentious and Catholic cousin, Mary Queen of Scots, imprisoned, and beheaded. Pushing the metaphor, one might say that the fiery colony of England would have only one monarch, and the English workers bearing the banner of the protestant allele carried out the Queen's will to have no other queen but protestant Elizabeth. Previously a sister, Mary (called by the protestants, Bloody Mary) had been on the English throne and had killed many protestants. Protestant workers feared to let another Catholic queen rule the English hive.

Now this is only a crude metaphor because no one single gene codes for religious and political affiliations in human colonies. But there is a parallel. Humans have been genetically hard wired for family, kinship, and tribal affiliation. The social phenomenon of this kind of group behavior (such as competitive national games as football, territorial wars, genocide), probably all, is based on some kind of genetic human adaptation for tribal affiliation. Human will is a functional description of how humans decide upon a plan of purposeful action. For less complex organisms than humans, much of their will (as in the will of the fire ant) has been programmed in their DNA – hard-wired behavior, so to speak. This kind of hard-wired willful behavior in animal species has been traditionally called "instinct."

Instincts provide the primitive "hard-wired" biological instructions for attaining the basic purposes of animate beings.

In complex organizations of life, instincts are not sufficient to explain behavior, particularly in the human species. The activities of learning and planning add to and even go beyond humans' basic instinctive instructions for action. In the complex cognitive and social capabilities of humans, their purposes are not all purely instinctual but are also learned and deliberate. Instinct combines with learning and reason in the human, so that human will is a mixture of instinct and reason. The concept of *reason* needs to be added to any theory of *action* to describe how actions can be deliberately chosen by some sentient being as a result of a prior analysis of an action as a *means and end.*

The point is this. Taken all together, the terms – of *function, intention, instincts, action, reasoning* – provide a complex set of related ideas for describing living phenomena outside of and beside the set of ideas of *mechanism.* (Mechanism is what atoms do, but Mechanism and Function together is what fire-ant societies do.) This set functional ideas we will call the *Scientific Paradigm of Function.*

As the idea of scientific method is the central concept in modern epistemology, so is the concept of a scientific paradigm central to modern ontology.

Physics uses the scientific paradigm of Mechanism but not that of Function. Biology uses both the paradigms of Mechanism and Function. In general, all the different fields of science differ by the different scientific paradigms used as intellectual frameworks for description and explanation. In later chapters, we will review how modern science uses four *Scientific Paradigms: Mechanism Function, System, and Logic.*

## *Illustration: National Science Foundation*

Now let us turn to science administration at the macro-level of a nation. The science administration infrastructure of a modern nation has government agencies that fund research.

Funded research determines principally what science is performed.

The US government, the National Science Foundation selects and funds about one-third of the nation's scientific projects and centers in US universities. NSF was established by US law in 1954 to fund both "basic research and applied research," which is to say "science" and "engineering" at American universities. In the 1950s, Vannevar Bush, an MIT engineering professor and research administrator of the war-time Office of Strategic Defense, conceived of the mission for such an agency and advocated in a famous paper: Science – The Endless Frontier (Bush 1950). As conceived by Bush and implemented by a US law signed by President Harry S. Truman, the mission of NSF was and continues to be to fund scientific and engineering research (basic and applied research) across the scientific and engineering disciplines in universities.[2]

NSF's headquarters is located in Arlington, Virginia, USA. There are about 1,700 employees – about one half Ph.D. scientists and engineers who administer the research divisions and programs of NSF. NSF receives research proposals from professors at American universities and processes these proposals, using peer review to judge quality. In 2007, NSF received about 40,000 proposals for research projects, of which about 10,000 were funded as research grants (http:www.nsf.gov, 2007). NSF's proposal selection process is depicted in Fig. 1.3.

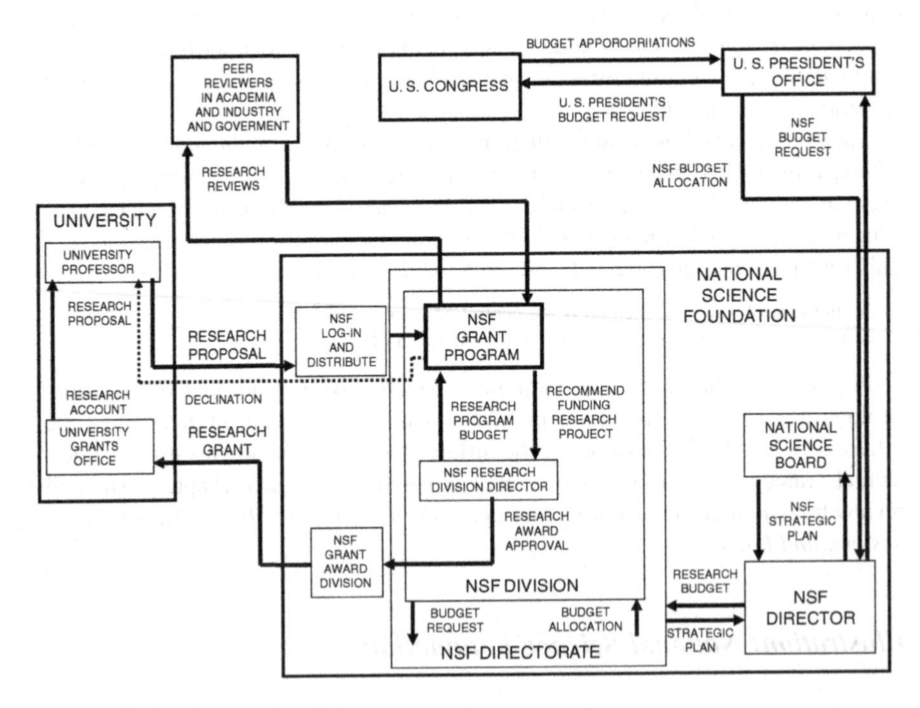


**Fig. 1.3** National science foundation award system

The organization of NSF is in research directorates, which closely follows the organization of research in universities. In 2009, the NSF research directorates were organized as scientific disciplines: biological discipline (Biological Sciences), computer discipline (Computer & Information Science & Engineering), engineering disciplines (Engineering), earth science discipline (Geosciences), mathematics and physics and chemistry and materials disciplines (Mathematical & Physical Sciences), and social science discipline (Social, Behavioral, & Economic Sciences). The other directorates are educational or administrative: (Education & Human Resources) (Budget, Finance & Award Management), and (Information & Resource Management).

Each research directorate was divided into divisions (which approximate disciplinary specialties) and then into research programs (which approximate disciplinary sub-specialties). Research proposals for research projects are submitted to an appropriate research program. In each research program, NSF has defined the focus and scope of research issues within which proposals could be funded and the criteria of scientific quality to be used by peer reviewers to judge for NSF the quality of submitted proposals. Then the high-quality proposals are funded from research program budgets.

When a research proposal reaches NSF (electronically submitted) the proposal is logged in and electronically distributed to the appropriate NSF research program. In that research program is an NSF research program officer with a doctorate in the field of the program. The professional responsibilities of the program officer are:

1. Read the proposal and identify the proper kinds of scientific expertise to review the research quality of the proposal
2. Select peer reviewers in academic, governmental, or industrial research positions who have published in the areas of expertise and are qualified to judge the research quality
3. Electronically forward a copy of the research proposals with a request to review the research proposal with a proper regard to confidentiality of information in the proposal
4. Receive the reviews and upon the advice in the review decide whether or not the proposal should be funded (graded as an excellent proposal, very good, good, fair, or poor proposal)
5. Allocate a grant budget to those of the excellent and very good proposals the program officer decides to fund and can fund within the program budget
6. In some programs, a program officer will use an external panel (selected by the program officer to offer funding advice) which will meet under the auspice of the program officer and will offer funding advice by ranking in terms of quality the reviewed research proposals, and then the program officer on this advice allocates funds to proposals (projects or centers) within the program budget constraint
7. Forward recommended proposals for grants to the division director overseeing the program

8. Upon approval by the division director, the proposals to be funded as research grants are sent to the NSF Division of Grants in the Directorate of Budget, Finance & Award Management
9. The Division of Grants then notifies the professor and university administration (for each approved proposal) that a research grant based upon the proposal will be awarded for a certain amount to begin by a certain date;
10. After formal notification of the award, the university administration establishes a research budget line upon which the professor can draw to fund and begin the research project;
11. As research progresses in the project, the professor sends any required progress reports to NSF and the university administration sends requests for reimbursement of research spent on the grant.

Thus the direct participation in the progress of science by an NSF program officer (science administrator) lies in determining the (1) areas for science research and (2) criteria of scientific quality.

To obtain money to fund research, NSF is budgeted in appropriation bills passed by the US Congress and signed by the US Director. For the budget process, each year the NSF directorates submit budget plans to the Director and National Science Board. In the NSF authorization law, both the Director and members of the Board are appointed by the US president. The Director and Board share equal responsibilities in governing NSF. A preliminary budget request agreed upon by both Board and Director is submitted to the President's Office of Management and Budget (OMB). An OMB examiner looks at the budget request and discusses it with the Director of OMB to set a budget target for NSF for the next budget year. This target provides an upper bound for the NSF budget. The Director and Board then revise the NSF budget to be within the OMB target and resubmit the budget to OMB.

This budget process of targets and budget submissions occurs between OMB and all federal agencies. Then in January, OMB assembles all the budget requests to the President from all agencies – and submits the whole budget package to the US Congress in February as the President's Budget Request (for the next Federal government fiscal year). From February to October, various authorization and appropriation committees in Congress review the budget requests and then Congress passes 13 appropriation bills to fund the Federal government for the next fiscal year.

When (in or after October), the appropriation bill in which NSF is included has been passed, then the Director of NSF assigns budgets to the NSF directorates – who in turn assign budgets to NSF divisions and programs within divisions. The program research budget funds research proposals as research grants.

The basic grant document is written and evaluated as a "research proposal." A scientific proposal consists of arguments on the where, how, and why research should be performed. The "methodological how" is expressed in science as a *research strategy* in a research proposal. The "administrative why" is expressed in *issues of research* issues of the proposal. Together, *research strategy and issues* are both necessary to gain funding for the research – a *research grant.* These research grants support graduate students to gain doctorates in science and engineering.

The research-proposal process begins at an American university with a professor having an idea for a research project, writing it as a research proposal and submitting it for approval to the university. The chairman of the professor's department and dean of the department's school then must approve (sign-off) on the proposal, and it next goes to the university's grant administration office. When all proper administrative rules are satisfied in the proposal, the grant administration office then submits the professor's research proposal to NSF. The reason for this formal procedure is that an NSF grant is formally awarded to the university for its administration – with the professor as principal investigator (project manager) responsible for the performance of the research funded by the grant.

The micro-level of research administration is the level at which funded research is performed.

The macro-level of research administration is the level at which research is selected to be funded.

## Two Levels of Procedures in Science

There are thus two different organizational procedures in science for planning research and funding research, which exist as micro- and macro-levels of activities in science. We sketch this in Fig. 1.4. On the (1) micro-level plane of science activities, we still have the four categories focused upon method (epistemology), paradigm (ontology), research (project administration), or application (technology). Now also on the lower (2) macro-level plane of science activities, we can identify four categories focused upon research strategy, research quality, research management, or technological progress.

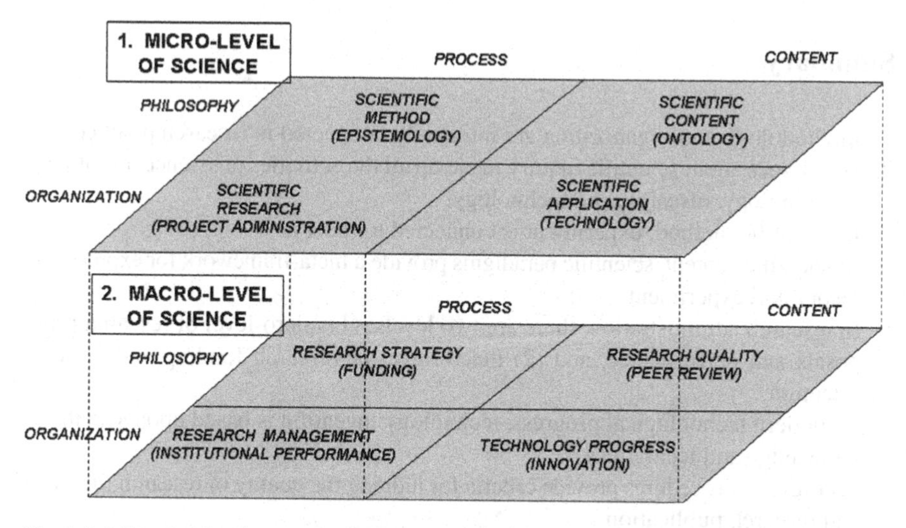


**Fig. 1.4** Micro-level and macro-level of science activity

We have discussed the upper micro-level plane as the activities of science, which involves scientific method, scientific content, scientific research, and science application as:

1. Modern epistemology
2. Modern ontology
3. Research-project administration
4. Technology research

Next, we can also see that a macro-level of science involves:

1. Procedures about scientific activity that focus upon both process and philosophy are procedures for guiding *research strategy – funding of science.*
2. Procedures about scientific activity that focus upon both process and organization are procedures for guiding *research management – performance of institutions of science.*
3. Procedures about scientific activity that focus upon both content and philosophy are procedures for guiding *research quality – peer review.*
4. Procedures about scientific activity that focus upon both content and organization are procedures for guiding *technology progress* (*innovation*).

> Science is a methodologically based process and also an organized procedure of research.
>
> The methodological process is one of asking questions about nature (research issues) – with answers derived from experiments and theory constructed upon the experimental observations, empirically grounded theory.
>
> Organizational research procedures provide funding for and contexts of performance for the methodological process of scientific inquiry.

## Summary

1. Methodology and organization are intimately connected in research practice.
2. Four issues about scientific inquiry make up all the activities of science: ontology, epistemology, research, and technology.
3. In scientific method, experiment is connected to theory.
4. In scientific content, scientific paradigms provide a meta-framework for expressing theory and experiment.
5. In research administration, there are two levels: (1) micro-level of research proposals and performance, and (2) macro-level of research funding and project selection.
6. In modern technological progress, technology invention is based upon scientific knowledge and techniques.
7. Peer review procedures provide criteria for judging the quality of research funding and research publication.

# Notes

[1] There are several biographies of Ernest Rutherford, among which are by those by Heilbron (2003), Pasachoff (2005) and by Reeves (2008).

[2] An history of the early years of NSF is (Lomask, NSF 76-18).

# Chapter 2
# Origin of Scientific Method

## Introduction

We have emphasized that scientific method is a methodological approach to the process of inquiry – in which empirically grounded theory of nature is constructed and verified. To understand this statement, it is useful to go back in time to see how the method evolved. The origin of modern scientific method occurred in Europe in the 1600s: involving (1) a chain of research events from Copernicus to Newton, which resulted (2) in the gravitational model of the solar system, and (3) the theory of Newtonian physics to express the model.

There were many important intellectual precursors to science. For example, alchemy was a precursor to the modern scientific discipline of chemistry, but it was not science. Alchemy was a confusion of practices and un-grounded theory. In medieval Europe, the fundamental stuff of the universe was viewed as air, earth, fire, water – alchemy. But now in modern Europe, the fundamental stuff of the universe is energy and mass, atoms and molecules, fields and particles – chemistry and physics.

As another example, the modern science of mathematics has important historical roots in Egyptian and Greek and Arab geometry and algebra. But algebra and geometry were not integrated until 1619, when Renes Descartes created the modern mathematical topic of analytic geometry. Nor was the modern topic of calculus created until in 1693, when Newton added to analytic geometry the ideas of a differential calculus of infinitesimals. (And about the same time and independently, Leibnitz contributed the ideas of integral calculus.) Then the modern discipline of mathematics intellectually grew in the 1700s, as mathematicians built upon a modern analytical foundation of geometry, algebra, calculus, vectors, and (later) set theory.

What is essentially different between the civilizations before and after the origin of science in the 1600s is a very different conception of nature. Before, nature was merely a manifestation of a super-nature – the supernatural and unobservable – the world of religion. Afterward, nature now is only what is observable in the world. Nature is thought about, described, and explained through experiments and theory and scientific paradigms. No longer do we live in a world of superstition and magic. We live in a modern world of science and technology – without magic.

F. Betz, *Managing Science*, Innovation, Technology, and Knowledge Management 9,
DOI 10.1007/978-1-4419-7488-4_2, © Springer Science+Business Media, LLC 2011

So before Isaac Newton's grand synthesis of mechanics, there was not science – at least not as we now know it. Modern science is both method and paradigms. Newton synthesized scientific method as theory-construction-and-modeling-upon-experimental-data. And Newton created the first scientific paradigm – *Mechanism.*

## Scientific Method

Science began in that intellectual conjunction of the research of six particular individuals: Copernicus, Brahe, Kepler, Galileo, Descartes, Newton. Why this particular set of people and their work? For the first time in history, all the component ideas of scientific method came together and operated fully as *empirically grounded theory*:

1. A scientific model that could be verified by observation (Copernicus)
2. Precise instrumental observations to verify the model (Brahe)
3. Theoretical analysis of experimental data (Kepler)
4. Scientific laws generalized from experiment (Galileo)
5. Mathematics to quantitatively express theoretical ideas (Descartes and Newton)
6. Theoretical derivation of an experimentally verifiable model (Newton)

### *Nicolaus Copernicus*

Nicolaus Copernicus (1473–1543) was what we would now call a *theoretician*, but he thought of himself as a "natural philosopher." He proposed an idea (actually a revival of an ancient idea) that the universe should be modeled with the sun as a center and not the earth – sun-centric versus earth-centric system.

Nicolaus Copernicus (1473–1543) was born in the city of Toruń, then in the Kingdom of Poland. Copernicus entered the Kraków Academy in 1491. Four years later he went to Italy to continue his studies, in law and in medicine at the University of Bologna and at the University of Padua. His uncle was a bishop in the Catholic Church, supported him and expected him to become a priest. While in Italy, he met an astronomer, Domenico Maria Novara da Ferrara and became his assistant for a time, making his first astronomical observations. Copernicus finished his studies at University of Padua and received a doctorate in canon law in 1503. He then returned to take a position at the Collegiate Church of the Holy Cross in Breslaw, Silesia. Just before his death 1543, he published his work, *De revolutionibus orbium coelestium*,[1]

Nicolaus Copernicus (http://en.wikipedia.org; Ncolaus Copernicus 2007)

Copernicus's model challenged an older and then widely accepted model of an earth-centered system – which had been refined by the Egyptian, Ptolemy (90–168 AD) of Alexandria. Ptolemy wrote scientific treatises, three of which were influential upon later Islamic and European thought: an astronomical treatise (Almagest), Geography, and "Four Books" astrology.

Claudius Ptolemy, by a Medieval Artist (http://en.wikipedia.org; Ptolemy 2007)

The Ptolemaic model had the Earth as center and the sun and planets circling the Earth. But it had awkward aspects – such as the planet of Venus showed an apparent retrograde motion, going forward most of the time but sometimes going backward. To account for this appearance, Ptolemy had put the planet upon a small circle upon a bigger circle around the Earth. This was to model the apparent "retrograde" motion of the planet Venus as seen from the earth. This was theoretically not elegant. It was neither simple nor direct in explanation. Copernicus argued that if all the planets were upon circles around the sun, the model became elegant – elegant in the manner of – simpler and without added complexity.

## *Tycho Brahe*

Copernicus's work stimulated new observations by the astronomer Tycho Brahe. Brahe wanted to determine which model was correct by direct astronomical observations. Now we could call Brahe an experimental scientist (in contrast to the theoretician Copernicus).

The importance of Brahe to Copernicus is that Brahe would use observations to ground theory – to place a theoretical model upon an empirical foundation – empirically grounded theory.

The greatly improved *precision* of Brahe's measurements over previous measurements of planetary positions enabled the breakthrough in astronomy. This precision of measurement provided data accurate enough to determine between two theoretical models of the planets which in fact was real: the Earth-centric (Ptolemy) or the Sun-centric (Copernicus) model?

In historical perspective, we can view Brahe as a great experimental scientist – because he understood that it was the precision of measurements that was the key to determining which model was correct in reality. This understanding by an experimenter as to what experimental data is critical to theory construction or validation is the mark of a great experimental scientist.

This is a key process in scientific method – precise experimental verification of a theoretical model of nature – by improved scientific instruments.

Tycho Brahe (1546–1601) was born in Denmark. His father was a nobleman. His uncle raised him, and in 1559, he went to the University of Copenhagen to study law. He turned his attention to astronomy after a predicted eclipse in 1560. Over the course of his life, he built several observatories, and constructed measuring instruments larger and much more precise than previous instruments. These were astrolabes, ten times larger than previous astrolabes. His measurements the planetary motion of Venus, Mars, and Jupiter were an order of magnitude more exact than older measurements of planetary motion.[2]

Tycho Brahe (http://en.wikipedia.org; Tycho Brahe 2007) Astrolabe

## *Johannes Kepler*

Brahe made many, many astronomical measurements and, in 1600, hired a mathematician, Johannes Kepler, to analyze all the data. To analyze means to abstract the underlying form of the data and to generalize the form, so that data from additional new observations would fit that form. Analysis of data is the connection of observation to theory.

Kepler moved his family from Austria to Poland and began working for Brahe. But Brahe died unexpectedly on October 24, 1601. Brahe had been the imperial mathematician to the court of Emperor Rudolph II; and Kepler was appointed as Brahe's successor. Kepler continued working on analyzing Brahe's measurements. By late 1602, Kepler found a law that nicely fit the planetary data – planets sweep out equal areas of their orbits in equal times. Here was a law of nature (the mind of God in Kepler's view). It was a phenomenological law – a law of nature which nature follows – and also a quantitative law!

Kepler understood that this law was a property of elliptical orbits. Copernicus's model had used circular orbits. But Kepler saw that, in reality, planets followed elliptical orbits. By the end of the year, Kepler completed a new manuscript, *Astronomia nova*, describing the elliptical orbits. But this was not published until 1609 due to legal disputes with Brahe's heirs over ownership of Brahe's data. (This was an early dispute over what today we would call "intellectual property").

This quantitative formulation of a law-of-nature was a major step toward scientific method.

Scientific method consisted not merely of qualitative observations of nature, but also of quantitative measurements and quantitative laws depicting the underlying form of the measurements – physical laws of a natural phenomenon.

Phenomenological laws are regular patterns of relationship observed as occurring in phenomenon of nature.

Johannes Kepler (1571–1630) was born in Germany. In 1589 he entered the University of Tubingen as a theology student but was soon to excel in mathematics. His love of astronomy was long standing, and he cast horoscopes as an astrologer. Learning of the Ptolemaic model and the Copernican model, he liked the Copernican model. Kepler than took a position as a teacher of mathematics and astronomy at a Protestant school in Graz, Austria (which later was to become the University of Graz). Kepler published his first astronomical work in 1595, Mysterium Cosographicum, in which he defended the Copernican system. He was at the time interested in geometric forms (polygons) which might be used to fit the astronomical data. But his intellectual breakthrough was not to occur until he gained access to Brahe's data. Kepler could not have created his theory of planetary orbits as ellipses without the extreme precision of Brahe's measurements.[3]

Johannes Kepler (http://en.wikipedia.org; Johannes Kepler 2007)

## *Galileo Galilei*

Just before Kepler's publication of Astronomia nova, the telescope was invented in 1608 in the Netherlands. Learning of this invention, Galileo Galilei in Italy made a telescope that same year with three power magnification. He used it to observe the moon and planets. He was the first to observe the moons of Jupiter, a large planet with four moons circling it. This was a clear analogy to Copernicus's solar model, with the sun the center of planetary orbits – as was Jupiter the center of its moons' orbits. Galileo published his first astronomical observations in March 1610 as Sidereus Nuncius. The double impact of Kepler's elliptical orbits and Galileo's moons-of-Jupiter established for the astronomical community then the realistic superiority of the Copernican model. The Ptolemaic model went into the dustbin of intellectual history.

Galileo went on to establish the first scientific laws of physics. He performed experiments about motion and gravity and inferred new physical theory based upon experimental results. He pioneered the scientific method of doing quantitative experiments whose results could be generalized in mathematical expression. After Kepler's mathematical analysis of Brahe's measurements, Galileo's physical laws provide a second historical example of modern scientific method.

Galileo Galilei (1564–1642) was born in Pisa Italy. He entered the University of Pisa to study medicine, but instead studied mathematics. In 1589 he was appointed to the chair of mathematics in Pisa. In 1592, he moved to the University of Padua, where he taught geometry, mechanics and astronomy. Here he made significant progress in the physics of motion. After Galileo published his account of the moons-of-Jupiter in 1610, he went to Rome to demonstrate his telescope and advocate the Copernican solar model. He was then admitted to a prestigious academy in Rome, Accademia dei Lincei.

But in 1612, some Catholic priests opposed the idea of a sun-centered universe. In 1614 he was denounced by Father Tommaso Caccini (1574–1648) as a heretic. Galileo was called back to Rome from Padua to defend himself. In 1616, Cardinal Roberto Bellarmino ordered him not to teach Copernican astronomy. But he was free to return to Florence.

Later in 1632, Galileo published a book which compared the two views of the universe, *Dialogue Concerning the Two Chief World Systems*, making the holders of the earth-centric model to appear as fools. This offended the Pope in Rome, who thought Galileo was making fun of him. In October of that year, Galileo was ordered to appear before the Holy Office in Rome. There he stood trial for heresy. As a judgment, he was required to recant his belief in the Copernican solar model, and he was ordered to be imprisoned. This was commuted to house arrest, and he was allowed to return to his house near Florence.

For the remaining 16 years of his life, Galileo remained under house arrest. Fortunately, he used his time to write what would become his most famous book, *Two New Sciences*. This book would establish the laws of physical motion. It was the work upon which later Isaac Newton would build his revolutionizing physical theory. Galileo died in 1642.[4]

Galileo Galilei (http://en.wikipedia.org; Galileo Galilei 2007)

Cristiano Banti's 1857 painting Galileo facing the Roman Inquisition

Scientific method was exemplified in Galileo's approach – physical experiments on observable objects, measurements of relationships, analysis of measurements, formulation of theory as phenomenological laws of relationship between objects.

## *Rene Descartes*

The next step in the emergence of the scientific method was to improve the language of quantitative analysis – the invention of analytical geometry and of calculus and their application to the expression of physical theory. And this was due principally to Descartes and Newton. Rene Descartes was a contemporary of Galileo and made a very major contribution to advancing mathematics. He conceived of analytical geometry – adding algebraic expressions to the classical geometry of Euclid. As shown in Fig. 2.1, Descartes proposed to describe a space with basis vectors, $X$, $Y$, $Z$, so that every vector was at right angles to each other. Then any point in the space could be described by three numbers $(x, y, z)$ as projections onto these vectors.

What Newton would add is another time dimension $t$. Motion of a particle in that space could then be described as the succession of points occupied by that particle as time $t$ elapsed. At time $t_1$, the particle would be at position $(x,y,z_1,t_1)$ and then proceed to position $(x,y,z_2,t_2)$ at time $t_2$ and so on. This analytical geometry would provide the critical mathematical representational basis for physics and for Newton's calculus. Without analytical geometry and calculus, modern physics would not have been possible.

**Fig. 2.1** Three-dimensional geometric space

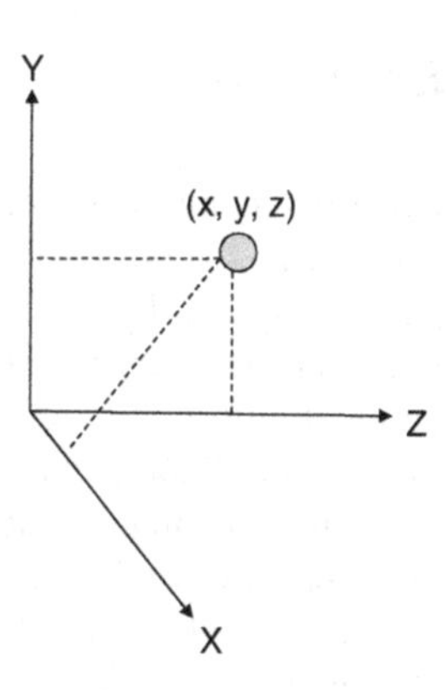


Renes Descartes (1596–1765) was born in France in 1596, and as a young man attended the University of Poitiers, graduating in 1616 with a Baccalureat and License in law. He did not practice law and entered court service in the Netherlands. There he met Isaac Beeckman who interested him in mathematics. In 1619, he was traveling in Germany and thinking about using mathematics to solve problems in physics. He had the idea to combine Euclidian geometry with algebra and created a new mathematical topic, "analytical geometry." This allowed the representation of space as a three-dimensional coordinate system, with any point in the space describable as projected distances along each Cartesian coordinate.[5]

Rene Descartes (http://en.wikipedia.org; Rene Descartes 2007)

As a historical footnote, Euclidean geometry derives from Euclid's "Elements." Euclid was a Greek philosopher of Alexandria living around 300 BCE. Algebra derives from Muhammad ibn Musa al-Khwarizmi (780–850). He was a Persian mathematician, who wrote on the systematic solution of linear and quadratic equations and is considered to be the "father" of algebra. His book, *On the Calculation with Hindu Numerals*, was translated into Latin in the twelfth century as *Algoritmi de numero Indorum*. The English word "algebra" is derived from the Arabic "al-jabr," one of the operations to solve quadratic equations. The English word "algorithm" derives from "algoritmi," the Latinization of al-Khwarizmi's name.

Euclid of Alexandria (http://en.wikipedia.org, of Alexandra, 2007)

Muhammad ibn Musa al-Khwarizmi (http://en.wikipedia.org, Muhammad ibn Musa al-Khwarizmi 2007)

## *Isaac Newton*

Descartes' combination of geometry and algebra enabled a quantitative description of space. This spatial description was essential to describe position and motion of particles in space. This would allow Newton to combine Galilean physics with that Cartesian geometry (as Descartes work is now called) and also with Kepler's astronomical ellipses to create a dynamic model of the solar universe. After all that time – from Plato and Aristotle – down through Augustine and Bacon – and down through Copernicus, Brahe, Kepler, Galileo, and Descartes – then finally the stage of history was set for Newton and his grand scientific synthesis of mechanism.

Isaac Newton (1643–1727) was born in England. He entered Cambridge University at the age of 19. He was engaged to Anne Storey. But she married someone else, and Newton never married. At Trinity College in Cambridge, most of the teachings were still those of Aristotle. But Newton read Descartes and Galileo and Copernicus and Kepler[6]

In 1665, he began to think about infinitesimal quantities and changes in velocities, and how to calculate with them in Cartesian space. This was the beginning of his development of calculus. In 1665, he obtained his degree. But then Cambridge University closed because of a Great Plague in England (which killed about one fifth of London's population, perhaps a bubonic plague). Newton went home and for the next year and a half worked on calculus and gravitation. Newton did not publish his calculus until 1693. But by then an independent invention of calculus had been made by Leibnitz which he published in 1684. Newton had approached calculus as differentials (which he called fluxions). Leibnitz had approached calculus as integration. (Of course, both differentiation and integration are essential to calculus.)

From 1670 to 1672, Newton lectured on optics. He thought light was be composed of particles (but had to also associate light as waves to explain diffraction of light). In 1675, Newton suggested that ether might exist to transmit forces between particles. Then in 1679, Newton returned to his work on mechanics. In 1687, Newton published *Philosophiae Naturalis Pricipia Mathematica.* This is the principle work which established physics on a quantitative basis (and now called the Newtonian Mechanics). It contained the three universal laws of motion:

1. Law of Inertia – The motion of a body is constant unless acted upon by an external force.
2. Law of Force – The effect of an external force upon a body is to change its acceleration, proportional to the body's mass: $F = ma = m\, \mathrm{d}v/\mathrm{d}t$.
3. Law of Action–Reaction – For every action (force) upon a body, there is an equal and opposite reaction (reactive force).

In Newton's calculus, the equation of force and motion ($F = m\, \mathrm{d}v/\mathrm{d}t$) is now called 'a "differential equation" – an equation containing differentials of calculus in the mathematical expression. Now there are standard ways in calculus to solve many differential equations – that is to find the kinds of algebraic solutions that fit a differential equation. Newton would not only pose the first differential equation but also solved it. The modern mathematical topic of calculus consists of Newton's formulation of differentials and of differential equations and the solution of a differential equation.

The next issue to Newton for his new physics was the quantitative expression for gravitational force, as diminishing according to the inverse square of distance. Newton then formulated the gravitation force (1) as proportional to the product of the masses attracted by the force of gravity between them and (2) decreasing in force as the square of the distance: $F_g = gMm/r^2$ (where $g$ is the gravitational constant as 32 ft/s/s and $M$ and $m$ are the quantities of the two masses which are attracted by gravity).

Newton set his differential force equation ($F = ma$) equal to the gravitational force to obtain a differential equation of motion for planets around the sun: $m\,\mathrm{d}v/\mathrm{d}t = gMm/r^2$. Newton solved this differential equation (using his new mathematical methods of calculus) to find that the solutions to this equation are (in analytical geometry form) the quantitative formulae which describe either an ellipse or a parabola (Fig. 2.2).

And thus was represented the solar system explained by a mechanics of universal gravitation. Using Galileo's physics, Newton's quantitative model derived Copernicus's solar model which fit with Brahe's astronomical measurements and Kepler's elliptical planetary orbital laws. These ideas all come together in the differential equation of motion of planets orbiting the sun bound by gravity!

This is scientific theory! It was created by the methods of science – observation and measurement, analysis, theory construction and prediction. It was then that discipline of modern physics began!

Isaac Newton (http://en.wikipedia.org; Isaac Newton 2007)

In historical perspective, the origin of science can be seen as a kind of systems problem – (a) getting all the pieces of the system together in one era and (b) getting all those pieces to be coordinated and integrated. This is what Newton did. He put it all together and created the theoretical paradigm of Newtonian mechanics. After

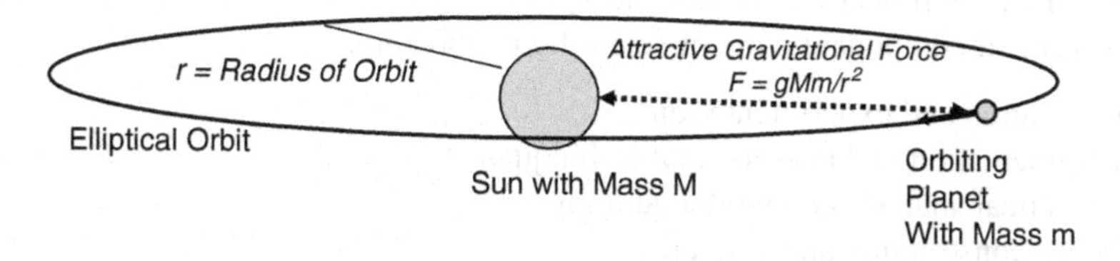


**Fig. 2.2** Copernican solar system planetary model

Newton's synthesis of mechanics, there arose, in the 1700s, the modern scientific disciplines of physics and chemistry and mathematics.

As footnote in science history, there was a disputed claim about who first discovered the quantitative law of gravity, Hooke or Newton? This created bitterness between the two contemporaries. Thomas Hooke (1635–1703) experimentally demonstrated it had the quantitative form of diminishing with the square of the distance. Apparently independently, Newton also had formulated the law of gravity as diminishing by the square of the distance. However, Hooke insisted upon acknowledgement by Newton that Hooke had first discovered the law. Newton refused since he believed he had not learned it from Hooke. This was particularly bitter to Hooke since he never received fame equal to Newton. Newton wrote that he independently discovered the square of the distance form of gravity from (1) reconsidering Kepler's previous work (Kepler's third law having the wrong proportion for diminution of gravity with distance), and (2) obtaining information from Flamsteed and Halley about the length of orbits of Jupiter and Saturn, Newton then concluded that the gravitational force between two massive objects diminished as the square of the distance between them increased.

In the organization of science, scientific credit and prestige goes to the person first to discover or explain a natural phenomenon.

## Scientific Method as Empirically Grounded Theory

The development of the disciplines of science, physics, chemistry, and biology did begin after Newton's synthesis of mechanical theory. After Newton published his seminal work in 1686, the next two centuries (the eighteenth and nineteenth centuries) saw the development of the disciplines of classical physics and chemistry and mathematics and biology. Further major theoretical developments continued to occur in physics and chemistry and biology in the twentieth century. Science is still progressing. Notably in the twentieth century and in addition to great continuing progress in the physical and life sciences occurred the founding of social sciences and computer science. And in all this progress in science, the process by means of which scientific knowledge has been obtained is now called the "scientific method." The heart of scientific method is the grounding theory construction based upon experimental data (Fig. 2.3). This means to base theory on experimental facts – construction and validation of theory upon empirical results. This distinctive approach of science can be called "empirically grounded theory."

The critical component parts of scientific method are:

1. Observation and experimentation
2. Instrumentation and instrumental techniques
3. Theoretical analysis and model building
4. Theory construction and validation
5. Paradigm development and integration.

**Fig. 2.3** Empirically grounded theory

Together these five techniques define the scientific method. All these components must be present and integrated. This is what happened first in the seventeenth century in Europe – when all the pieces of scientific method came together:

1. Copernicus (a theoretician) proposed a theoretical model which could be experimentally tested against another model of Ptolemy (an ancient theoretician).
2. But existing astronomical measurements of annual planetary motions were not accurate enough to determine which model more exactly fit the data, and Brahe (an experimentalist) constructed larger astronomical measuring instruments to obtain more accurate data.
3. Brahe hired Kepler (a mathematician) to analyze his measurements to determine if they fit a Copernican model; and Kepler found that an analytical pattern of elliptical planetary orbits did exactly fit the data.
4. Galileo (an experimentalist and theoretician) experimented with motion of physical bodies and induced three laws of motion (theories) giving mechanical behavior.
5. Descartes (a mathematician) invented new mathematics, analytical geometry, to extend Euclidian geometry.
6. Newton (a mathematician and theoretician and experimenter) invented differential calculus to extend analytical geometry to apply this to the description of spatial motion and also discovered the quantitative form of the gravitation force and applied all this to derive the Copernican solar model in the physical framework of Galileo's laws of motion.
7. After Newton's grand synthesis of mechanical theory, the new scientific disciplines of physics and chemistry were begun, describing material behavior in the new Newtonian mechanics.

> Central to the scientific method is the construction of theory of nature based upon and validated upon experimental observations of nature.
>
> Knowledge which is not "empirically grounded theory" is not scientific.
>
> One calls the research approach of experiments-on-nature as an "empirical approach" in scientific research – empiricism.
>
> One calls the research approach of theory-construction about nature a "theoretical approach" in scientific research – theoretical.

## Vienna Circle's Logical Positivism

However, not all philosophers of science have recognized this full complexity – about how scientific theory is constructed in the practice of science. An example in the early twentieth century was the school of philosophy of science called "logical positivism." They had two positions: (1) that all objects in science must be observable and (2) scientific theory is merely logically induced from experiment. Their first position corresponds with actual historical events in science, but the second does not. Let us briefly review this school because it is still mentioned in books about research methodology.

In Vienna in 1907, Philipp Frank and Hans Hahn and Otto Neurath began holding meetings in Vienna coffee shops about the philosophy of science – hence the name for the group became "Wiener Kreis" – Vienna Circle. Frank was a theoretical physicist in classical physics (Newton's mechanics and Maxwell's electromagnetism). Hahn was a mathematician. Neurath had studied sociology, economics, and philosophy. These meetings evolved into a philosophical school called "logical positivism." Morris Schlick joined the meetings and organized the group into the Ernst Mach Society. Many others joined the group, including: Gustav Bergmann, Rudolf Carnap, Herbert Feigl, Kurt Fodel, Tscha Hung, Victor Draft, Karl Menger, Richard von Mises, Marcel, Natkin, Theodor Rdakovi, Rose Rand, Moritz Schlick, and Friedrich Waismann. Also Wittgenstein and Karl Popper would attend later meetings.[7]

Central to the philosophy of the group were two philosophical positions: (1) experience is the only source of knowledge and (2) logical analysis is the way to solve philosophical problems. The first position was an opposition to any traditional philosophy that held there was an additional reality, metaphysical reality, behind the reality of the physical world. The Vienna Circle attacked metaphysics as any source of knowledge. The Vienna circle philosophically attracted logicians – particularly influenced then by the work of Russell and Whitehead on the logical foundations of mathematics.

Ernest Mach (1838–1916) was born in Chirlitz (now part of the Czech Republic). In 1860, he obtained a doctorate in physics from the University of Vienna. In 1867, he became a professor of Experimental Physics at the Charles-Ferdinand University in Prague. Mach photographed and described shock-waves in air. Mach also advocated a philosophy of phenomenalism, recognizing sensations as the ground of reality.

(http://en.wikipedia.org; Ernest Mach 2009)

In 1929, Hans Hahn and Otto Neurath, and Rudolf Carnap wrote a "manifesto" for the Ernst Mach Society, *The Scientific Conception of the World: The Vienna Circle*. Their central theme was to abolish any metaphysics as a contender to physics for the source of reality. They wrote: "No room remains for a priori synthetic judgments. That knowledge-of-the-world is possible rests, but upon the principle of a material being ordered in a certain, natural way – and not on human reason impressing any form on the material. The kind and degree of this order cannot be known beforehand... Only step by step can the advancing research of empirical science teach us in what degree the world is regular. The method of induction, the inference from yesterday to tomorrow, from here to there, is of course only valid if regularity exists... However, epistemological reflection demands that an inductive inference should be given significance only insofar as it can be tested empirically." Hahn, Neurath, Carnap, 1929. "The Scientific Conception of the World: The Vienna Circle" (http://gnadav.googlepages.com/TheScientificConceptonoftheWorldeng.doc).

One sees in this manifesto, the three central assumptions of the logical positivism:

1. Experiment is the foundation (base, ground) of knowledge.
2. Regularity in the world (logical order) must be discovered and not presupposed philosophically (metaphysically).
3. Theory is constructed directly by induction from experiment.

Rudolf Carnap later wrote: "I will call metaphysical all those propositions which claim to represent knowledge about something which is over or beyond all experience, e.g., about the real Essence of things, about Things in themselves, the Absolute, and such like... (Traditional metaphysics) pretended to teach knowledge which is of a higher level than empirical science. Thus they were compelled to cut all connection between their (metaphysical) propositions and experience; and precisely by this procedure they derived them of any sense" (Carnap, Rudolf. 1935. Philosophy and Logical Syntax). (http://www.philosophy.ru/edu/ref/sci/carnap.html, 2009)

And because of that emphasis on induction-as-scientific-method, the Vienna Circle emphasized the role of logical analysis in philosophy. If the traditional role of metaphysics was excluded from modern philosophy (replaced by scientific method), then what was left for the modern role of philosophy? The logicians in the school, such as Rudolf Carnap, agreed that modern philosophy should become "logical analysis." Carnap wrote: "The function of logical analysis is to analyze all knowledge, all assertions of science and of everyday life, in order to make clear the sense of each such assertion and the connections between them. One of the principal tasks of the logical analysis of a given proposition is to find out the method of verification for that proposition" (Carnap 1934) and (http://www.philosophy.ru/edu/ref/sci/carnap.html, 2009).

The Vienna Circle (1) emphasized experience as the source of knowledge but (2) supposed that inductive inference is the way theory is constructed. Moreover, this latter methodological assumption – of simply inducing theory – led to an overly simple view of scientific method: (1) formulating a hypothesis (theory) and (2) then validating or invalidating that hypothesis by an experiment. However, these are two separate assumptions. Given the first assumption (that experience is the source of knowledge), the second assumption (about scientific method as simply inductive inference) need not necessarily follow.

The philosophical position we have taken in this book is to agree with the proposition that all science is based upon experience.

But philosophically we disagree that theory construction is simply and necessarily made only by simple induction from experiment.

Instead, one finds in historical cases of scientific progress a circularity in logic (induction and deduction) between theory construction and experiment.

Rudolf Carnap (1891–1970) was born in Wuppertal, Germany and attended the University of Jena but had to serve in the German Army for 3 years. In 1917–1918, he enrolled in the University of Berlin (where Albert Einstein had just been appointed as a professor of physics). But then Carnap returned to finish at the University of Jena. He wrote a thesis on an axiomatic approach to space and time, which the physics department said was too philosophical and the philosophy department said was pure physics. Carnap then wrote another thesis on space and time, but taking a traditional Kantian approach on this as transcendental aesthetics, which was published in 1922 in the German philosophy journal of Kant Sudien. In 1926, Carnap was introduced to Moritz Schlick, who offered Carnap an appointment as a professor in the University of Vienna. In 1931, Carnap moved to the University of Prague as a full professor of the German language. He wrote his book there of *Logical Syntax of Language* (Carnap 1934). After the Nazi government began in Germany in 1933, Carnap foresaw the future and emigrated to the United States in 1935. (That next year in 1936, Moritz Schlick was murdered in Vienna.) From 1936 to 1953, Carnap taught at the University of Chicago and then moved to the philosophy department of University of California at Los Angles in 1954 (http://en.wikipedia.org; Rudolf Carnap 2009).

Moritz Schlick (1882–1936) was born in Berlin and studied physics at Universities of Heidelberg, Lausanne, and Berlin. In 1904, he wrote his physics dissertation under Max Planck. But he changed from physics to logic and wrote his habilitation essay in 1910 on "The Nature of Truth According to Modern Logic." In 1915, he published a philosophy/physics paper about Einstein's special theory of relatively. In 1922, Schlick obtained an appointment as a professor of inductive sciences at the University of Vienna. Schlick joined the Vienna group meetings and organized regular meetings as the Ernest Mach Society. With the rise of Nazism in Germany, many members of the Vienna Circle began emigrating. On June 22, 1936, Schlick was shot and killed by a former student, who thought him a Jew. The student was tried and sentenced and then paroled – as a cause celebre for the anti-Semites in Vienna (but Schlick was not Jewish). The student became a member of the Austrian Nazi Party after the "Anschuss," in which the Germany Army marched into Austria.

(http://en.wikipedia.org; Mortiz Schlick 2009)

Hans Hahn (1879–1934) was born in Austria. He attended the Technische Hochschule in Vienna and then studied mathematics in Universities in Strasbourg, Munich and Göttingen. He wrote mathematical papers in functional analysis, topology, set theory, the calculus of variations, real analysis, and order theory. In 1905, he obtained a teaching appointment to the University of Vienna and became a professor of mathematics in 1921 (http://en.wikipedia.org; Hans Hahn 2009).

Otto Neurath (1882–1945) was born in Vienna. He entered the University of Vienna and obtained a doctorate from the Department of Political Science and Statistics. He taught political economy at the New College of Commerce in Vienna. After World War I, Neurath directed museums (institutes) for housing and city planning. In the 1920s, he joined the Vienna Circle and was an author of its manifesto. In 1937, Hitler's Germany annexed Austria, and he fled to Holland and then to England, dying there of illness in 1945.

Otto Neurath (http://en.wikipedia.org; Otto Neurath 2009)

As a footnote for those familiar with the Vienna Circle, I have not made use of Wittgenstein's association with the group. While he was valued by members of the group for his attacks on metaphysics, he made no significant contribution to the understanding of scientific method. For example, Rudolf Carnap commented on Wittgenstein: "I, as well as my friends in the Vienna Circle, owe much to Wittgenstein, especially as to the analysis of metaphysics. But on the point just mentioned ("Whereof one cannot speak, thereof one must be silent.") I cannot agree with him. In the first place he seems to me to be inconsistent in what he does. He tells us that one cannot state philosophical propositions, and wereof one cannot speak, therefore one must be silent; and then instead of keeping silent, he writes a whole philosophical book. Secondly, I do not agree with his statement that all his propositions are quite as much without sense as metaphysical propositions are. My opinion is that a great number of his propositions (unfortunately not all of them) have in fact sense; and that is the same is true for all propositions of logical analysis" (Carnap 1934).

Also in the present epoch of methodology, what significantly was been passed along from the Vienna School as scientific method was a technique of hypothesis testing urged by a late-comer to the school, Karl Popper. Popper focused upon the emphasis in logical positivism was of the role of "induction" in theory construction. In 1928, Popper joined the logical positivists' movement in Vienna and formulated his idea that scientific theory could not be verified but only falsified.

Popper argued that if theory is only induced from experiment, then theory can never be completely validated – never be certainly true.

Logically, it is correct to say that a theory constructed only inductively and directly from observation can never be absolutely true – only probably true. The reason for this is that the validity of directly induced theory depends logically always only upon the last instance of observation. There is no methodological guarantee that a future observation may not occur which contradicts the theory. No finite number of experiments can ever provide methodological certainty in a purely inductive method – only a probability of its truth. For this logical reason, Popper assumed that scientific theory could never be verified but only falsified – *because Popper believed all scientific theory is only constructed inductively.* Accordingly, Karl Popper proposed a simple methodology for science which consists only of formulating a hypothesis and testing the hypothesis for falsification. However the methodological issue is – from whence is a good hypothesis formulated?

While hypothesis-testing is a useful research technique, it is also important to establish why and how and where-from a good research hypothesis is formulated.

Karl Popper (1902–1994) was born in Vienna; and in 1928, he earned a doctorate in philosophy at the University of Vienna. In 1934, he published, *The Logic of Scientific Discovery*. He was forced to leave Austria after Hitler's takeover of the country in 1937, when he migrated to New Zealand. After the Second World War, Popper obtained a professorship at the London School of Economics, where he remained until he retired.[8]

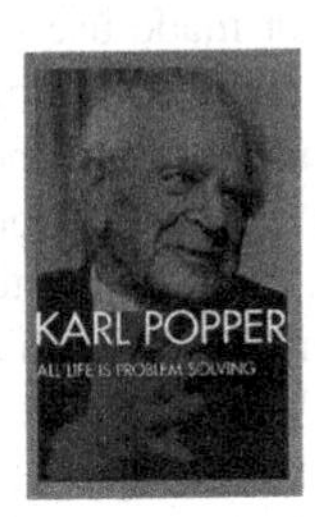


(http://en.wikipedia.org, Karl Popper 2009)

## *Illustration: Inference in Newton's Gravitational Solar Model*

How does Popper's prescription for scientific method as a simple kind of "hypothesis-testing" stand up in historical comparison of how science has really progressed? Not so validly.

For example in Popper's view – if the theory of the solar system had merely been inferred inductively from watching the sun rise repeatedly, what is the proof that it will always so arise? The sun's always rising is only a probable inference – a likelihood – but not a certainty. However, suppose that instead of merely an inductively inferred hypothesis of the sun always rising in the morning, one also has Newton's quantitative gravitational model of the solar system. Newton's theory predicts this as an orbiting Earth – and with great precision and certainty. The reason in the "certainty" of the scientific theory lies in the verified solar model – *in the elliptical orbits which are solutions to the gravitational-force model.* In elliptical orbits, the earth always circles the sun (while spinning). Unless new force disturbs the earth's orbit and spin, such as an asteroid hit, then the earth will continue to circle and spin mostly as it has.

Let us briefly review the idea of "inference" in logic – inductive inference and deductive inference. Historically, the idea of "logic" originated as a formal structure of a language, the "grammar" of the language. Grammar consists of the forms of sentences constructed from terms (nouns) and relations (verbs) and their order in a sentence. The order of a sentence is to assert something about a subject – a propositional phrase ascribed to the subject. Sentence order specifies the order of terms and relations into a meaningful predication of a subject. Sentences also have different modalities: declarative, inquisitive, and imperative. The logic of sentence-structure-and-order provides a given language its grammar (sentential form).

Also we need to remind ourselves that reasoning is an operation of the mind that both constructs and relates mental objects, concepts. Linguistic reasoning operates with language. Language not only provides modes of expressing experience in sentences but also as *connections between experiences* as *inferences.*

Inference is the form of reasoning about things as linguistic objects – relating ideas.

Inference is a proper form for making valid arguments from premises. Philosophically, there are two directions for inference:

1. Inductive reasoning – in which statements of particular facts are generalized into general ideas about the facts,
2. Deductive reasoning – in which particular statements of facts are deduced from general statements of theory.

Thus inference may proceed either from particular statements to general statements (induction) or vice versa (deduction). For example, in the classical dialogs of Plato about his teacher, Socrates, Plato emphasized logic's importance of probing at the assumptions of an intellectual position. This is the so-called Socratic method of questioning a person about assumptions. The direction of the inference lies in going from the conclusions to the assumptions of the argument – inductive inference.

In contrast, Aristotle's syllogism inference was deductive, going from general premises to particular instances. The famous example of a syllogism often repeated in philosophy courses is: all humans are mortal, and Socrates is a human, and therefore Socrates is mortal. (And which certainly was a correct inference – since historically Socrates was forced by his Athenian community to drink the lethal hemlock).

Let us now look at the construction of Newtonian theory of mechanics from this perspective of logical analysis – induction and deduction. Testing out the logical positivists' ideas about philosophy in empirical cases of the history of science – is in a form of "keeping with the scientific spirit." Certainly, philosophical theories about scientific method should be empirically grounded in the progress of science – as are all other scientific theories so grounded. This is only logically self-consistent. And in

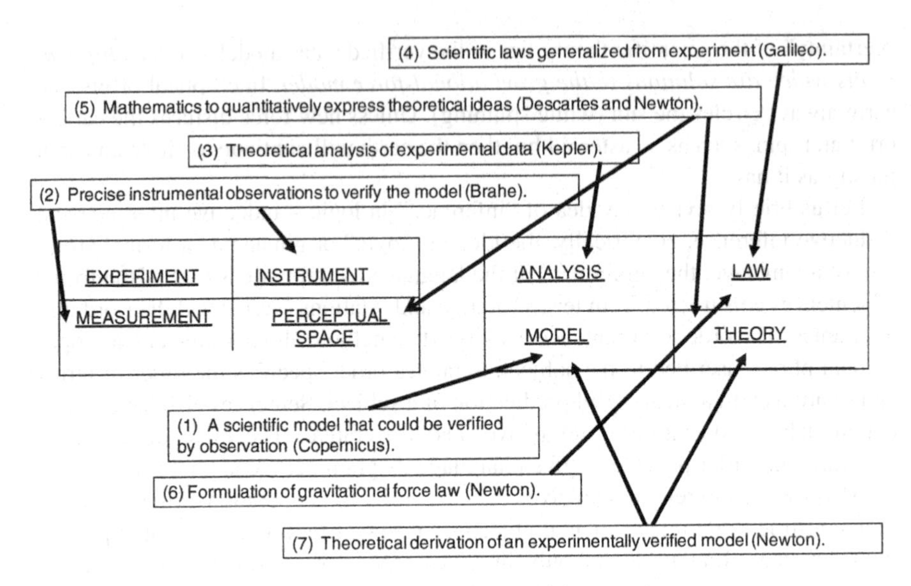


**Fig. 2.4** Empirically grounded theory for Copernican model

this spirit, we saw that Newton did not directly induce his mechanical theory directly from Brahe's measurements of planetary motions (Fig. 2.4).

1. Copernicus provided a scientific model that could be verified by observation – *deductive logical approach.*
2. Brahe developed instruments and made more precise measurements to verify the model – *inductive logical approach.*
3. Kepler made a theoretical analysis of experimental data, developing a phenomenological law about planetary motion – *inductive logical approach.*
4. Galileo performed physical experiments and formulated scientific laws generalized from the experiments – *inductive logical approach.*
5. Descartes integrated geometry and algebra and Newton created differential calculus to provide new mathematics for describing and modeling physical events – *deductive logical approaches.*
6. Newton formulated a phenomenological law of gravitation as a force varying inversely with the square of the distance – *inductive logical approach.*
7. Newton theoretically derived Copernicus's solar model as a consequence of his newly formulated mechanics – *deductive logical approaches.*

In Fig. 2.5, we sketch the views of the different participants in construction of the scientific theory of the solar system.

Copernicus imagined that if earth circled the sun, then the calculations for an astronomical almanac could be simplified from a theory that the sun circled the earth. Copernicus proposed a *theory* of a sun-centric planetary system – deductive inference. Brahe decided to put Copernicus's theory to an empirical test by improving

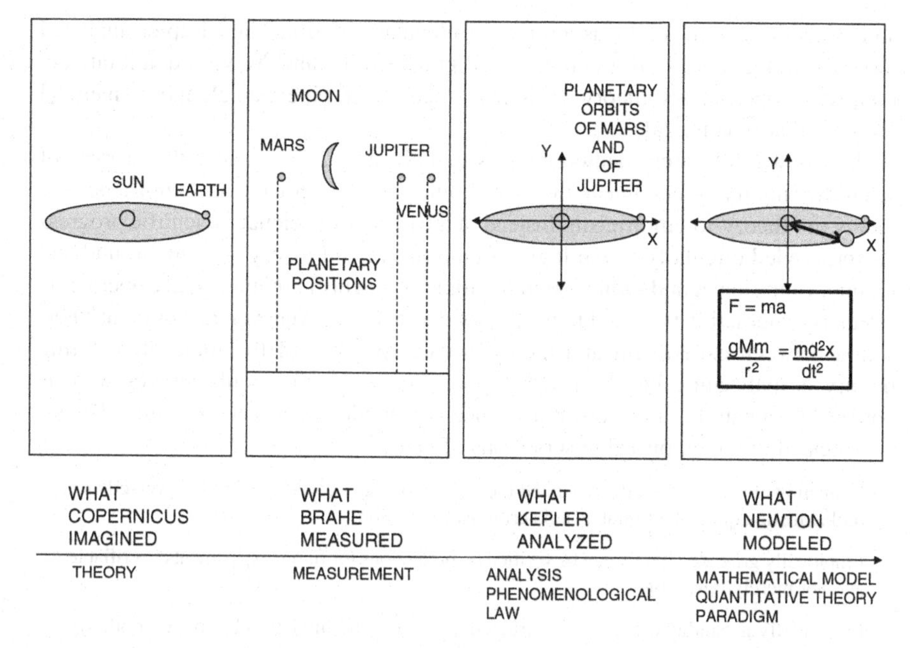


**Fig. 2.5** Measurement and theory of solar system

upon the astronomical measurements of the appearance of the planets, Mars, Jupiter, Venus, throughout the year, *measuring* their planetary positions – inductive inference. Brahe hired Kepler to try to fit these planetary data to circular orbits around the Sun. The data didn't fit a circular orbit but did fit an elliptical orbit. Kepler *analyzed* the data, proposing a *phenomenological law* (Kepler's law) about planetary motion – inductive inference. Newton developed new mathematics (differential calculus) – deductive inference. Newton also proposed a *theoretical law* for gravitation (varying as the inverse square of the distance) – inductive inference. Newton *modeled* the solar system and created a *theory* of mechanics (Newtonian mechanics) – deductive inferences.

## Circularity Between Empircism and Theory in Scientific Method

In the historical example of Newton's model of the solar system, we saw the empirical research technique of measurement; and this was combined with the theoretical techniques of analysis, modeling, and theory to create this dramatic progress in physical science – combinations of both inductive and deductive inferences. Most historical instances of progress in scientific theory have shown the use of both induction and deduction in the construction of theory based upon experiments. In temporal sequence, Copernicus proposed a theoretical structure for the geometry, and Brahe improved the measurement of planetary motions, and Kepler analyzed the measurements to find a pattern of elliptical orbits, and Newton quantitatively modeled Copernicus's theory in a new theoretical kinematical and dynamical physical theory (Fig. 2.4).

This is why research methodology is complicated. It is because the process of scientific inquiry is not linear in logic, going directly either from empiricism-to-theory nor theory-to-empiricism. Instead in the history of science, scientific progress has proceeded circularly in the logic of empiricism and theory – going around and around. Empiricism and induction in the logic of scientific inquiry really operate by circularly interacting with deductive theory construction. Yet even in a circular interaction between experiment and theory, a basic premise of the logic of scientific inquiry is that nature be observable. Or conversely, science only studies what is observable in nature. Empiricism in science is grounded in observing nature. Theory is grounded in the empirical observations of nature.

> Deductions from empirically grounded theory are logically certain (even if expressed as a probabilistic equation and not as a deterministic equation).
>
> Empirically grounded theory can be verified by predictions for future experiments –verifiable and not merely falsifiable.
>
> Empirically grounded theory is constructed not simply by induction but by a circularity of induction and deduction – in experiment and theory and prediction.

Yet due to the influence of the logical positivist's, much of subsequent philosophy of science taught in the twentieth century for the social sciences (but not for the physical or life sciences) became that of a simple methodological idea – theory construction consists of only a "hypothesis testing" as a research technique. This view of science was widely shared and many empirical studies in the social sciences followed the format of (1) formulating a hypothesis, (2) conducting an experimental sample, and (3) statistically analyzing the probability of the truth or falsity of the hypothesis.

A methodological weakness of the simple Popperian-research-approach occurs when researchers do not address the issue of from where or how should hypotheses come – which hypotheses are worth the effort of testing for falsification? In contrast to Popper's assumption about only being able to falsify scientific theory, the historical evidence in science is different. Most often in scientific history, it has been a scientific model rather than a hypothesis which has provided the basis for scientific verification. Newton's quantitative model of the Copernican solar system was one important

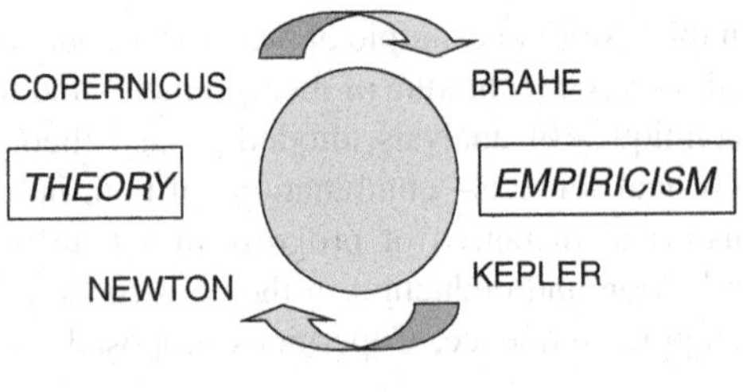

**Fig. 2.6** Circularity in historical interactions between research techniques

example. Watson and Crick's model of DNA is another example (which we later review). Generally speaking, a hypothesis which is not derived from a scientific model of a phenomenon has often been insignificant.

> A significant scientific hypothesis worth testing experimentally should be derived from a scientific model.
>
> For empirical scientific research, modeling is more likely a fruitful research activity for researchers than merely hypothesis-making-and-testing. What we will do in this book is learn how to manage this circularity in scientific method (between induction and deduction) through the interactions between experiment and analysis and modeling and theory.

## Summary

1. Science began when the scientific method of inquiry was established as the systematic way of understanding nature – basing theory construction and validation upon experimental data.
2. Empirical and Theoretical techniques in scientific method enable the construction of empirically grounded theories of nature.

## Notes

[1] There are many books and biographies about Copernicus, such as Bienkowska (1973).
[2] An interesting account on the relationship of Brahe to Keler is Ferguson (2002).
[3] There are many books and biographies about Kepler, such as Dreyer (2004) and Stevenson (1987).
[4] There are many biographies of Galileo, such as Langford (1998).
[5] There are many biographies of Des Cartes, such as Keeling (1968).
[6] There are many books and biographies on Newton, such as Tiner (1975).
[7] There are many books about the Vienna Circle, such as Kraft (1953) and Sakar (1996).
[8] Popper's principle methodological book is Popper (1934).

# Chapter 3
# Research Funding

## Introduction

We turn from the method to the organization of science. As we have emphasized, science is expensive. Proposals for scientific research projects are written by scientists and submitted to funding programs – which in turn award some of the projects as research grants. Research grants support the scientists, professors and graduate students, and their equipment to perform the specified research in the proposal. Research proposals describe the issues of a research to be performed and the methods and team for performing the research.

Research programs define goals for research support, areas of support, and criteria for funding. Research programs may provide funding for research projects external or internal to the funding organization. The largest external-funding research programs occur in government agencies and in philanthropic organizations. Research programs which primarily fund projects internal to the organization are in industrial or governmental laboratories.

*The reason to begin with understanding a research funding program is that – a research project which is perceived by a program as outside its planned scope will never get funded by the program – no matter how good the proposal.*

## *Illustration: US National Institutes of Health*

Government agencies now fund most of all scientific research. In the USA the expansion of research support for science occurred after the Second World War. Then new science and technology policies by the US government expanded and greatly increased the government support of research in American universities. Particularly, research support grew rapidly in the Department of Defense and in the

F. Betz, *Managing Science*, Innovation, Technology, and Knowledge Management 9,
DOI 10.1007/978-1-4419-7488-4_3, 

National Institutes of Health, in the National Science Foundation, and in the Department of Agriculture. In the second half of the twentieth century, government support of science in USA overwhelmed the research support in universities by private philanthropy.

As an example of a government research program, we look at the US government's National Institutes of Health (NIH). The US Federal government R&D responsibility for advancing medical research is embodied in this agency (NIH is an agency within the US Department of Health and Human Services). NIH funds research both within its own governmental research centers (institutes) and externally to US universities and hospitals and small businesses. The main campus of its institutes is in Bethesda, Maryland, just on the northern border of the capital of USA in Washington, DC. The mission of NIH is: "science in pursuit of fundamental knowledge about the nature and behavior of living systems and the application of that knowledge to extend healthy life and reduce the burdens of illness and disability" (http://www.nih.gov, 2007).

In 2007, NIH identified its primary mission goals to foster discoveries in biology and applications to medical and health technologies. For this, NIH supported research focused upon:

- The causes, diagnosis, prevention, and cure of human diseases
- The processes of human growth and development
- The biological effects of environmental contaminants
- The understanding of mental, addictive, and physical disorders
- Directing programs for the collection, dissemination, and exchange of information in medicine and health (http://www.nih.gov, 2007)

In this list, we see that the missions of NIH focused first on the *public welfare* for health ("causes, processes, biological effects, understanding of disorders") and then upon *research infrastructure* for health ("collection, dissemination, and exchange of information in medicine and health").

The history of the NIH began in 1798, when the government of the new nation of USA established a Marine Hospital Service for seamen. In 1836, the US Army established a Library of the Office of Surgeon General; and in 1887, the government established a Laboratory of Hygiene (to deal with infectious diseases). In 1902, a National Advisory Health Council was established, and the name of the Marine Hospital Services was changed to Public Health and Marine Hospital Services (later becoming in 1912, the Public Health Service). In 1930, the Hygienic Laboratory was renamed the National Institute of Health (NIH). In 1937, a National Cancer Institute was established. In 1946, a Division of Research Grants was added within NIH to fund external medical research. From 1948 through 1997, many new institutes for different health problems were established. In 2007, there were many NIH institutes and centers, with a total budget of $28 billion dollars and 18,727 employees. The institutes can be grouped into four categories of research focus: health problems, systems, biological nature, and medical technology.

NIH Institutes

*Health Problems*

National Cancer Institute (NCI)
National Eye Institute (NEI)
National Heart, Lung, and Blood Institute (NHLBI)
National Institute on Alcohol Abuse and Alcoholism (NIAAA)
National Institute of Allergy and Infectious Diseases (NIAID)
National Institute of Arthritis and Musculoskeletal and Skin Diseases (NIAMS)
National Institute on Deafness and Other Communication Disorders (NIDCD)
National Institute of Dental and Craniofacial Research (NIDCR)
National Institute of Diabetes and Digestive and Kidney Diseases (NIDDK)
National Institute on Drug Abuse (NIDA)
National Institute of Mental Health (NIMH)
National Institute of Neurological Disorders and Stroke (NINDS)
National Institute of Neurological Disorders and Stroke (NINDS)

*Systems: Biological, Environmental, Institutional*

National Institute of Child Health and Human Development (NICHD)
National Institute on Aging (NIA)
National Institute of Environmental Health Sciences (NIEHS)
National Institute of General Medical Sciences (NIGMS)
National Institute of Nursing Research (NINR)

*Biological Nature*

National Human Genome Research Institute (NHGRI)

*Medical Technology*

National Institute of Biomedical Imaging and Bioengineering (NIBIB)
(http://www.nih.gov, 2007)

## *Focus of Government Research Programs*

One can generalize, from his classification of NIH institutes, that there are four important criteria, which can provide guides to formulating research strategies in government programs. These are researches focusing either upon problems or on

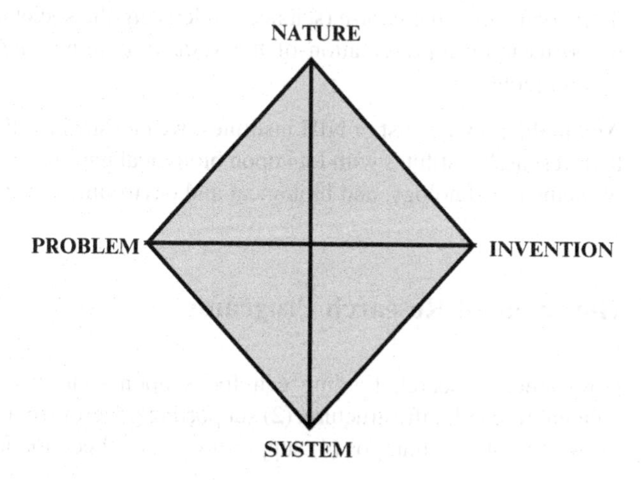


**Fig. 3.1** Dimension of research strategy

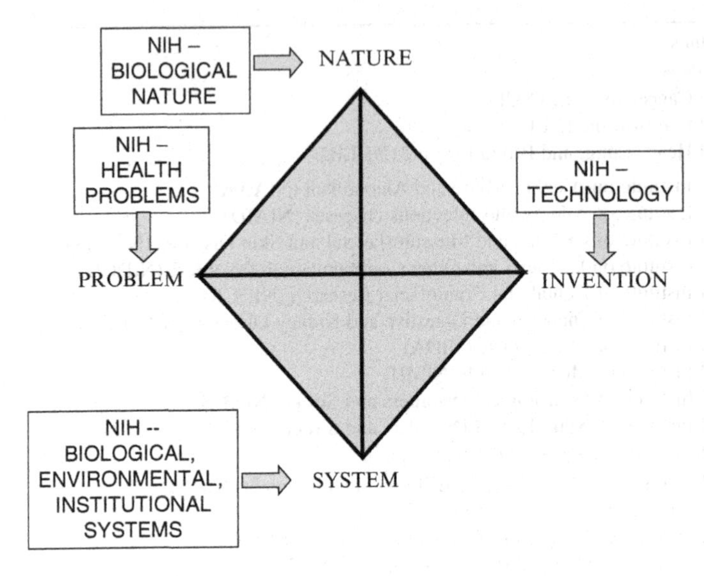


**Fig. 3.2** Research strategy for government programs

systems or on nature or on technology. As shown in Fig. 3.1, research strategy in government research programs can be explained in terms of different research foci upon: (1) particular societal problems, (2) nature underlying societal problems, (3) technology invention to solve problems, or (4) systems underlying problems.

This diagram shows how research planning in government research programs can be explained by different research opportunities for a government mission:

1. A focus upon research about a *problem* in a society
2. A focus upon research for invention to improve *technology* for dealing with the problems
3. A focus upon the *nature* (science) underlying the societal problems
4. Focus upon representation of the *systems* (engineering) operating in societal problems

And in the previous list of NIH institutes, we can see how historically NIH evolved their research institutes with foci upon biological nature, health problems, invention of medical technology, and biological and environmental systems (Fig. 3.2).

## Government Research Programs

Government research funding can focus upon four missions in (1) supporting national research infrastructure, (2) supporting government functions, (3) supporting areas of public welfare, or (4) supporting national economic development.

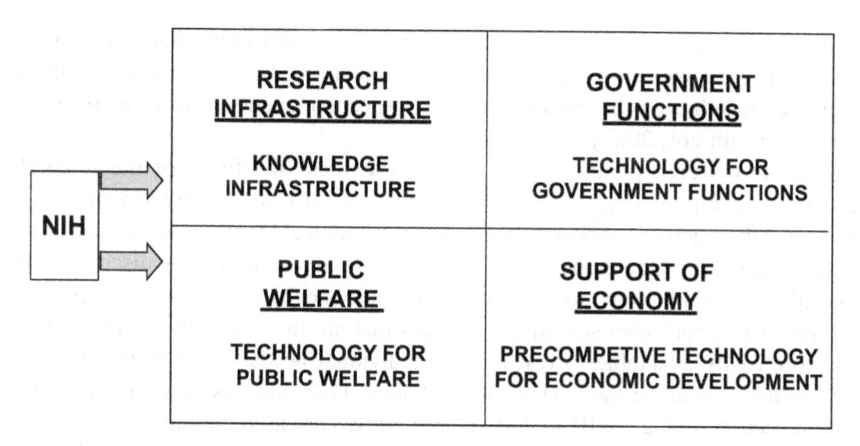


**Fig. 3.3** Science and technology knowledge issues

Research in medicine (such as in the US NIH) contributes to the public welfare and to the national infrastructure for public health. As another example, military force is a function of national governments, and so research and research labs focused upon weaponry technology is aimed at the government as customer. Since a national defense industry develops to commercialize the technology as weapon systems, some of the weapons may be sold to other friendly nations. As a third example, government funding for science in university research (such as in US NSF) contributes to the national educational infrastructure and to the public welfare when scientific progress advances technology.

As shown in Fig. 3.3, government science and technology issues can focus on providing research funding for (1) developing and maintaining a nation's knowledge infrastructure, (2) developing technology for use in government functions, (3) developing technology for use in areas of public welfare, or (4) developing technologies prior to use by specific industries (pre-competitive technologies). Also we can see that the NIH mission had focused upon the two left columns of: (1) supporting *research infrastructure* in medical research and (2) developing medical technology for the *public welfare* of health.

## Research Planning in NIH

Strategic planning for the NIH Institutes was done each fiscal year. For example, in 2007, NIH planned for next fiscal budget year in terms of strategic program initiatives which NIH called a Roadmap for Medical Research. And underlying their roadmap was the idea of the human body as a *system*: "The human body is dauntingly complex. To truly revolutionize medicine and improve human health, we need a more detailed understanding of the vast networks of molecules that make up our cells and tissues, their interactions, and their regulation. We also must have a more precise knowledge of the combination of molecular events leading to disease" (http://www.nih.gov, 2007).

Strategic research planning must not only express the focus of the research (human body) but also find scientific opportunities for research on the focus. NIH envisioned the scientific opportunities as arising in molecular biology: "To capitalize on the completion of the human genome sequence and recent exciting discoveries in molecular and cell biology, the research community needs wide access to technologies, databases, and other scientific resources that are more sensitive, more robust, and more easily adaptable to researchers' individual needs." (http://www.nih.gov, 2007) With a focus on the system of the human body and research opportunities from molecular biology, NIH in 2007 identified five research initiatives:

1. Building blocks, biological pathways, and networks
2. Molecular libraries and molecular imaging
3. Structural biology
4. Bioinformatics and computational biology
5. Nanomedicine (http://www.nih.gov, 2007)

## *Building Blocks, Biological Pathways, and Networks*

"Complex elements – from individual genes to entire organs – work together in a feat of biological teamwork to promote normal development and sustain health. These systems work because of intricate and interconnected pathways that enable communication among genes, molecules, and cells. Scientists are still working to discover all of these pathways and to determine how disturbances in them may lead to disease.... Ultimately, scientists hope to be able to completely map an organism's protein and metabolism networks, and create models to help predict the human body's response to disease, injury, or infection." (http://www.nih.gov, 2007) *Here we see that this NIH technology-research roadmap was in the direction of funding research on biological nature to better understand natural systems* (Fig. 3.4).

## *Molecular Libraries and Molecular Imaging*

"The Molecular Libraries initiative will provide public sector biomedical researchers access to small organic molecules that can be used as chemical probes to study the functions of genes, cells, and biochemical pathways in health and disease. The component is also expected to facilitate the development of new drugs by providing early stage chemical compounds to researchers so they can find successful matches between a chemical and its target and thus help validate new targets with potential

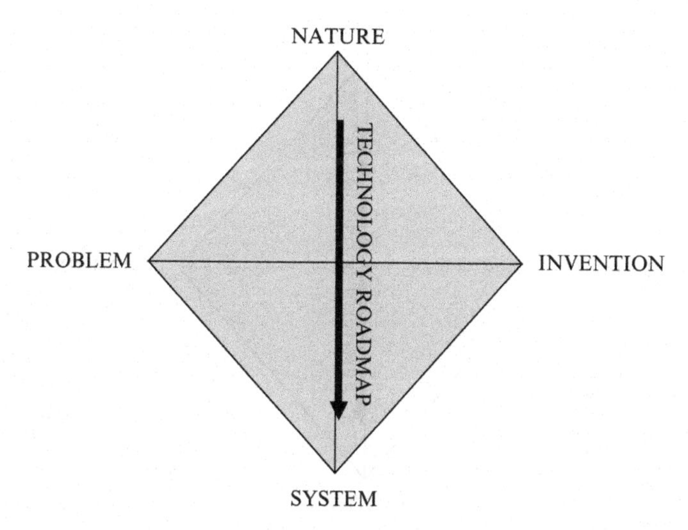


**Fig. 3.4** Nature to system technology roadmap

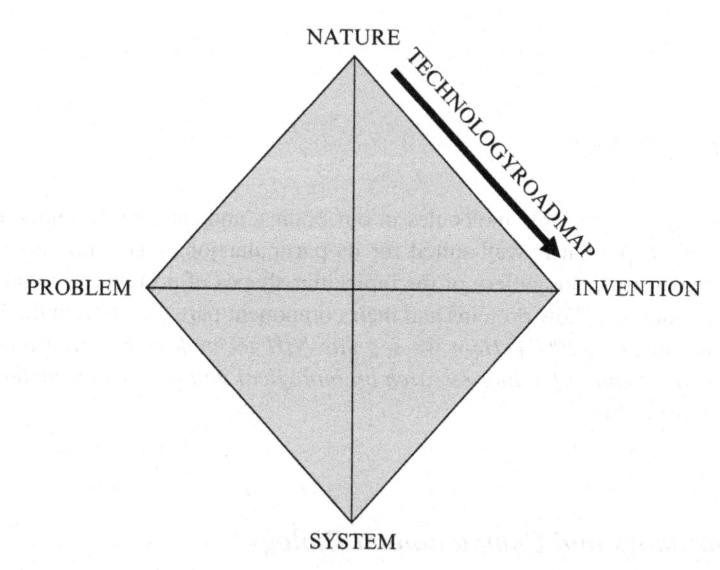


**Fig. 3.5** Nature to invention technology roadmap

for therapeutic intervention. The ultimate goal is to enable a detailed molecular understanding of cell and tissue function in normal and disease states, which may lead to greater power to diagnose and treat disease" (http://www.nih.gov, 2007). *Here we see this NIH technology-research roadmap was in the direction of funding research on biological nature to invent new medical treatments (therapeutic interventions)* (Fig. 3.5).

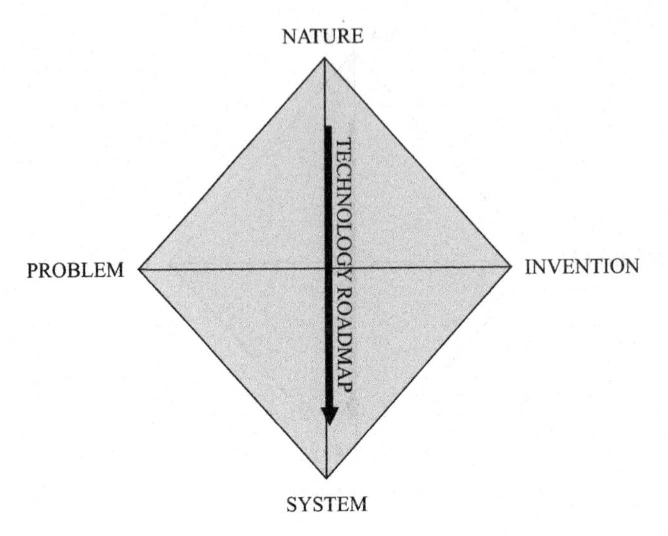


**Fig. 3.6** Nature to system technology roadmap

## *Structural Biology*

"Proteins are indispensable molecules in our bodies, and each has a unique three-dimensional shape that is well-suited for its particular job.... This initiative is an effort to create a picture gallery of the molecular shapes of proteins … to advance our understanding of how proteins and their component parts function in the body" (http://www.nih.gov, 2007). *Here we see this NIH technology-research roadmap was in the direction of funding research on biological nature to better understand natural systems* (Fig. 3.6).

## *Bioinformatics and Computational Biology*

"Scientists are using computers and robots to separate molecules in solution, read genetic information, reveal the three-dimensional shapes of proteins, and take pictures of the brain in action. These techniques generate huge amounts of data, and biology is changing fast into a science of information management. Researchers need software programs and other tools to analyze, integrate, visualize, and model these data." (http://www.nih.gov, 2007) *Here we see this NIH technology-research roadmap was in the directions of funding research for instrumental inventions to improve the study and modeling of biological nature* (Fig. 3.7).

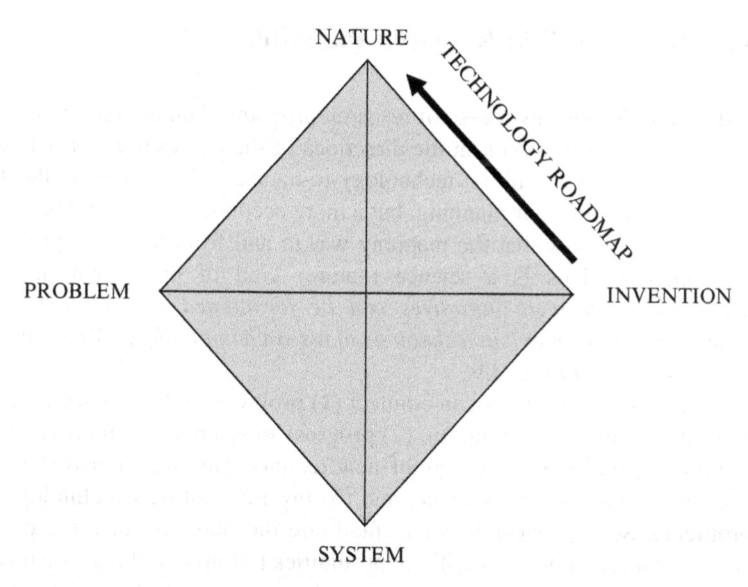


**Fig. 3.7** Invention to nature technology roadmap

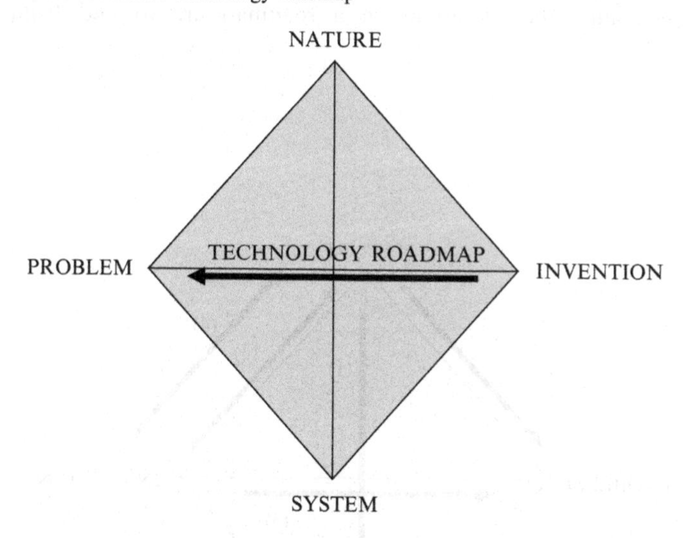


**Fig. 3.8** Invention to problem technology roadmap

## *Nanomedicine*

"A long-term goal is to create materials and devices at the level of molecules and atoms to cure disease or repair damaged tissues, such as bone, muscle, or nerve ..." (http://www.nih.gov, 2007). *Here we see this NIH technology-research roadmap was in the direction of funding research for technological inventions to improve medical solutions to health problems* (Fig. 3.8).

## Science "Roadmaps" in Research Planning

In the NIH example, we saw a case of research program planning based primarily upon science, and this was seen in the directions of the connections of nature-to-system or nature-to-problems or technology-to-nature. NIH has used the term technology-roadmap for their planning, but a more accurate term should have been science-roadmap. We saw that the mapping was to and from nature to problems-systems-technology. This is a science (nature) kind of directional mapping. *Research program strategic initiatives can be formulated through connecting research in science to research in technology to research in problems. We summarize this strategy approach in* Fig. 3.9.

Strategic program initiatives can connect (1) progress in the science of nature for improved systems representation, (2) progress in science for improved problem analysis, (3) and (4) inventions of new technologies for improved instrumentation and techniques for science, and (5) invention of new technologies to solve problems. Science-roadmaps can facilitate the planning of research program initiatives based upon scientific opportunities to improve the understanding of nature underlying societal problems and applications. In science-based research programs, the directions in a roadmap are to and from nature (science).

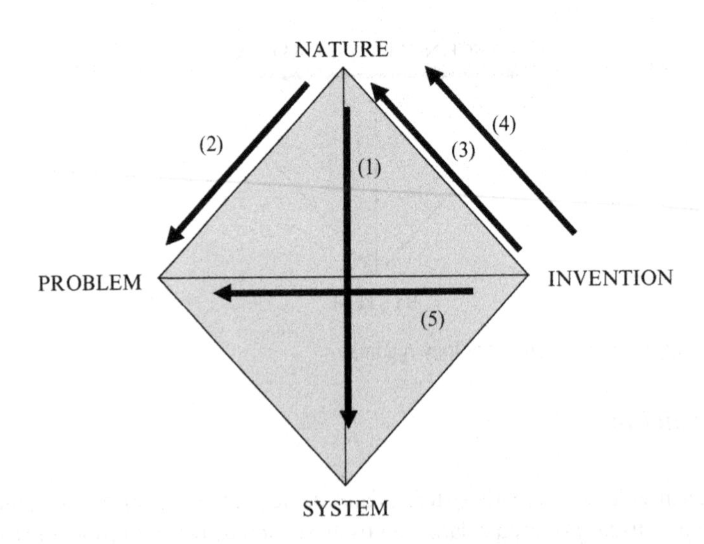


**Fig. 3.9** Science agenda roadmaps

## Peer-Review in Research-Funding Procedures

Government research budgets determine the areas and levels of scientific research to be performed in the nation. How these budgets are allocated to individual research projects is then determined at the macro-level by the research-project selection procedures of a government agency. For quality in research, the project selection should occur through a peer-review, methodological procedure. In any public research funding program – such as in US's NSF or Turkey's TUBITAK or the European Union's European Research Commission (ERC) – selection criteria are publicly stated and then adhered to in the review procedure.

Generally in government research grant programs, the criteria for scientific quality are merit and significance. For example, this express of quality is NSF's official peer review criteria:

> **Criterion 1: What is the intellectual merit of the proposed activity?**
> "How important is the proposed activity to advancing knowledge and understanding within its own field or across different fields? How well qualified is the proposer (individual or team) to conduct the project? (If appropriate, the reviewer will comment on the quality of prior work.) To what extent does the proposed activity suggest and explore creative and original concepts? How well conceived and organized is the proposed activity? Is there sufficient access to resources?"
>
> **Criterion 2: What are the broader impacts of the proposed activity?**
> "How well does the activity advance discovery and understanding while promoting teaching, training, and learning? How well does the proposed activity broaden the participation of under-represented groups ( gender, ethnicity, disability, geographic, etc.)? To what extent will it enhance the infrastructure for research and education, such as facilities, instrumentation, networks, and partnerships? Will the results be disseminated broadly to enhance scientific and technological understanding? What may be the benefits of the proposed activity to society?"
>
> (NSF 99-172, OFFICE OF THE DIRECTOR, September 20, 1999, National Science Foundation Merit Review)

As a second example, in 2007 Turkey's TUBITAK used the following three criteria for peer reviewing the selection of proposals:

1. Intellectual/Scientific/Professional Merit
   Theoretically sound, Novel, and possibly Significant Contribution
2. Achievability of the Research
   Sound method, Appropriate and Complete team, Proper equipment
3. Expected Impact of Anticipated Outcome
   In science or in Technology or in Industry or in Society (http://www.tubitak.tr, 2007)

The quality and range of science in a nation are strongly influenced by the procedures of research administration – at both the macro and micro levels of science. The macro-level of the organization of research is focused upon selecting and funding research. The micro-level of the organization of research is focused upon gaining funding and performing and using research.

This issue of macro-level quality of science and technology policy is at the heart of modern economic development. Historically over the last 200 years, economically competitive nations have been those advanced in science/technology. A major modern problem about research is why are some nations in the world today advanced in science and technology and others lagging? The answer lies in the great differences between nations in:

1. *Quality* of institutional procedures of science
2. *Quantity* in the level of funding of science and technology research
3. *Application* as national capability to transform science technology into economic utility

At the micro-level of research organization, research is formulated in a project expressed as a project proposal (2), as sketched in Fig. 3.10.

The project proposal describes the goal of the research as an issue in science or technology. Proper formulation of a research issue is a first stage of scientific methodology – requiring both an issue formulated as cognizant of the relevant research literature and a sound methodological approach. The research project proposal is essential because it is then submitted to a funding organization (governmental or industrial or philanthropic organization). Research is expensive and requires funding for salaries of the researchers, purchase of research equipment, maintenance of research laboratory facilities, and support of research administration.

Within a research funding organization, research proposals are reviewed for quality (5). The funding program (4) selects an external group of relevant scientific

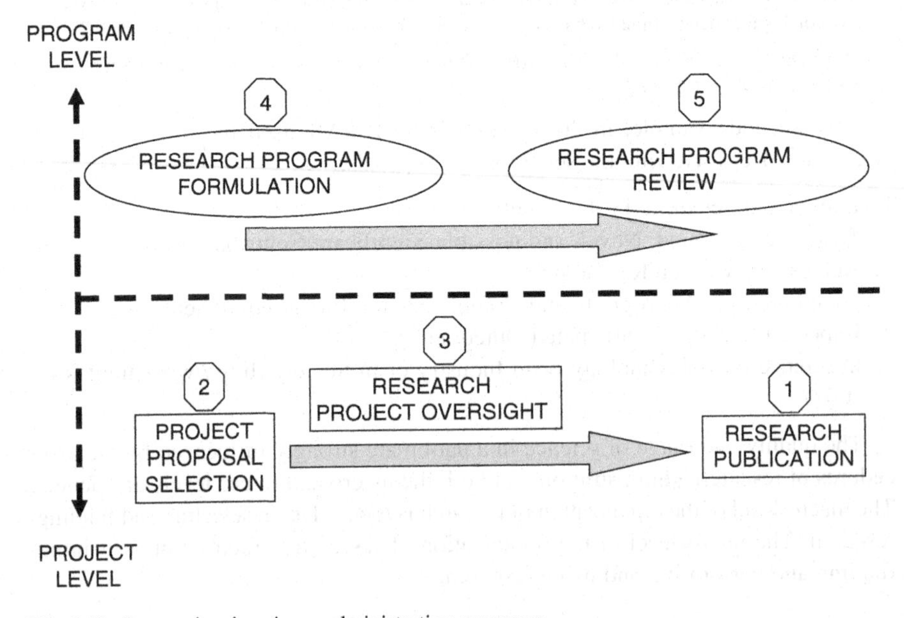


**Fig. 3.10** Peer review in science administration processes

and technical peer reviewers to read the submitted research proposals and asks them to grade each proposal in quality as excellent, very good, fair, or poor. The funding agency then selects only among the proposals rated excellent or very good for funding within their research budget constraints. Proper peer review is an organizational procedure vital to ensure the quality of the research methodology of a proposal.

Once a research project is funded, its research should be performed (3) in a proper research setting (university, government, or industrial research laboratory), with qualified researchers (doctoral researchers and doctoral graduate students) having access to proper research equipment necessary for the methodology. Moreover, a proper administration of the research setting must account that the research funds were spent as intended upon researchers and equipment of the project.

Finally, proper research project reports should be sent to the funding agency on the progress and completion of the project. Moreover, (1) results of the completed projects should be published in the scientific and technical and professional journals (with possible delays for the filing of patents on intellectual property created in the project). Scientific and engineering and professional societies publish journals that archive progress in science and engineering and technology. In their journals, these societies also use peer review to judge whether or not to publish a scientific or technical article as a real contribution to progress in knowledge. Modern research is performed in discrete projects, which are managed both organizationally and methodologically for quality.

For example, research in Stanford University is performed annually at the level $200 million dollars, with research funding from US government agencies (such as NSF, DARPA) and philanthropic organizations (such as the Rockefeller Foundation) and industrial firms (such as IBM or Google). These are the research "facts-of-life" – in nations advanced in science – and for new researchers desiring an academic career. Upon obtaining a first university appointment, a newly-minted assistant professor must realize the importance of two faculty promotion criteria: (1) publishing scientific papers and (2) obtaining research grants. Both have become critical accomplishments for academic tenure and promotion in research universities, worldwide. Also for research careers in government and industry, there are two essential criteria: (1) research publications and (2) research applications. Thus it is essential for any modern career in research to understand how research performance (methodology) connects with research funding and application (organization).

*This is why we are using the term "managing science" – in order to focus on the interconnection between methodology and organization*

## *Illustration: Empirical Evidence for the Theory of Plate Tectonics: Tharp and Heezen*

A nice illustration of a scientific research project, funded by a government research grant to a university, was the project of Tharp and Heezen – who together first empirically established the definitive scientific evidence for the theory of plate tectonics. Plate tectonics theory explains geological activity of the surface of the Earth – as a crust which has been divided into plates which shift with respect to one another. The first direct empirical evidence to substantiate this theory resulted from their ocean-floor mapping project at Columbia University.

Stephen Hall described Tharp: "With her red hair and tailored tweed suits, Marie Tharp reminded a lot of people of Katharine Hepburn, and not just because of her beauty and self-assured style.... Tharp was an English and music major as an undergraduate before obtaining a master's degree in geology at the University of Michigan – a degree possible only because the geology department there opened its doors to women after World War II began. She possessed a keen mind, a garrulous personality and, by the time she arrived in New York in 1948, another degree, in mathematics. All those qualities help explain why she was hired by the geology department at Columbia University. Not as a scientist, mind you, but as a technical assistant. She eventually ended up working with a hulking, quick-tempered graduate student from Iowa named Bruce Heezen. Against considerable odds, Tharp and Heezen rewrote 20th-century geophysics" (Hall 2006, p. 45).

Marie Tharp (1920–2006) was born in Ypsilanti, Michigan. Her father was a map maker for the US Dept of Agriculture, and her mother was an instructor in German and Latin. She attended Ohio University, graduating in 1943 with degrees in English and music. Then she attended the University of Michigan, gaining a master's degree in geology. She worked as a geologist for an oil company and also attended the University of Tulsa, gaining an undergraduate degree in mathematics.

Bruce Charles Heezen (1924–1977) was born in Vinton, Iowa. He attended Iowa State University and received a B.A. degree in 1947. He went to Columbia University for graduate work, attaining a master's degree in 1952 and a doctorate in geology in 1957.

Their project began in 1952 when Heezen thought about doing something with a pile of ocean measurement data. These were charts of sonar measurements of ocean depths, taken by many ships passing through the Atlantic Ocean. This was a large stack of charts with no coordination in the measurements, as different ships took different ocean routes. Heezen gave the stack to Tharp with the instruction: "Here, do something with these" (Hall 2006, p. 45).

Tharp began plotting the ocean depths in a two-dimensional chart; but she quickly began imagining a three-dimensional chart with the ocean floor and its hills and mountains and canyons showing across the Atlantic floor – an ocean-floor landscape. Moreover, since the measurements were taken on different uncoordinated paths, Tharp had to imagine many interpellations between the missing details of the ocean floor features. And as the sea-floor landscape took shape, Tharp saw a surprising feature. There was a mountain ridge that ran north-to-south down the middle of the Atlantic (the Mid-Atlantic Ridge), which researchers had noticed before.

But now in this detail, Tharp saw a new feature: "But as Tharp's careful drafting made clear, there was also a valley that ran down through the middle of the mountain range. It was a hugely important geophysical feature; this rift valley marked a dynamic seam in the crust of the planet, the boundary of huge continent-size plates where new portions of crust rose from the interior of the earth to the surface like a conveyor belt and then, in a geological creep known as drift, moved outward in both directions from the mid-ocean ridge" Hall 2006, p. 45).

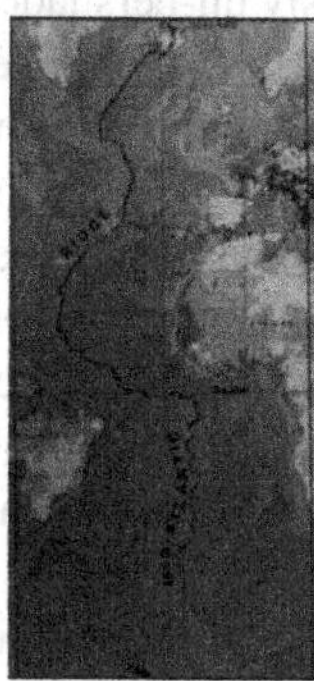

MID-ATLANTIC RIDGE Mid-Atlantic Rift (http://en.wikipedia.org, Mid-Atlantic Ridge 2007)

Tharp was the first scientist to see this feature, which meant that Tharp had discovered a new thing in nature – discovery as a focus of science research. At the time, what was significant about this discovery was that it provided direct empirical evidence for a geological theory that had been around in geology for a long time: "The idea that vast tracts of the earth's crust moved across the surface, known as continental drift, was unpopular at the time. Most geophysicists were – fixists – who believed the planet's surface was static, and Tharp later remarked that a scientist could be fired for being – a drifter – in the 1950s. But she was the first to see the signature of plate tectonics on the surface of the earth…" (Hall 2006, p. 45).

At first Heezen was skeptical of the idea that this could be a rift-valley which empirically confirmed the plate tectonic theory: "Heezen was the first of many scientists who rudely dismissed it. 'Girl talk,' he said. 'It cannot be. It looks too much like continental drift.' (Hall 2006, p. 45) But as Tharp's chart-plotting continued, Heezen became enthusiastic about the evidence: 'Almost on impulse,' Tharp said, she and Heezen decided to make the mapping project even more ambitious. They would create a 'physiographic diagram' of the ocean floor – a kind of map that shows a landscape as it would appear from a low-flying airplane. Their first map, published in 1957, showed the North Atlantic. Over the next 20 years, they would map the underwater landscape of all the world's oceans. Rift valleys were a feature of every ocean floor" (Hall 2006, p. 46).

If a rift valley was a feature of every ocean floor, the pattern became clear. The middle of the Atlantic Ocean's floor was spreading east and west along a north-south fault. The Pacific Ocean floor was being subsided with America's continents drifting west and Asia's continents drifting east. This created a ring-of-fire around the Pacific Ocean – where the continental plates were being pushed over the Pacific Ocean floor – all around the ocean.

The importance of the discovery was that it provided direct empirical evidence of sea-floor spreading, as a consequence of continental drift. In 1912, Alfred Wegener had proposed that the present continents had once been a gigantic single land mass which had broken up.

Tharp's and Heezen's seascape maps were not initially received favorably by all: "Maurice (Doc) Ewing, the brilliant and autocratic director of what is now the Lamont-Doherty Earth Observatory at Columbia, remained famously un-persuaded by the growing evidence of continental drift and began to clash with Heezen over both ideas and ego. Heezen had become a tenured professor, but Ewing did what he could to thwart the mapping project. He refused to share important data about the sea floor with the map makers – data that Heezen's graduate students sometimes surreptitiously exported to Tharp and her assistants. He stripped Heezen of his departmental responsibilities, took away his space, drilled the locks out of his office door and dumped his files in a hallway. Most important, Ewing blocked Heezen's grant requests and, as Fox said, "was essentially trying to ruin Bruce's career" (Hall 2006, p. 46).

Moreover: "Ewing could not fire Heezen. But in the middle of the 1960s, he did fire Tharp. But Tharp continued the project from her own home in South Nyack, New York. The US Navy and the National Geographic Society continued to provide support for Tharp and Heezen to continue the mapping project. In the 1970s, Tharp and Heezen hired Heirich Berann, an alpine landscape painter from Austria to make visual maps of the ocean's floors. The *World Ocean Floor* was published in 1977 and looked as if someone had pulled the plug on a globe-size bathtub – draining all the water from the world's oceans and revealing hidden features of the earth's surface. The map showed a continuous, 40,000-mile-long seam running across the world's surface, like the stitching on a throbbing geophysical baseball" (Hall 2006, p. 46).

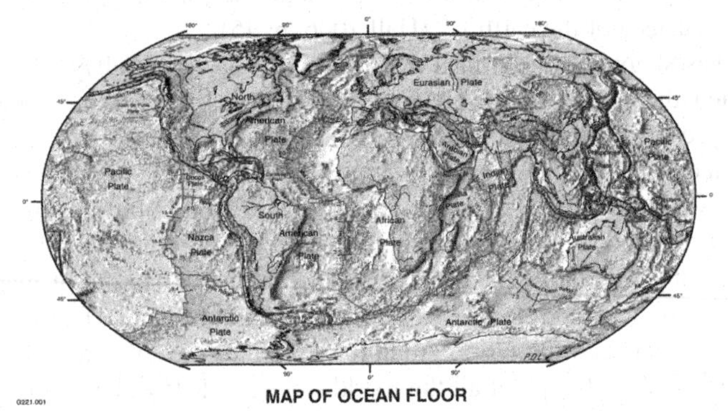


Plate Tectonics Map (http://en.wikipedia.org, Plate Tectonics 2007)

On Tharp's and Heezens' long and productive professional relationship, Hall wrote: "Although she and Heezen fought like cats and dogs over the accuracy of the map, the adversity with Ewing united them as fiercely as a wedding pact. They dined and drank together like husband and wife, and Heezen received his graduate students at her house. Yet according to people who knew them, their unusually intense relationship was platonic" (Hall 2006, p. 46).

In 1977, with the printer's proofs of new maps in his possession and sailing on a research vessel near the coast of Iceland, Heezen suffered a fatal heart attack. Marie Tharp never married and continued to live in her house until her death in 2006. Now the theory of continental drift is an established and universally accepted scientific theory in geological science. The reason Heezen and Tharp succeeded, despite the opposition by a jealous chairperson, was the continuing research funding for the project by the Navy office.

*We can see in this illustration that (1) empirical discovery of patterns in nature (as features, properties, relations) and (2) how theory about nature as empirically-grounded theory (in theory construction and validation) provide the research issues in a science research project and (3) how sustained funding by a research agency is necessary for the initiation and continuation of a research project.*

## Format of Research Proposals

As we have emphasized, scientific research is a philosophical form of inquiry – asking and answering a question about nature, using scientific method. A research proposal expresses this approach. In a proposal, the question asked about nature is the *research issue* of the proposal. The way the question will be answered is the *scientific method* of the proposal. The value of the question will be its *research impact*.

A scientific research proposal for a research grant should clearly pose:

1. A scientific or technical vision of a *research issue*
2. How the research can be performed in *research methodology*, and
3. Potential use of the research results as *research impact*

The format of a research proposal should first address the four criteria of research vision, research methodology, research team, and research impact:

1. *Research Issue – Vision*
   (a) What is the frontier-of-science and/or state-of-art-in-technology in the specific research field?
   (b) What are the issues or challenges or barriers to scientific/technical progress in the field?
   (c) Which of these issues/challenges/barriers (research issues) will the research project address?
2. *Research Methodology*
   (a) What research techniques will be used to address these?
3. *Research Team*
   (a) What kinds of skills will be required in the methodology.
   (b) Which members of the research team will provided these.
4. *Research Equipment*
   (a) What research instruments will be needed to perform the research?
   (b) How will these be accessed or purchased?

5. *Research Impact*
   (a) How would successful results of the proposed project contribute to the advancements of knowledge and/or used on important problems?
   (b) How will such progress be communicated or implemented to and for whom?
6. *Research Budget*
   (a) What funding amount is requested for the project and for how long?

## University Procedures in Research-Proposal Writing and Submission

The procedures in writing and submitting research projects from university researchers use both scientific and technical journals and university science department organization to select and guide future research, as sketched in Fig. 3.11.

The scientific techniques of (1) observation and experimentation and instrumentation, (2) analytical techniques, (3) theory and model construction are the key foci of research projects. They are used to (4) discover and understand nature. These ideas are used procedurally to formulate (5) research issues for a research proposal (9). But the formulation of the research issues (9) also require information input from (7) scientific journals to identify if the issue is novel and significant to future scientific progress. In addition, information input also must come from (6) funding program topics to find a funding program in whose domain the research issue

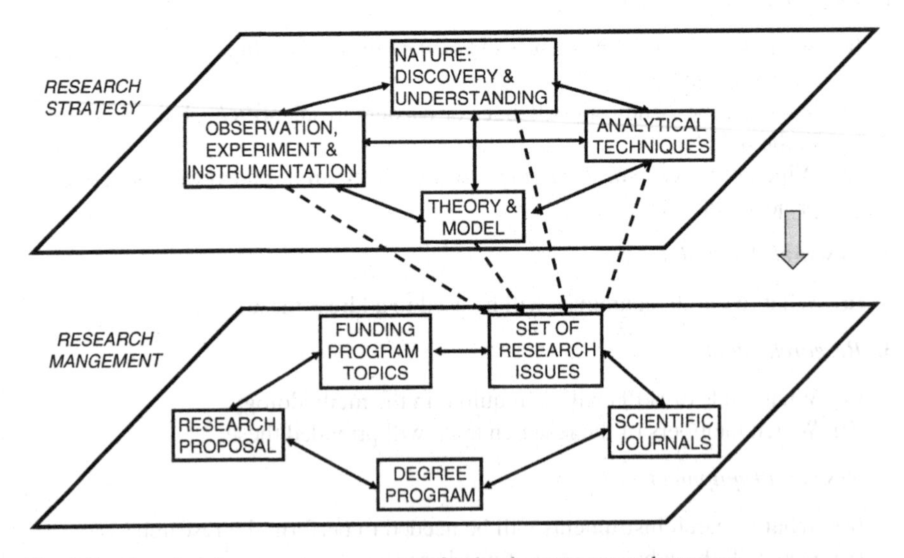


**Fig. 3.11** Procedure in research management for development research strategy

recognizably falls. And information input must also come from (8) a university graduate degree program so that the research issue could provide a thesis topic for a doctoral student who would perform the research tasks on the proposed research project. Thus, the procedures in formulating a research proposals require effective processes for gathering information from (7) scientific journals and about (6) funding-research proposals and about (8) university degree programs.

An essential feature is the literature review in (7) scientific journals of all published references to the frontiers of science (as the latest published research results from recently complete research results) are recorded in the journals and conference proceedings. Thus understanding the frontier of science in a scientific area requires a reading of the most recent scientific literature in the area.

In the disciplines of science, the procedures of journal publication and on-line Internet publication of journals are central to science administration. The journals define their focus upon a specific area of science, a small piece of the totality of science. More than one journal may cover the same area or overlapping areas. Journals are usually published by scientific societies, which are voluntary organizations. If there are more than one society covering an area of a scientific field, then there may be more than one journal in the area. For example, there often are national scientific societies and international scientific societies whose interests overlap scientific areas.

Scientific journals use scientific members of the society to review the quality of a proposed article from the perspective of methodological soundness and novelty. This peer review of publications tries to ensure scientific merit and newness of contribution to knowledge. It is the archival function of a scientific journal is to place on permanent record the results of research projects that have been performed in the scientific field in recent years. Since research normally takes about 3 years to perform, at least a prior year to get funded, and about a year to get published. The research ideas in a current issue of a scientific journal are usually 5 years old. Because of this time delay between scientific conception, achievement and communication, scientific researchers also maintain a more current mode of communication among themselves in the form of annual scientific conferences and informal discussion networks.

Accordingly, the research proposal in defining the research issue of the proposed project should identify all the relevant recent articles in scientific journal that together describe the frontier of science (or state-of-art of technology) in the area of the research and the challenges to advancing that knowledge which the proposed research will address. The most significant and relevant scientific journal articles and recent conference presentations need to be identified in the research proposals.

Once the research issue of the proposal has been described and justified by the review of relevant scientific literature, the next section of the research proposal should describe how the research will be performed, the research methodology. Basic to all scientific methodology are the procedures in science for observing and experimenting on nature, analyzing the results of experiments and constructing theory and testing theory through prediction of the results for new experiments.

Scientific method defines the approach of any research strategy. In a research project, the reality of nature and problems involving nature (technical and social) are queried as a research issue in the project. The method of query can be through:

1. Providing new observations/measurements/experimentation of nature and/or new scientific instrumentation
2. Providing new analytical techniques or use of analytical techniques on new areas
3. Providing new theory and/or new models of objects or use of existing theory/ models on new areas

The set of scientific issues must be recognized as appropriate by a research funding program in a funding agency for the proposal to be seen as valid and timely. Again the use of the scientific journals in defining frontiers of scientific research are fundamental both the scientist proposing the research and to the program officer in the funding agency to which the proposal is submitted. And as we saw in the NSF funding chart, the funding agency program officer will send out the research proposal to scientific peer reviewers to judge the appropriateness of the research issue and the soundness of the scientific methodology to be used in the proposal.

Next the proposal must describe the research team of the proposal, as to their appropriate specialties and qualifications to perform the research. Peer reviewers used by the funding agency will also judge the appropriateness and capability of the proposed research team.

Any appropriate research equipment and instrumentation needed to perform observations, measurements, or experimentation in the project must also be described in the research proposal. And software or computer capability needed to develop or use theory or modeling in the research project must be described. And these too will be judged by the funding agency peer reviewers as to appropriateness and quality to do the proposed research tasks.

In some scientific projects, the research issue is defined not only by the state-of-knowledge (frontier of science or state-of-art of technology) but also by some important problem (industrial technology or government technology or societal problem). In such a case, then the research proposal needs not only to address the research issue but the importance of the use of the research results and how they can be implemented.

Finally, the research proposal needs to specify a duration and research budget for performing the research. Normal research proposals for scientific research at universities specify a duration of 3 years. This period corresponds with the time normally required to do a doctorate dissertation. In universities, scientific research is performed by graduate students working under a professor as a thesis supervisor. Scientific workers as doctoral graduate students need financial support for living and tuition expenses. The research grant will provide the funds needed to support the graduate students working on the project. The number of graduate students on a research project should be determined by the number of distinct tasks in the research, each of which could lead to a doctoral thesis. The product of the (research assistant salary plus tuition) times the number of students equals the labor cost of the research. In addition to the "research labor" cost on the project, there is an overhead charge by the university for providing space and other support functions to the project.

*A research proposal summarizes a possible contribution to science as a discrete future research project.*

Before submitting any research proposal to a funding program, one needs to first understand what the real research focus and boundaries are in the current operation of a program (current program officer's perception of research scope for the program). As program officers are changed to administer a research program, the program's priorities in research support also change. Some ways to understand a program's research priorities include: (1) looking up a program's recent list of research grants, (2) visiting the current program officer, (3) learning about a program officer's resume and publications.

*Be creative in formulating a good research proposal but also know thy program officer to anticipate its relevance to program priorities.*

## Summary

1. Government research programs can be organized around research issues: (a) for supporting the infrastructure or science, (b) for addressing specific societal problems, (c) research for the invention of technologies to address societal problems.
2. Research visions for desirable new research goals may be expressed in terms of (a) a research area and (b) desirable achievements from future research in the area as research goals.
3. Research project proposals summarize research to be performed in a finite period of time by a particular research team.
4. The format of a research proposal includes discussions of (1) research issue, (2) research methodology, (3) research team, (4) research equipment, and (5) research impact.

# Chapter 4
# Research Techniques

## Introduction

We saw how scientific method was developed in the 1600s, culminating in Newton's grand synthesis of classical mechanics. And we saw that scientific method is a specific set of research techniques – empirical and theoretical – aimed at experimentally observing nature and constructing and validating theory based upon such experimental data. And this is called "empirically grounded theory." We next look at the research techniques which provide the intellectual tools for implementing the research method of science.

## Research Techniques

Figure 4.1 lists the different methodological techniques in the empiricism and theoretical aspects of scientific research.

For example in the previous case of Copernicus to Newton, we can identify the kinds of research techniques that were used, as shown in Fig. 4.2.

1. Copernicus provided a scientific model that could be verified by observation
2. Brahe developed instruments and made more precise measurements to verify the model
3. Kepler made a theoretical analysis of experimental data, developing a phenomenological law about planetary motion
4. Galileo performed physical experiments and formulated scientific laws generalized from the experiments
5. Descartes integrated geometry and algebra and Newton created differential calculus to provide new mathematics for modeling physical events
6. Newton formulated a phenomenological law of gravitation as a force varying inversely with the square of the distance
7. Newton theoretically derived Copernicus's solar model as a consequence of his newly formulated mechanics

F. Betz, *Managing Science*, Innovation, Technology, and Knowledge Management 9,
DOI 10.1007/978-1-4419-7488-4_4,  

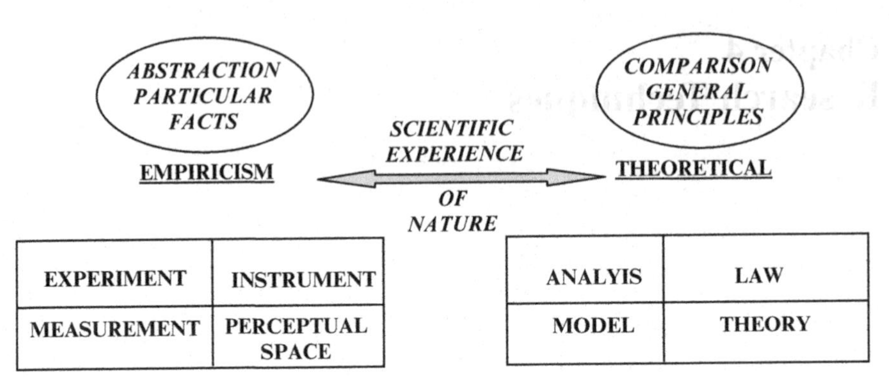


**Fig. 4.1** Empirically grounded theory – methodological techniques in the empiricism and theoretical aspects of scientific research. *Experiment* is the controlled observation of nature – experiencing nature through the human senses aided by scientific instruments. *Instrument* is a device which provides a means of extending the range and sensitivity of human sensing of nature. *Measurement* is an instrumental technique in observing nature that results in quantitative data about the phenomenal thing. *Perceptual Space* is conceptual framework within which a natural phenomenon is described. *Analysis* is inferring a mathematical pattern in the quantitative data of a set of experiments. *Phenomenological Law* is a generalization of relationships observed between natural objects. *Model* is a symbolic simulation of the processes of a phenomenal object. *Theory* is a symbolic description and explanation of a phenomenal field of objects

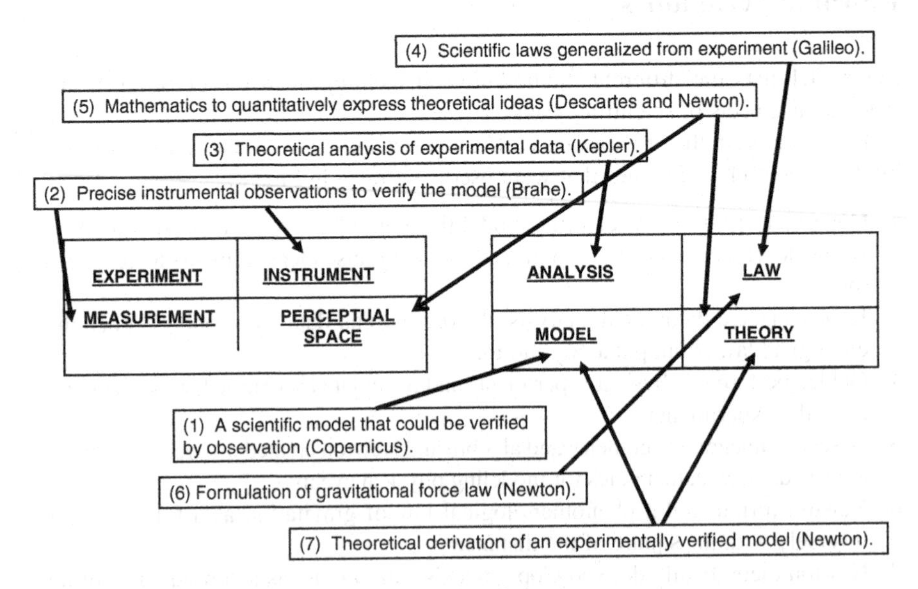


**Fig. 4.2** Empirically grounded theory – research techniques

A taxonomy is a set of categories based upon distinctions that define the categories. This taxonomy classifies the kinds of research techniques used in a history of scientific progress. Moreover, the taxonomy provides a guide to the techniques the methodology of science comprises.

> The method of science consists of the interaction of empirical and theoretical research techniques to discover natural phenomena and to construct empirically-grounded theory of nature.

## *Illustration: Discovery and Modeling of DNA*

To see the different research techniques in a historical case, we will review the scientific history of the discovery and modeling of DNA.[1] This occurred over a century of research, from the 1850s to the 1950s. From this sequence of research, great scientific progress in biology occurred – which resulted in establishing the new specialty of molecular biology. And this occurred as a scientific issue focused upon a basic research issue: How is life reproduced?

The first methodological issue in any research is to formulate a question about nature. Copernicus's question about nature was "does the sun revolve around the earth or the earth around the sun?" Questions about basic nature – questions of science – consist of *inquiries* into nature. What things exist in nature? How do such things interact? In the origin of molecular biology, the "phenomenal thing" studied in nature is "life." A generational interaction between the phenomenal things-of-life (organisms) is reproduction, or "heredity." The scientific answer to the question of heredity required many events of scientific research, which were performed over a long period of time:

1. Investigating the structure of the cell
2. Isolation and chemical analysis of the cell's nucleus, DNA
3. Establishing the principles of heredity
4. Discovering the function of DNA in reproduction
5. Discovering the molecular structure of DNA

### The Structure of the Cell

For the early history of DNA, Portugal and Cohen began the story in the 1800s with investigations of the structure of biological cells (Portugal and Cohen, 1977). Scientists used a new invention, the microscope, to look at bacteria and cells. Cells were seen to be the constituent modules of living beings. The observation of the cell by the microscope was a basic discovery about an underlying form of living nature which occurred in the late 1700s. The technique in scientific method of using the microscopic (technology) for observation of nature discovered a content of nature, the cellular structure of life.

> The new observational instrument (microscope) enabled a new discovery about nature. This activity was one in the structural category of 'scientific method'.

The microscope itself arose from optical technology in the eighteenth century and was used as an instrument in science. All instruments in science are technologies. In the taxonomy of all the activities of science, the importance of the category of technology (applications of science) is, to science itself, the use of technology in scientific instrumentation. The technology of the microscope was invented in the 1600s, and Hans Lippershey, Hans Jenssen, and Zacharias Jenssen (Dutch eyeglass makers) are given historical credit for the optical lens development for the first telescopes which then later became inverted into microscopes.

In 1838, Christian Ehrenberg was the first to observe the division of the nucleus when a cell reproduced (Portugal and Cohen 1977). Ehrenberg (1795–1876) was born in Germany and completed his doctoral work at the University of Leipzig in 1818. He collected specimens of plants and animals on an expedition to the Middle East in 1820–1825. In 1827, he was appointed professor of medicine at Berlin University. In 1829, he accompanied Wilhelm von Humboldt on an expedition to eastern Russia at the Chinese frontier. On his return, he concentrated on studying microscopic organisms.

Christian Ehrenberg (http://www.widipedia, Christian Ehrenberg 2007)

In 1842, Karl von Nageli observed the rod-like chromosomes within the nucleus of plant cells. Nageli (1817–1891) was born in Switzerland. He studied at the University of Zurich and then in Jena, focusing on the microscopic study of plants. He became a professor at the University of Zurich and later at the Universities of Freiburg and Munich.

Thus by the middle of the nineteenth century, biologists had seen that the key to biological reproduction of life involved chromosomes which divided into pairs when the nucleus of the cell split in cell reproduction. The first significant scientific event toward the discovery of DNA had been the description of the structure of a biological cell, using the microscope to look at bacteria and cells. Then "cells" were thought to be theoretically the constituent modules of all living beings.

We can now use the case of the scientific discovery and modeling of DNA to understand some of the terms we use in "research techniques." *Experimental observations* in biology were extended by the *Instrument* (microscope) to a smaller scale of *Perceptual Space*. This provided observations of units of life that were abstracted in the idea of a natural biological thing called a "cell" (empiricism) and generalized for all organisms into a new underlying form (*Theory*). All this was called biological

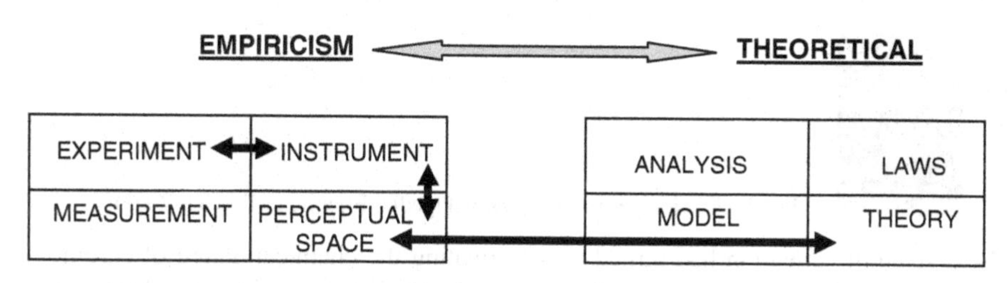


**Fig. 4.3** Research techniques: perceptual space to theory

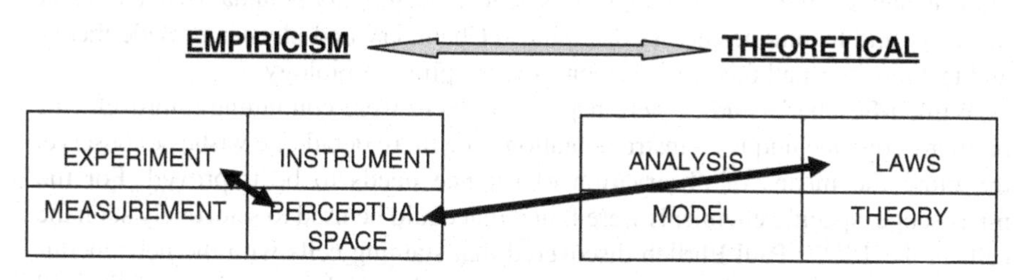


**Fig. 4.4** Research techniques: perceptual space to laws

"cellular theory." Also, we noted that in 1838, Christian Ehrenberg was the first to observe the division of the nucleus when a cell reproduced, and in 1842, Karl von Nageli observed the rod-like chromosomes (*Thing*) within the nucleus of plant cells (Fig. 4.3).

Biologists had seen that the key to the cellular *Perceptual Space* in the biological reproduction of life involved chromosomes, which divide into pairs when the nucleus of the cell splits in cell reproduction. Thus, further biological *Experiments* were discovering the process (*Laws*) of cellular reproduction (Fig. 4.4).

## Discovery and Chemical Analysis of DNA

We next describe the scientific progress in the chemical analysis of DNA. In 1869, a chemist, Friedrich Miescher, reported the discovery of DNA, by precipitating material from the nuclear fraction of cells. He isolated the nucleus of a cell and determined that it chemically contained phosphorus, nitrogen, and sulfur. He called the material "nuclein." Subsequent studies showed that it was composed of two components, nucleic acid and protein (and later nucleic acid would be called DNA) (Portugal and Cohen 1977). Miescher (1844–1895) was born in Switzerland and studied medicine at the University of Basel, obtaining his medical degree in 1968. He turned to physiological chemistry and was appointed as a professor at the University of Basel. He and his students studied the chemistry of the nucleic acid but still did not understand its specific cellular function.

Friedrich Miescher (http://www.wikipedia, Friedrich Miescher 2007)

Next scientific attention had turned to investigating the chemical nature of chromosomes. Science is organized into disciplines, the techniques and knowledge in one discipline may be used in another discipline. This is one way that the organization of science interacts with the philosophy of science: disciplinary boundaries and transfer of theory and techniques across disciplines. Chemistry and physics provide theory and techniques (methods) to the scientific discipline of biology.

While Miescher's studies were happening, there were continuing improvements in microscope techniques – instrumentation. The more detail one wishes to observe, the more the means of observation of science needs to be improved. For the microscope, specific chemicals were found that could be used to selectively stain the cell. In the 1860s, Paul Ehrlich discovered that staining cells with the new chemically derived, coal-tar colors matched with the chemical composition of the cell components. This is an example of how technology contributes to science, for the new colors were products of the new chemical industry. Underlying progress in technology is progress in science. But also progress in technology can provide instrumental bases for progress in science. This is the interaction of *scientific method and scientific application.* Ehrlich (1854–1915) was born in Prussia and obtained a medical degree at the University of Leipzig. He joined the Institute for Infectious Diseases in Berlin in 1886. 1896, he became director of the Institute of Serum Research and Examination in Berlin. In 1908, he received the Nobel Prize for Medicine for developing new drugs, as he developed drugs to treat sleeping sickness and syphilis.

Paul Ehrlich (http://www.wikipedia, Paul Ehrlick 2007)

From Ehrlich's staining of cells, new techniques for *Instrument* use were improving the *Experimental observations* in the microscope level of the *Perceptual Space* to better reveal the structure (*Model*) of the cellular object (*Thing*) (Fig. 4.5).

Then in 1873, A. Schneider described the relationships between the chromosomes and various stages of cell division. He noted two states in the process of mitosis. Mitosis is the phenomenon of chromosome division resulting in the separation of the cell nucleus into two daughter nuclei. So, the cell division process of "mitosis" continued to be described from experiments as a process (a natural process).

In 1879, Walther Flemming also studied cell division and introduced the term "chromatin" for the colored material found within the nucleus after staining.

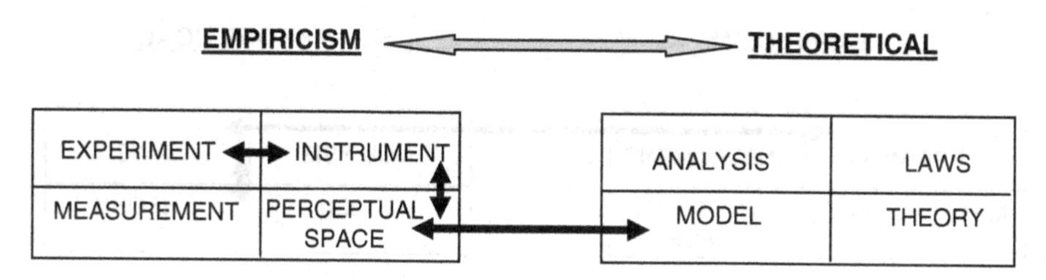


**Fig. 4.5** Research techniques: perceptual space to models

He suggested that chromatin was identical with Miescher's nuclein. Flemming (1843–1005) was born in Germany and graduated in medicine at the University of Rostock in 1868. In 1876, he became a professor of anatomy at the University of Kiel. He used aniline dyes to identify the chromatin and investigate the process of cell division and the distribution of the chromosomes to the daughter nuclei, calling the process as "mitosis."

Walther Flemming (http://www.wikipedia, Walther Flemming 2007)

At this time, studies of nuclear division and the behavior of chromosomes were emphasizing the importance of the nucleus. But it was not yet understood how these processes were related to fertilization. In 1875, Oscar Hertwig demonstrated that fertilization was not the fusion of two cells but the fusion of two nuclei – *scientific method interacting with scientific content – epistemology with ontology*. Oscar Hertwig (1849–1922) was born in Germany and studied at the University of Jena. In 1988, he became a professor of anatomy at the University of Berlin. He discovered fertilization in sea urchins and recognized the role of the cell nucleus in inheritance. He found that fertilization included the penetration of spermatozoon into an egg cell.

Oscar Hertwig (http://www.wikipedia, Oscar Hertwig 2007)

Thus, Schneider had described the relationships between the chromosomes and various stages of cell division. Then, Hertwig demonstrated that fertilization was not the fusion of two cells but the fusion of two nuclei. Thus, *Experiments* were

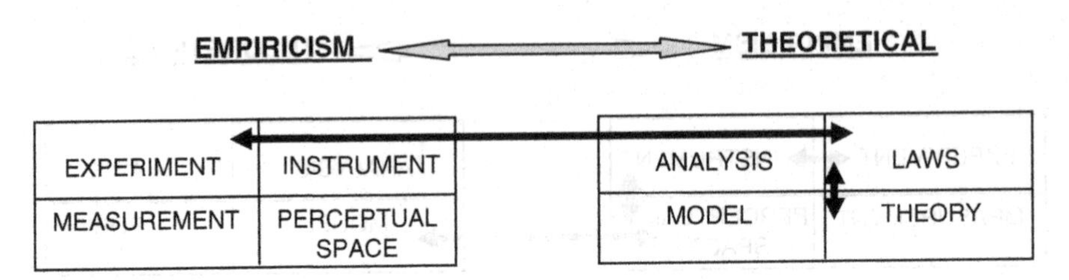


**Fig. 4.6** Research techniques: experiment to laws and theory

showing another "Law" in the theoretical topic of cellular theory (*Representation*) of cellular reproduction as involving the fusion of two nuclei (Fig. 4.6).

Meanwhile, the study of nucleic acid components was progressing. In 1879, Albrecht Kossel began publishing on nuclein. Over the next decades, he (along with Miescher) was foremost in the field of nuclein research. Kossel (1853–1927) was born in Rostock as son of the then Prussian consul. He attended the University of Strasbourg to study medicine and graduated later from the University of Rostock in 1878. Kossel worked in physiological chemistry, on the chemistry of tissues and cells. He began studying the cell nucleus in 1880. He received a Nobel Prize in Medicine in 1910 for his research in cell biology.

Albrecht Kossel (http://www.wikipedia, Albrecht Kossel 2007)

Together, the research studies of Miescher and Kossel, along with Pheobus Levine, finally laid the clear basis for the determination of the chemistry of nucleic acids (Portugal and Cohen 1977). Pheobus Levine (1869–1940) was born in Russia and studied medicine at the Imperial Military Medical Academy, receiving a medical doctorate in 1891. In 1893, he moved with his family to New York in the United States. In 1905, he was appointed head of the biochemical laboratory at the Rockefeller Institute of Medical Research. He identified components of DNA. In 1909, he had discovered ribose and dexoyribose in 1929.

Back as early as 1914, Herman Emil Fisher had attempted the chemical synthesis of a nucleotide (component of nucleic acid, DNA). Emil Fisher (1852–1919) was born in Germany and earned his doctorate in chemistry at the University of Strasbourg in 1874. In 1881, he was appointed Professor of Chemistry at the University of Erlangen, later moving to the University of Wurzburg and the University of Berlin. He received the Nobel Prize in Chemistry in 1902. He studied purines and sugars and proteins. He separated and identified individual amino acids, the constituents of proteins.

Herman Emil Fisher (http://www.wikipedia.org, Herman Emil Fisher 2007)

But real progress was not made in synthesis of nucleotide until 1938 – two decades later. Chemical synthesis of DNA was an important scientific technique necessary to understand the chemical composition of DNA. The length of time it took to establish the chemical basis of nucleic acids was long, from 1869 to 1938.

In the organization of science (science administration), the timelines for scientific progress are measured in decades, not months or years.

One of the problems was that DNA and RNA were not distinguished as different molecules until 1938. This is an example of the kinds of problems that scientists often encounter.

Nature is often more complicated than originally observed – scientific content (ontology) interacting with scientific method (epistemology).

By the end of the 1930s, the true molecular size of DNA had been determined. In 1949, C. E. Carter and W. Cohn found a chemical basis for the differences between RNA and DNA. And finally by 1950, DNA was known to be a high-molecular-weight polymer with phosphate groups, linking deoxyribonucleosides between 3 and 5 positions of sugar groups. The sequence of bases in DNA was then still unknown. Thus, by 1950, the detailed chemical composition of DNA was finally determined but not yet its molecular geometry. Almost 100 years had passed between the discovery of DNA and determination of its chemical composition. Many *Experiments* had to be done to finally create the chemical *Analysis* of DNA (Fig. 4.7).

## The Principles of Heredity

From 1900 to 1930, while the chemistry of DNA was being sought, the foundation of modern genetics was also being established. Understanding the nature of heredity began in the nineteenth century with Darwin's epic work on evolution and with

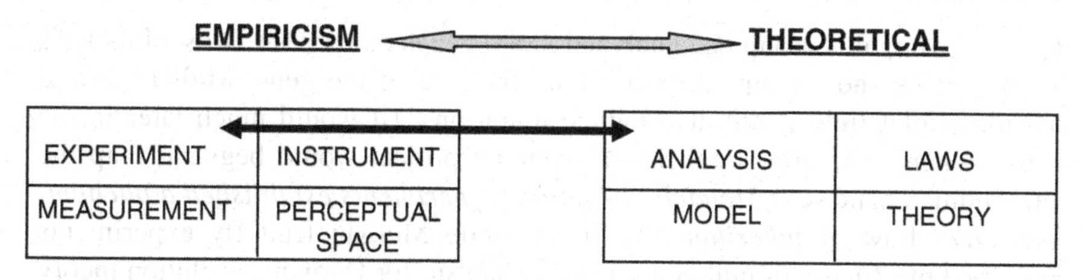


**Fig. 4.7** Research techniques: experiment to analysis

Mendel's pioneering work on genetics. This established a new specialty area in biology – the ontological understanding of heredity. Gregor Mendel (1822–1884) was born in Austrian Silesia (now in the Czech Republic). He studied at the Philosophical Institute in Olomouc, and in 1843 entered the Augustinian Abbey of St. Thomas in Brno. He was christened Johann but took the name Gregor on entering monastic life. He conducted his genetic studies in the monastery's garden. He cultivated pea plants and showed that one in four plants showed regressive genetic characteristics. He read a paper on his studies at meetings of the Natural History Society of Brunn in Moravia in 1865. He became an abbot in 1868. His pioneering studies have earned him the honor of being called the "father of modern genetics."

Gregor Mendel (http://www.wikipedia, Gregor Mendel 2007)

Modern advances in genetic research began in 1910 with Thomas Morgan's group researching the heredity in the fruit fly, Drosophila meanogaster. Morgan demonstrated the validity of Mendel's analysis and showed that mutations could be induced by x-rays, providing one means for Darwin's evolutionary mechanisms. Morgan introduced physical instrumentation methods into the study of heredity – epistemology interacting with ontology. Thomas Hunt Morgan (1866–1945) was born in the United States and received a doctorate from Johns Hopkins University in 1891. He taught at Bryn Mawr. In 1900, the work of Mendel was rediscovered by scientists. In 1904, Morgan moved to Columbia University, where he began genetic researches on the fruit fly. He demonstrated that the genes are carried on chromosomes, which are the mechanical basis for heredity. In 1928, Morgan moved to The California Institute of Technology to establish a then new division of biology. He was awarded the Nobel Prize in Physiology or Medicine in 1933.

Thomas Hunt Morgan (http://en.wikpedia.org, Thomas Hunt Morgan 2007)

By 1922, Morgan's group had analyzed 2,000 genes on the four Drosophila fly's chromosomes and attempted to calculate the size of the gene. Muller showed that ultraviolet light could also induce mutations. (It would much later in the 1980s, when an international human genome project would begin to map the entire human gene set.) *Mendel's pea plant experiments established a phenomenological* Law *of inheritance of traits* while Morgan fruit fly experiments developed the *Theory* of mutations as a mechanism for Darwin's evolution theory (Fig. 4.8).

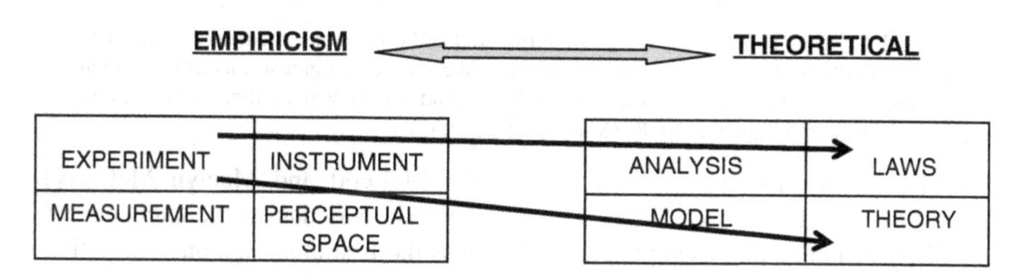


**Fig. 4.8** Research techniques: experiment to laws & theory

## Function of DNA in Reproduction

While the geneticists were showing the principles of heredity, the mechanism of heredity had still not been demonstrated. Was DNA the transmitter of heredity, and if so, how? Other scientists had been studying the mechanism of the gene, with early work on bacterial reproduction coming from scientists using bacterial cultures.

R. Kraus, J.A. Arkwright, and O. Avery with A.R. Dochez had demonstrated the secretion of toxins by bacterial cultures (Portugal and Cohen, 1977). This had raised the question of what chemical components in the bacterium were required to produce immunological specificity. The search for the answer to this question then revealed a relationship between bacterial infection and the biological activity of DNA.

Also as early as 1892, there had occurred the identification of viruses and their role in disease. In 1911, Peyton Rous discovered that rat tumor extracts contained virus particles capable of transmitting cancer between chickens. In 1933, Max Schlesinger isolated a bacteriophage (virus) that infected a specific kind of bacteria. Next, scientists learned that viruses consisted mainly of protein and DNA. In 1935, W. M. Stanley crystallized the tobacco mosaic virus, which encouraged scientists to study further the physical and chemical properties of viruses.

In 1928, Frederick Griffith showed that a mixture of killed infectious bacterial strains with live noninfectious bacteria could create a live infectious strain. He called this a transformation but did not yet understand it was a transfer of genetic material as DNA. In his experiment, a laboratory mouse was injected with a non-virulent strain of bacteria and the mouse lived. A second mouse was injected with a virulent strain of bacteria that killed the mouse. Then, a heat-killed batch of the virulent strain (bad-but-dead bacteria) was injected into a mouse and that mouse lived. So that only live, virulent strains of bacteria would kill a mouse. Next a heat-killed virulent strain of bacteria (bad-but-dead) was mixed with a live non-virulent (good-and-alive) bacteria and the mix was injected into a mouse. The mouse died. The only way the mouse could have died was if there was a transfer of genetic material from the bad-but-dead to the good-and-alive bacteria, so that some of the good-and-alive bacteria became the bad-and-alive bacteria. This experiment proved that the virulence characteristic in the bacteria was controlled and transmitted by genes in the bacteria.

Frederick Griffith (1879–1941) was born in England and trained as a medical doctor. After the First World War, he tried unsuccessfully to make a vaccine against pneumonia which was infectiously prevalent after the war. In the Second World War he died in air raid on London. (http:en.wikipedia.org, Frederick Griffith 2007)

In 1944, Oswald Avery (along with Colin McLeod and Maclyn McCarty) showed that the Griffith's transformation in bacteria was due to the exchange of DNA between dead and living bacteria. This was the first clear demonstration that DNA did, in fact, carry the genetic information.

This is an example of scientific method, whereby a relationship is established between a mechanism (DNA) and a biological process (infection).

Oscar Avery (1877–1955) was born in Novia Scotia, Canada, and his family moved to the United States. In 1904, he received a medical doctoral degree from the College of Physicians and Surgeons in New York. He joined the faculty of the Rockefeller Institute in 1923. (http://en.wikipedia.org, Oscar Avery 2007)

Also, in parallel research by 1940, work by George Beadle and Edward Tatum further investigated the mechanisms of gene action and showed that the genes control cellular mechanisms – and through control of the cellular production of substances. It is the genes that control the production of enzymes. These enzymes are needed for chemical synthesis in the cell. Thus, the genes control the production of enzymes which in turn control the cellular production of proteins.

The scientific stage was now set to understand the structure of DNA and how DNA's structure could transmit heredity. Beadle and Tatum (along with Joshua Lederberg) would share the Nobel Prize in Physiology or Medicine in 1958.

George Beadle (1903–1989) was born in the United States in Nebraska. He received a PhD from Cornell University in 1931. Beadle worked as a post-doc for Thomas Hunt Morgan at Cal Tech. He was a professor at Harvard University and then Stanford University, where he worked with Tatum. (http://en.wikipedia.org, George Beadle 2007)

Edward Tatum (1909–1975) was born in Colorado and attended the University of Chicago and received a PhD in biochemistry from the University of Wisconsin in 1934. In 1937 he worked at Stanford University with Beadle. Then he moved to Yale University in 1948 but in the same year returned to Stanford. Then in 1957, he moved to Rockefeller Institute. (http://en.wikipedia.org, Edward Tatum 2007)

Joshua Lederberg was born in 1925 in New Jersey, U. S. He received a PhD from Yale University in 1947, working under Edward Tatum, when they showed that the bacterium E coli entered a sexual phase during which it could share DNA through bacterial conjugation. (http://en.wikipedia.org, Joshua Lederberg 2007)

We see that while the geneticists were examining the function (paradigm) of heredity, other biologists were examining the mechanism (paradigm) of heredity, showing that mechanism of DNA was the transmitter of the function of heredity. Along the timeline of scientific progress, again the scientific method of experiment was connecting theory across the two scientific paradigms in biology of mechanism and function. Griffith's experiment demonstrated that DNA passed the infectious function from a dead bacterium to a different live one. Beadle, Tatum, and Lederberg demonstrated that genes control cellular processes and genes could be

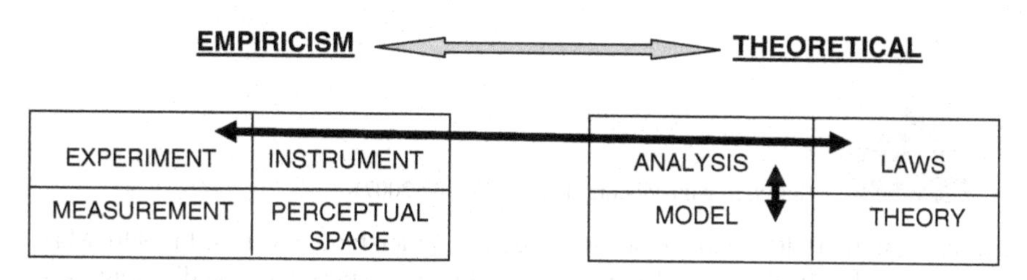


**Fig. 4.9** Research techniques: experiment to laws, analysis & model

sexually exchanged between bacteria. In this way, a mechanism was connected to a function in biology.

Biological explanation requires both a mechanism and a function (connected together) connected paradigms of mechanism and function. Thus the two kinds of *Experiments*, one on genetics and another on DNA transmission of gene information, together established that the process (*Law*) of heredity transmission through the *Model* of DNA exchange (Fig. 4.9).

## Structure of DNA

We have reviewed the long and many lines of research necessary to discover the elements of heredity (genes and DNA) and its function (transmission of heredity). Yet before scientific application of this knowledge (technology) could use this kind of information, one more scientific step was necessary – understanding the mechanism.

This step was achieved by a group of scientists that were to be later called the "phage group." From the scientists in this group, the discovery of the molecular form of DNA would then create the modern scientific specialty of molecular biology. (Judson 1979)

Historically, bacteriophages (viruses) had been studied in biology since 1917 when they were discovered by Felix d'Herelle. Felix d'Herelle isolated phages (viruses) by (1) infecting a nutritional medium with bacteria, (2) infecting the bacteria with phages, and (3) filtering the medium through a porcelain filter in which only the small phages pass through.

Felix d'Herelle (1873–1949) was born in Montreal, Canada. At age six, after the death of his father, his mother moved with him to Paris. At seventeen years of age, he began traveling through Europe and met and married Marie Caire. Then he moved his new family to Canada. He worked for the Canadian government studying fermentation and distillation of maple syrup to schnapps. He next worked as a bacteriologist for Guatemala and then Mexico and then returned to France. In France, he worked in the laboratory of the Pasteur Institute. There he extracted bacteria harmful to locusts that were plaguing the Mexican production of sisal. During World War I, d'Herelle remained in France and produced medication for the allied military. In 1917, d'Herelle announced the discovery of 'an invisible antagonistic microbe of the dysentery vacillus' – a phage (virus).

(http://en.wikpedia.org, Felix d'Herelle 2007)

It was not until 1940 when viruses were used to understand genetics. In 1940, Max Delbruck and Salvador Luria began collaborating on experiments on different bacteria and bacteriophages and showed that a given bacteria could only be infected by a specific strain of phage.

Max Delbruck (1906–1981) was born in Berlin, Germany. He received a PhD in physics in 1930 from the University of Gottingen. He became interested in biology and moved to the United States in 1937, taking work as an assistant in the Biology Division at Cal Tech under Morgan. In 1939, he coauthored a paper in biology on viruses (or bacteriophage or "phage"). He moved to Vanderbilt University to teach physics but continued genetic research with Salavdor Luria at Indiana University. They showed bacterial resistance to virus infection is caused by random mutation. For this work, Delbruck and Luria would win a Nobel Prize in Physiology of Medicine in 1969 (sharing it with Alfred Hershey).

Max Delbruck (http://en.wikpedia.org, Max Delbruck 2007)

Salvador Luria (1912–1991) was born in Turin, Italy. He received a medical degree from the University of Turin in 1936. He took classes on radiology at the University of Rome and read Delbruck's paper. . In 1938, he received a fellowship to study in the United States and intended to work with Delbruck. But Mussolini's Fascist regime banned Jews from academic research fellowships. So Luria went to Paris in 1938. Then when the Nazi German army invaded France in 1930, Luria fled to Marseilles and immigrated to the United States. In 1940 in New York, he visited Enrico Fermi at Columbia University. Luria knew Fermi (who had emigrated to the United States two years earlier) from his time of study at the University of Rome. Fermi helped him obtain a fellowship from the Rockefeller Foundation for work at Columbia.

At Columbia University, Luria met Delbuck and Hershey, and they began collaboration on experiments at Cold Spring Harbor Laboratory on Long Island, New York. This collaboration continued even after Delbruck returned to Vanderbilt at the end of the summer. (Eventually, this collaboration would win them all a Nobel Prize.) From 1943 to 1950, Luria taught at the University of Illinois, and his first graduate student would be James Watson. Later in 1950, Luria moved to Indiana University, where he discovered that specific enzymes could cut DNA in E. coli genes at specific places in the DNA – restriction enzymes. In 1959, he moved to Massachusetts Institute of Technology.

Next, the story of the discovery of the molecular structure moves on (and dramatically so) to Luria's first student, James Watson.[2] Watson had been educated in the ontological and epistemological approach of the phage group. As a graduate student, Watson knew that the great goal (ontological issue) was to discover the structure of DNA. He received a post doctoral fellowship to do research in Europe – looking for a way to pursue his research goal. First, he went to Copenhagen. That summer, he attended a scientific meeting in Italy where Maruice Wilkins talked about his x-ray diffraction studies on the DNA molecule.

James Watson was born in 1928 in Chicago, Illinois, USA. He attended the University of Chicago, graduating in zoology in 1947. In 1946, he had read Ewin Schrödinger's book, *What is Life?,* reorienting him from ornithology to genetics. He went to the University of Illinois for graduate work, having read of Luria and Delbruck's research on viral mutations (Phage Group). In 1948, Watson began research work under Luria and met Delbruck during Watson's first trip to Cold Spring Harbor Laboratory that summer. Watson's research projects was to use x-rays to inactivate bacterial viruses (phages), and he obtained his PhD in Zoology from Indian University in 1950. He graduated with a PhD under Luria in 1951.

James D. Watson (http://en.wikpedia.org, James D. Watson 2007)

Meanwhile as Watson was in Europe, Linus Pauling in 1951 published a molecular model of a protein in the geometric form of a helix. Pauling was using x-ray crystallography and molecular model building.

Linus Pauling (1901–1994) was born in Portland, Oregon, USA. In 1917, Pauling attended Oregon Agricultural College (now Oregon State University) and graduated in chemical engineering in 1922. He then went to California Institute of Technology, on obtaining a PhD in physical chemistry and mathematical physics. Then he traveled to Europe on a Guggenheim Fellowship to study under the German scientists who had created quantum mechanics: Arnold Sommerfeld at the University of Munich, Niels Bohr in Copenhagen, and Erwin Schrodinger in Zurich. Pauling became acquainted with the new quantum mechanical theory, which was the key to atomic binding into molecules – the physical basis of chemistry.

Linus Pauling (http://en.wikpedia.org, Linus Pauling 2007)

In 1927, Pauling became an assistant professor in theoretical chemistry at Cal Tech. In the 1930s, Pauling began publishing work on the nature of the chemical bond and wrote a pioneering text on the subject. (Later in 1954, he would receive a Nobel Prize in Chemistry for this work.) Also in the 1930s, Pauling began working on crystalline structures of biological molecules, using x-ray pictures of protein molecules made by William Astbury in England. In 1948, Pauling modeled hemoglobin in which atoms are arranged in a helical pattern. In 1951, Pauling modeled amino acids and peptides, proposing both an alpha helix model and beta sheet (flat) model to structure these. This was the "helix" model which was to stimulate Watson.

*Watson saw himself in a race against Pauling to be the first to model DNA!* From Pauling's paper, Watson saw a research strategy for how to model DNA. First, he would need to obtain an x-ray crystallographic picture of the structure of DNA to infer its geometry. Then, he would have to build a chemical model of DNA in that geometric structure. And possibly, DNA might have a helical geometry, as did Pauling's peptide molecule. Since he had heard Wilkins report on the x-ray diffraction studies on DNA, Watson decided he had to go England. Watson asked Luria to arrange for his fellowship to allow him to work at Cambridge.

Scientists, as well as managers, perceive competition in their environments. Watson looked at research from a research management perspective – philosophical and organizational. He needed to beat scientific competitors to the goal and needed a proper research facility and team to win the competition.

Watson had two kinds of intellectual inputs (research techniques) from Pauling's work to put into his research strategy. First Watson had x-ray pictures of the crystalline structure of DNA molecule. Second Watson had a molecular modeling approach from Pauling, using quantum mechanics and showing a helix form. To use these inputs, he needed a both a collaboration with a physicist for the quantum mechanics and a collaboration with an x-ray experimentalist for x-ray pictures of DNA. These were to be provided by Francis Crick and Maurice Wilkins. Crick was in Cambridge, and Wilkins was in London.

Once at the Cavendish Lab in Cambridge University, Watson found a collaborator in Francis Crick. Crick was then still a graduate student working on a physics degree. He was older than Watson because his graduate studies had been interrupted by service in the Second World War. In 1951, Crick was assisting in the development of the mathematical theory of how a helical model would produce x-ray diffraction pictures. Late in 1951, Crick met Watson and they began working together on a molecular model of DNA – Watson's goal.

Crick (1916–2004) was born in Northhampton, England. In 1937, he obtained a B Sc degree in physics from the University College London. Crick began graduate work, measuring the viscosity of water at high temperatures, but the work was interrupted when a bomb fell upon the laboratory after World War II had begun. During the war, Crick worked for the British Admiralty Research Laboratory, designing magnetic and acoustic mines. In 1947, Crick turned to biology, working first in Cambridge at the Strangeways Laboratory and then in the Cavendish Laboratory. He had been influenced by the works of Linus Pauling (chemist) and Erwin Schrodinger (physicist). The electronic bonding of atoms into biological molecules were in the direction of research interesting Crick. Crick began working on x-ray crystallography in the Cavendish Lab.

Francis Crick (http://en.wikpedia.org, Francis Crick 2007)

But the x-ray diffraction pictures of DNA were not being taken in the Cavendish Lab at the University of Cambridge. Instead, they were being made in King's College London at the University of London. In 1946, a new biophysics laboratory had been established in King's College in Cambridge, and Wilkins began doing x-ray diffraction studies on proteins. Fortunately, Wilkins was a friend to Crick. With DNA samples from calf thymus, Wilkins was able to produce thin threads of DNA that contained highly ordered arrays of DNA, suitable for production of x-ray diffraction patterns.

Wilkins (1916–2004) was born in New Zealand. When he was six, his family moved to Birmingham, England. He studied physics at the University of Cambridge in St. John's College. Next he went to the University of Birmingham for graduate work and in 1940 received a PhD in physics for a dissertation on phosphor material. During the World War II, he first worked on improving radar screens in Britain and later on isotope separation in the Manhattan project in the United States.

Maurice Wilkins (http://en.wikpedia.org, Maurice Wilkins 2007)

But although Wilkins had begun the x-ray research, while he was away on holiday, the Director of the Biophysics Laboratory, John Randall, hired a new postdoc to continue the research. Her name was Rosalind Franklin. When Wilkins returned from holiday, he assumed that Franklin worked for him. But Franklin assumed she worked for Randall. This difference hampered good working relationships between Wilkins and Franklin – which would have unfortunate professional consequences.

Rosalind Franklin (1920–1958) was born in England. In 1938, she entered Newham College in the University of Cambridge, graduating in 1941. She began research in Cambridge under Ronald Norrish (a Nobel Prize chemist in fast chemical reactions. In 1942, she left to do war research for the British Coal Utilization Research Association, studying the porosity of coal. In 1945, she obtained a doctorate degree from the University of Cambridge in the physical chemistry of solid organic colloids, with special reference to coal. At the end of the war, Franklin went to Paris to work under Jacques Mering at the Laboratoire Central des Services Chimiques de l'Etat, learning x-ray diffraction techniques. She then wanted to return to England and accepted a position at King's College London offered by Randall to work the DNA project.

Rosalind Franklin (http://en.wikpedia.org, Rosalind Franklin 2007)

Earlier in field of science in the 1930s, x-ray crystallography had been developed as a new scientific instrumental technique. X-rays (high-energy photons) were sent through crystals and formed pictures on photographic negatives of the scattering of these x-rays from planes of atoms in the crystal's atomic structure. By looking at the x-ray diffraction picture and calculating, one could describe the crystalline structure in terms of planar structure of its atoms and spacing between planes of atoms. Just as the microscope was essential to observe the cell and its structure, x-ray crystallography was essential to observe the structure of DNA.

(An analogy of water waves, instead of photon waves, would be to look at the structure of a pier extending out into the ocean. By watching waves come in and under the pier, one could see new waves refracted around the pilings of the pier. Then one could see how the spacing of the pier's pilings produced smaller waves. Then watching these smaller waves, one can calculate backwards to measure the spacing between the pilings.)

But Franklin's research task was not turning out to be easy. It happened that there were two crystalline forms of DNA, and only one would yield a good picture. Moreover, the crystal in that picture would have to be oriented just right – to get a picture that would be interpretable as to its structure (seeing the helix of the crystal form just dead on down the axis of the helix). Finally, Franklin did get a good picture of the right form, a picture right down the center of a clearly helical form.

Creativity in experimental techniques to tease out the secrets of nature is the talent of an outstanding experimental scientist (and very different from the talents of a theoretical scientist).

Meanwhile Watson and Crick had begun their modeling research in Cambridge. Watson brought knowledge of biology and organic chemistry to their collaboration. Crick brought a knowledge of physics. Both skills were necessary. And from Pauling's research showing a helical form of a protein, Watson had conjectured that if DNA too had a helical structure, what kind of x-ray diffraction picture would a helical molecule make? Watson wanted to be prepared to interpret it from an x-ray, and he asked another young expert in diffraction modeling for a tutorial. He was told (if the picture were taken "head-on" down the axis of the helix) how to see the helix and how to measure the angle of the helix. Watson was thus equipped to interpret an x-ray picture of DNA, if he could only get his hands on one.

This is an example of scientific method as connecting observation to theory.

Crick also needed to know precisely how the chemistry was ordered in a DNA molecule. Fortunately, the one scientist who knew this was Erwin Chargaff. He visited England and lectured about it in Cambridge in early 1952. After Watson and Crick listened to his lecture, they realized they needed to learn more about the nucleotide biochemistry.

Erwin Chargaff (1905–2002) was born in Austria. In 1923, he attended the University of Vienna, receiving a doctorate in chemistry in 1928. He then traveled to the United States as a research fellow and did postdoctoral work in organic chemistry at Yale University. In 1930, he returned to Europe working as research assistant at the University of Berlin and then at the Pasteur Institute in Paris. In 1935, he immigrated to New York, as a research associate in the biochemistry department at Columbia University, becoming an assistant professor in 1938 and a full professor in 1952. He published studies in nucleic acids, using chromatographic techniques. Chargaff received the Pasteur Medal in 1949 and later the National Medal of Science in 1974. Chargoff had discovered that the relative proportions of amino acids that compose DNA. He had chemically determined that the amounts of adenine (A) and thymine (T) are roughly equal (A=30.9% and T=29.4% of the bases in DNA) and that the amounts of cytosine C) and guanine (G) are roughly equal (C=19.9% and G=19.9% of the bases in DNA). This is known as 'Chargaff's rules'.

Chargaff's rules were critically important to Watson and Crick's modeling efforts. In February 1953, Watson and Crick began to make the first of their geometric models of the structure of the DNA molecule using cutouts in white cardboard and paste. The atoms composing the bases, A, T, C, and G are all geometrically flat in a ring structure. Watson cut out hexagonal rings in the white cardboard to represent their flat ring structure. Then, he paired A and T and paired C and G, looking to see where hydrogen bonds might couple the paired rings.

In his model, Watson saw that each type of base (G, C, A, or T) on one strand of DNA must form a bond with just one type of base on the other strand – with hydrogen atoms providing the bond. According to the "Chargaff's rules," he assumed guanine would bond only to cytosine and ȧdenine would bond only to thymine. He assumed hydrogen atoms would form the bonds between the complementary bases on opposite strands.

Watson and Crick then thought that building a 3-D structure in a helical form of these bases might model a real DNA molecule. They cut pieces of thick paper (as symbols of chemical organic rings) and arranged them geometrically together as a helix model, holding the cutouts together by wires.

At first they had tried a triple-helix model, and it did not work. They were dismayed but persisted. They knew they had the chemistry right. They just needed to get the geometry right. Then Watson learned through the lab "grapevine" that Franklin had got a good picture. But Franklin had not shown it to any one yet. She wanted to first analyze the picture. It was her experimental picture, so she had the right to be the first to analyze it.

Watson feared that if he simply asked Franklin to see it she would not show it to him. So Watson did not ask. Franklin was as fierce a scientific competitor as Watson. They were both new postdocs seeking to build a career in science. Both she and Watson were looking for the structure of DNA. Watson went uninvited into her lab and looked at the picture on her desk – without Franklin's permission.

There it was! Watson saw it! It was clearly the pattern of a helix. The pattern was of a double helix! Quickly, Watson measured the pattern and rushed to Crick with the information on the angle of the helix.

Watson and Crick put their model together in the form of a double helix, two strands of amino acid chains, twisting about each other like intertwined spiral staircases. They used the angle for the helix, as measured from Franklin's picture. All the physical calculations and organic chemistry then fit together in the model beautifully! Without a doubt! This was it! This was the major goal of biology to determine the structure and mechanism of the DNA. It was a double-helix molecular structure of DNA!

This was progress both in scientific epistemology and ontology! The chemical methods of model construction and the x-ray picture of the molecular geometry together (scientific method) provided the evidence for the nature of DNA (ontology).

The Double Helix Structure of DNA (http://en.wikpedia.org, DNA, 2007)

The structure itself was informative about the dynamics of mitosis. DNA was structured as a pair of twisted templates, complementary to one another. In reproduction, the two templates untwisted and separated from one another, providing two identical patterns for constructing proteins, that is, reproducing life. In this untwisting and chemical reproduction of proteins, life was biologically inherited.

The DNA story proceeded – but not so nicely in terms of ethics. When Watson and Crick published their paper, they added Wilkins name onto it, because they had used research from his group. But they did not add Rosalind Franklin's name. That Wilkins allowed his name to appear without Franklin's name and that Watson and Crick did not offer to have her as a co-author – this has gone down in science history as poor professional conduct.

In 1962, Watson and Crick and Wilkins were awarded the Nobel Prize in Biology. Unfortunately, Rosalind Franklin was not so honored. The reason given by the Nobel committee spokesperson was that she had died before the prize was awarded. Yet at the time, the committee could have acknowledged and honored her major contribution (even without further dividing the prize money). Watson became director of the Cold Spring Harbor Laboratory – *science administration*. In 1988, Watson was appointed head of the Human Genome Project at the National Institutes of Health. Crick contributed to research to try to decode the genetic code and later worked on the idea of consciousness as a biological process.

But there was some good part of the story for Franklin. She felt she had not been treated well at Kings College, and she went to France. There she was well treated. She continued to do excellent research in picturing organic structures. There she had good friends and enjoyed Paris. Unfortunately, she was to die young of cancer. Some have conjectured that it was all that exposure to x-rays in her research that finally triggered the cancer.

This mention of improper acknowledgement of contributions to scientific progress is not meant to denigrate the major contributions of Watson, Crick, and Wilkins. They were great scientists, but they did fail to give proper acknowledgement to Franklin. She should have been made another co-author of the paper, as she did the actual x-ray research and not Wilkins. Proper acknowledgement of individual contributions to scientific progress is an essential feature to the "openness" of scientific method. If scientists are not given proper recognition, there is no personal incentive to publish openly and freely.

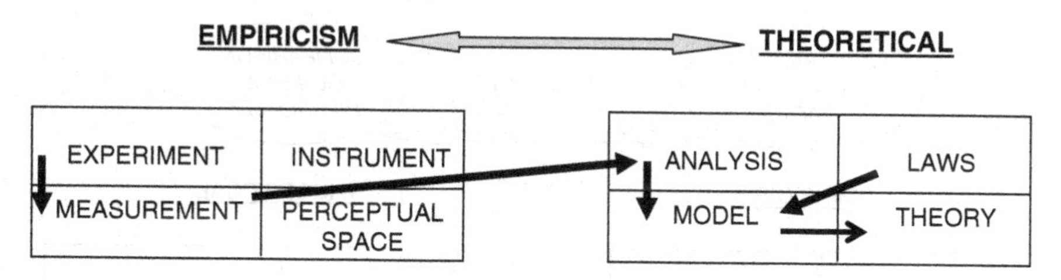


**Fig. 4.10** Research techniques: measurement to theory

> Science is a kind of managed activity: managed in research projects, managed in research journal publications, managed in professional recognition, and managed in research and teaching jobs.

In summary, nearly all the empirical evidence and theory ready for the construction of a geometrical and chemical model of DNA was not assembled until February of 1953, when James Watson and Francis Crick made the first of their geometric models of the structure of the DNA molecule using cutouts in white cardboard and paste. The atoms composing the bases, A, T, C, G were geometrically flat in a ring structure. Then, one more empirical piece was needed to get the geometry of DNA correct. And this came from Rosalind Franklin's x-ray pictures of DNA. Moreover, the structure itself was informative about biological process. It clearly indicated the molecular action of DNA in the mitosis of cell reproduction. In cell division, the DNA of the mother cell must divide and replicate into two sets of identical sets of DNA for each daughter cell. The two strands of DNA divide and then each strand duplicates itself (Fig. 4.10).

> In the molecular modeling of DNA, the Experiment to obtain geometrical x-ray Measurements of the DNA crystal were critical to establish the geometry between chemical components (Analysis) to Model DNA as a double helix. And the chemical laws were used to construct the model.
>
> The (1) discoveries of the cell and DNA, (2) constructing theories of genetics and mitosis, (3) demonstration of DNA as the mechanism of hereditary, and (4) modeling of the chemical molecule of DNA – all together formed the theories of molecular biology – the empirically grounded theory of the transmission of heredity in microbiology

## Empirical and Theoretical Research Techniques

We now return to the table of research techniques to expand on them (Fig. 4.11.)

### *Experiment*

Experiment is the controlled observation of nature – experiencing nature through the human senses aided by scientific instruments. Frederick Griffith's experiment

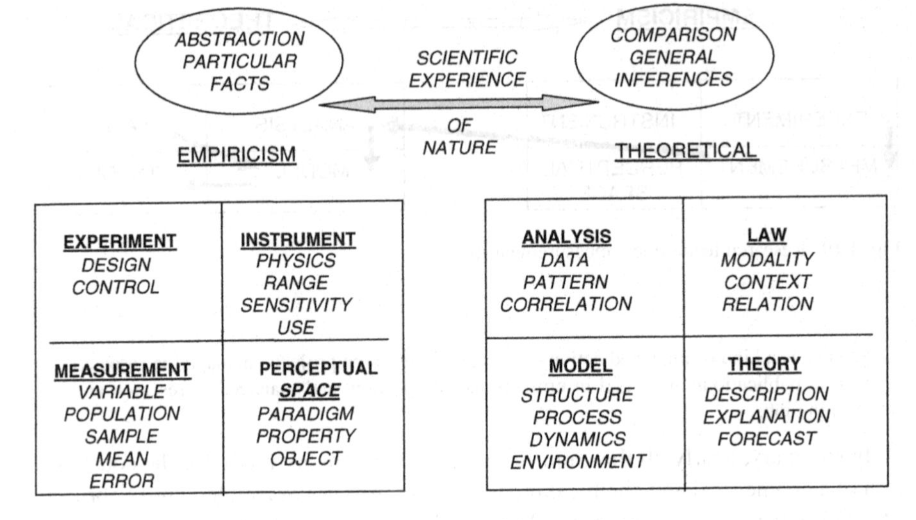


**Fig. 4.11** Empirically grounded theory – Table of research techniques

demonstrated that DNA was the transmitter of genetic information. Later Oswald Avery, MacLeod, and McCarty reproduced Griffith's experiment and chemically isolated the transmitting chemical, showing it was DNA carrying the genetic information. Two important aspects of experimental technique are design and control.

In the Experimental technique, Design and Control are critical features in the conception of the Experiment – determining what scientific question is asked and can be answered by the Experiment.

## *Instrument*

Instrument is a means of extending the range and sensitivity of human sensing of nature. In the example of the microscope as a scientific instrument, one can see that any physical instrument operates by physical interaction with nature. Microscopes use the physics of optical light waves or x-rays or electron beams to image small objects in nature. The type and range of the physics differs in different instruments. Optical microscopes and x-ray microscopes use different spectra of electromagnetic radiation. A given instrument will detect the physical interaction of nature with a given sensitivity.

In the research technique of Instruments, important research features in the Instrument are: Physics, Range, Sensitivity, Use.

## Measurement

*Measurement* is an instrumental way to observe nature expressed in quantitative terms of data. Measurement is also now a major topic in research methodology, treated from a statistical point of view. Statistics is a mathematical topic concerned with selecting an appropriate sample for a measurement of a population and then estimating the likely value of the measurement for the population from the sampled measurement. And also statistical methods enable estimating the error on the measurement of the population, depending upon the size of the sample.

> Measurement estimates the quantity of a Variable from statistical techniques about Populations and Sampling and estimates of the Mean value and Error of a Measurement.

## Perceptual Space

The perceptual space is the formal framework for observing and describing a natural thing, a phenomenal object. For example, the microscopic observation of the cell provided a small scale of perceptual space in which to view basic units of life, cellular structure. The perceptual space for the physical science consists of space and time. Perceptual spaces for the social sciences are different (a topic we explore in a later chapter).

> A natural thing which is first observed in nature is called a natural phenomenon.

When experimental measurements of the properties of a thing is described in a perceptual space, then the natural thing is called an "object" of nature.

## Analysis

*Analysis* consists of inferring a mathematical pattern in the quantitative data from experiments. To analyze means to abstract the underlying form of the data and to generalize the form, so that data from additional new observations would fit that form. Analysis of data is a first connection of empiricism to theory. Mendel's analysis of the pattern of inherited traits provided the idea of dominant and recessive traits in inheritance.

> Theoretical analysis of experiments provides the first step in constructing empirically-grounded theory by determining the qualitative and/or quantitative patterns in experimental Data.

## Laws

*Laws* are the expression of the process discovered in relationships between natural objects as relationships in a natural phenomenon, phenomenological laws. In the discipline of biology, phenomenological laws were called biological "principles" instead of "laws." The principles of heredity were established by Darwin's epic work on evolution and by Mendel's pioneering work on genetics in the 1800s. In 1910, Thomas Morgan's experiments demonstrated the validity of Mendel's principles and also showed that mutations could be induced by x-rays, providing a mechanism for Darwin's evolutionary mechanisms. Phenomenological laws can be of different modes, depending upon the discipline of science, such as causal laws in physics and prescriptive laws in social science (a topic we will later expand).

### *Model*

Model is a symbolic simulation of the processes of a phenomenal object. James Watson and Francis Crick constructed a double-helix model of the DNA as two connected and complementary strings of base sequences (A, T, C, G). The molecular action of DNA in the mitosis of cell reproduction was indicated as the two strands of DNA divide and then each strand duplicates itself.

> Models theoretically represent (describe and explain) objects (phenomenological things) observed by experiments of nature.

### *Theory*

Theory is a symbolic description and explanation of a phenomenal field of objects. The theory of genetic inheritance and the modeling of DNA and its process in mitosis together provides the theory for molecular biology.

## Scientific Inquiry as Both Inductive and Deductive

We can now resolve the role of logical analysis in scientific method by looking at the research techniques as they contribute to induction or deduction (Fig. 4.12).

Therein are indicated the intermediate research techniques interactions:

(a) Between 1-*theory* and 3-*experiment* is the step of 2-*models*
(b) Between 3-*experiment* and 5-*experience* is the step of 4-*instrument*

Also, we show the connections at a meta-level of the steps of:

(c) 11-*Paradigm* beneath 1-*theory*
(d) 6-*Law* beneath 2-*model*
(e) 7-*Analysis* beneath 3-*experiment*
(f) 8-*Measurement* beneath 4-*instrument*

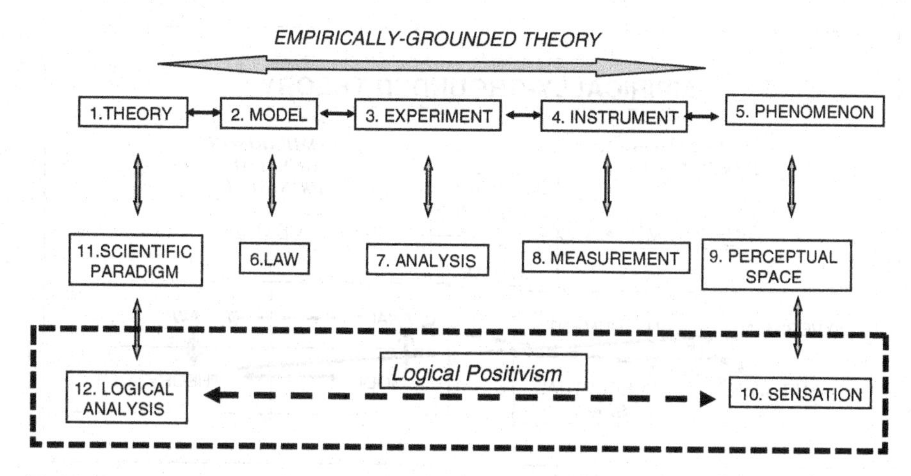


**Fig. 4.12** Scientific procedure of grounding theory on experiments

(g) 9-*Perceptual space* beneath 5-*experience-of-nature*

And finally, another meta-meta level in steps:

(h) 12-*Logical-analysis* beneath 11-*scientific-paradigm*
(i) 10-*Sensation* beneath 9-*perceptual space*

Moreover from this, we can see that the Logical Positivistists ignored all these interactions and assumed that (1) Theory was simply constructed as (12) Logical Analysis directly by induction from (10) Sensation.

But now we can summarize all the places wherein inductive and deductive processes occur in the connections between research techniques, as shown in Fig. 4.13.

In summary, scientific theory construction involves a circularity in inductive and deductive interactions between empiricism and theory over time, sometimes a long time.

1. Scientists pursue research that asks very basic and universal questions about what things exist and how things work. (In the case of genetic engineering, the science base was guided by the questions: What is life? How does life reproduce itself?) This is a deductive approach.
2. To answer such questions, scientists require new instrumentation to discover and study things. (In the case of genetic research, the microscope, chemical analysis techniques, cell culture techniques, x-ray diffraction techniques, and the electron microscope were some of the important instruments required to discover and observe the gene and its functions.) This is an inductive approach.
3. These studies are carried out by different disciplinary groups specializing in different instrumental and theoretical techniques: biologists, chemists, and physicists. (Even among the biologists, specialists in gene heredity research

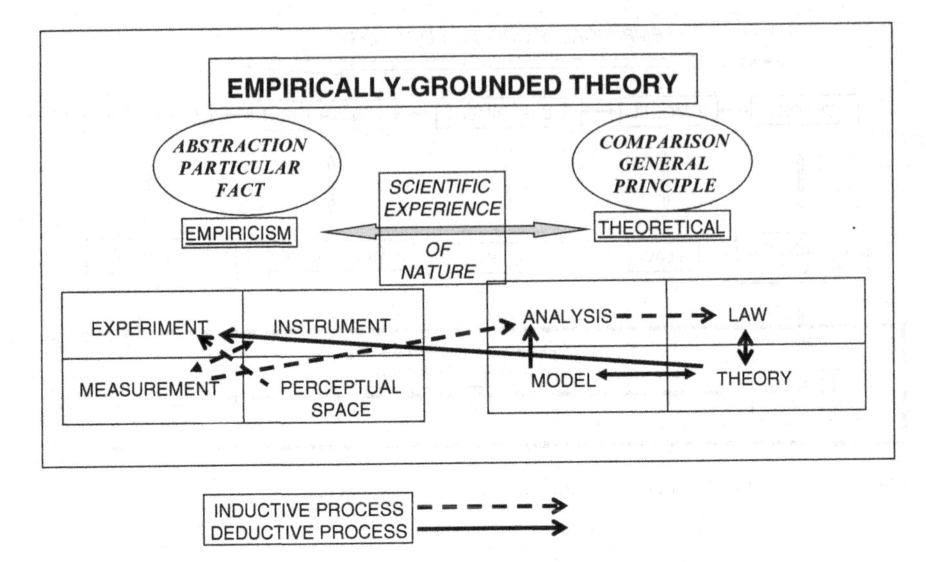


**Fig. 4.13** Inductive and deductive processes

differ from specialists in viral or bacterial research.) Accordingly, science is pursued in disciplinary specialties, each seeing only one aspect of the existing thing. Nature is always broader than any one discipline or disciplinary specialty. This is both inductive an deductive in approaches.

4. Major advances in science occur when sufficient parts of the puzzling object have been discovered and observed and someone imagines how to put it all together properly (as Watson and Crick modeled the DNA molecule). A scientific model is conceptually powerful because it often shows both the structure and the dynamics of a process implied by the structure. This is a deductive approach.
5. Scientific progress takes much time, patience, continuity, and expense. Instruments need to be invented and developed. Phenomena need to be discovered and studied. Phenomenal processes are complex, subtle, multileveled, and microscopic in mechanistic detail. (In the case of gene research, the instruments of the microscope and electron diffraction were critical, along with other instruments and techniques. Phenomena such as the cell structure and processes required discovery. The replication process was complex and subtle, requiring determination of a helix structure and deciphering of nature's coding.) This is circularity in both inductive and deductive approaches.

## Summary

1. Empirical research techniques include experimentation, instrumentation, measurement, and phenomenal objects.
2. Theoretical research techniques include analysis, laws, modeling, and theory construction.

# Notes

[1] The are many histories of molecular biology and DNA, such as Olby (1974), Portugal and Cohen (1977), Moreange (1998).

[2] Watson published his own account of the modeling of DNA, with the last edition of Watson (2001).

# Chapter 5
# Communities of Scientists

## Introduction

As we noted earlier, scientists organize themselves into scholarly societies, publishing research results in society journals and attending scientific conferences. These activities are peer-reviewed by the community for scientific quality. All that scientific information is for free, and scientists do not get paid for publication of scientific results. Instead, scientists find employment teaching in universities or doing research in industrial or in governmental labs. How do scientists in these different kinds of employment settings interact as research communities?

### *Illustration: The Royal Society*

The first step to organizing a research community is the formation of a scientific society. As an example, we will look at the British Royal Society, one of the oldest scientific societies.[1] On November 28, 1660, 12 members formalized meetings as a society, and those who were to become famous British scientists, participated in society: "The Royal Society was dedicated to the free flow of information and encouraged scientific communication. Then Robert Boyle (the famous British chemist) began the practice of reporting his experiments in great detail so that others could replicate them. This became the standard of scientific communication in the Royal Society" (http://en.wikipedia.org, Royal Society 2007). Their philosophical commitment was on experiments about nature as a basis of knowledge, the ground of knowledge,and the book that popularized the then new experimental approach toward nature was Sir Francis Bacon's *Novum Organum*.[2]

> Francis Bacon (1561–1626) was born in London. At the age of thirteen, he entered Trinity College in Cambridge in 1573. In 1576, he became a student barrister in the Honorable Society of Gray's Inn – one of the four professional associations, Inns of the Court, to which lawyers had to belong to practice law in the Royal Courts of Justice (High Court or Crown Court or Court of Appeal). In 1582, Francis Bacon began practicing law from Gray's Inn. In 1584, he entered parliament. In 1596, he was made a Queen's Counsel to Queen Elizabeth. After James I succeeded Elizabeth, Bacon continued as a counsel and was knighted in 1603. In 1608, Bacon was appointed to serve on James' Star Chamber,

F. Betz, *Managing Science*, Innovation, Technology, and Knowledge Management 9,
DOI 10.1007/978-1-4419-7488-4_5, 

which tried political libel and treason cases. In 1618, Bacon was appointed to the position of Lord Chancellor, a second highest ranking in the government.

But in 1621, a Parliamentary Committee charged him with corruption. Trying to maintain a lordly standard of living on the low salary of a judge, he had accepted bribes from plaintiffs in his court. He signed a confession and was sentenced to a fine and imprisonment and barred from holding any future official office. He was imprisoned only a few days in the Tower of London. But he was allowed to keep his titles. He devoted the rest of his days to study and writing. He died five years later of pneumonia.

In the Novum Organum, Bacon had written: "Man, being the servant and interpreter of Nature, can do and understand so much only as he has observed in fact or in thought of the course of nature. Beyond this he neither knows anything nor can do anything." (The New Organon, http://www.constitution.org;bacon;nov_org.htm) Bacon's book was a clear argument that the science should be based upon experiment, theory built on empirical grounds. This was the new 'spirit' of science that pervaded natural philosophers in the 1600s – the century in which science began.

Francis Bacon (http://en.wikipedia.org, Francis Bacon 2007)

Bacon's book expressed the new scientific perspective for knowledge-of-nature as scientific-theory-based-upon-experiment. In 1661, the Royal Society was charted by King Charles of England: "Moray first told the King, Charles II, of this venture and secured his approval and encouragement… the name, 'The Royal Society', first appears in print in 1661, and in the second Royal Charter of 1663 the Society is referred to as 'The Royal Society of London for Improving Natural Knowledge'" (http://www.royalsociety.org, 2007).

Meetings of the new society were held regularly: "The Society found accommodation at Gresham College and rapidly began to acquire a library. In 1662 the Society was permitted by Royal Charter to publish, and the first book it produced was John Evelyn's *Transactions*, edited by Henry Oldenburg, the Society's Secretary. Journal publication became an important activity of the Society. And one of the oldest scientific journals in continuous publication is the *Philosophical Transactions*." (http://www.royalsociety.org, 2007).

In 2007, the Royal Society published seven journals, which are peer-reviewed by qualified scientists before an article is published:

1. *Biology Letters* – "peer-reviewed journal publishing short, high-quality letters from across the biological sciences."
2. *Journal of the Royal Society Interface* – "cross-disciplinary journal highlighting research at the interface between the physical and life sciences."
3. *Notes & Records of the Royal Society* – "journal publishing articles in the history of science."
4. *Philosophical Transactions of the Royal Society A* – "publishes Theme Issues and Discussion Meeting Issues on key topics across the physical sciences."
5. *Philosophical Transactions of the Royal Society B* – "publishes Theme Issues and Discussion Meeting Issues on key topics across the life sciences."

6. *Proceedings of the Royal Society A* – "high-quality peer-reviewed research journal, publishing articles across a wide range of topics in the physical sciences."
7. *Proceedings of the Royal Society B* – "high-quality peer-reviewed research journal, publishing articles across a wide range of topics in the life sciences." (http://www.royalsociety.org, 2007)

The society is "… governed by its Council of Trustees, which is chaired by its President. The members of council and the president are elected from its fellowship. Fellows are elected annually by the existing fellowship for their substantial contribution to the improvement of natural knowledge including mathematics, engineering science, and medical science. Fellows must be citizens or ordinarily resident of the Commonwealth or Republic of Ireland, otherwise they may be elected as a foreign member" (http://www.royalsociety. org, 2007).

In the list of the Presidents of the Royal Society, one finds the names of some of the most famous British scientists:

Sir Isaac Newton (1703–1727) – Physicist, Newtonian mechanics and calculus
Joseph Banks (1778–1820) – Botanist, taking part in Cook's first voyage around the world
Sir Humphry Davy (1820–1827) – Chemist, discovered chlorine and hydrogen in acids
Sir Joseph Dalton Hooker (1873–1878) – Botanist and explorer in the Himalayas
Thomas Henry Huxley (1883–1885) – Biologist and comparative anatomist
William Thomson, 1st Baron Kelvin (1890–1895) – Physicist, thermodynamics
Joseph Lister, 1st Baron Lister (1895–1900) – Medicine, antiseptic treatment of wounds
John William Strutt, 3rd Baron Rayleigh (1905–1908) – Physicist, scattering of light
Sir Joseph John Thomson (1915–1920) – Physicist, discoverer of the electron
Sir Ernest Rutherford (1925–1930) – Physicist, alpha and beta radioactivity and atomic structure
Sir William Henry Bragg (1935–1940) – Physicist, x-ray spectrometer

The importance of the organization of the Royal Society for British science was in the *procedures for scientific communication* as: Royal Society meetings and scientific journals.

> Scientific societies are essential to the disciplines of science in the procedures by which they judge and record progress in a discipline, organizing scientific meetings, and publishing scientific journals.

## Scientific Societies and Peer Review

Scientific communication is an essential procedural feature of science, and for this purpose scientific societies were organized. As science evolved from the 1600s in Europe, the organization of a "scientific society" also evolved. The mission of a "scientific society" is to facilitate scientific communication, through scientific meetings and journals. The formation of a society is an important step in the emergence of a new discipline or subdiscipline.

Scientific societies are a subset of the idea of a "learned society," adding the scientific communities' particular procedure of peer review : "A learned society is a society that exists to promote an academic discipline or a group of disciplines. In a society, membership may be open to all and/or may require possession of some qualification, or may be an honor conferred by election, as is the case with the oldest learned societies such as the Royal Society of London (founded 1660), the Roman Accademia dei Lincei (founded 1603), the Académie Française (founded 1635), or the Sodalitas Litterarum Vistulana (founded 1488)" (http://en.wikipedia.org, Learned societies 2007). Organizationally, learned societies are incorporated as nonprofit organizations. They hold scholarly conferences for members to present new research results. They also sponsor academic journals in the discipline of the society. Learned societies are of key importance in the sociology of science.

> What distinguishes scientific societies from other learned societies is the dependence upon scientific method – empirically grounded theory – as a basis for progress in knowledge.

Scientific societies use peer review in decisions to publish articles in their journals. Scientific peers are defined as members of scientific societies. In publishing a scientific article in a science-discipline archival journal, the article is usually reviewed by three disciplinary peers for the editor of the journal. If the peer reviewers judge that the research results published in the article are valid, novel, and contribute to knowledge in the discipline, then reviewers recommend publication of the article. If there is disagreement on the quality, the editor of the journal can make an independent judgment on whether or not to publish or obtain additional reviews. If the peer review is generally positive but suggests revisions in the article, then the editor may send the article back to the author to make changes.

The journal editor then schedules publication of the article in a future issue of the journal based upon a quality consensus of the peer reviews and notifies the author of the article that it has been accepted for publication. For an academic author, the acceptance of research articles for publication in disciplinary journals is an important criteria for academic tenure and advancement. The three quality criteria in publication are thus validity, novelty, and contribution, so that progress in science (contribution) is built into the criteria for scientific publication. The peer review process generally works well within the boundaries of a subspecialty of an established discipline – because there is a general consensus in the discipline as to the boundaries of research efforts. This is to say (1) what view of nature belongs within the boundaries of a specialty and (2) what previous scientific literature is validated within the boundary.

> Peer review is the review of a scientific claim or proposal by equals of the scientist – scientific peers.

There are four procedures in science administration in which peer review provides the essential process in judging the quality of scientific research:

1. Peer review for publication of scientific articles
2. Peer review in selection of research projects to fund

3. Peer review in oversight of research project management
4. Peer review in reviewing research program performance

But the peer review process does not work so simply when:

1. There is no community consensus as to the proper boundaries of research focus, or
2. What view of nature should fall within a boundary, or
3. What key articles define the core scientific literature within a boundary

When there are such disagreements, then different "schools" arise within the specialty area and they argue with each other as to which school of thought is scientifically correct. Differing schools in a discipline or subdiscipline indicate a lack of consensus upon just what is the discipline. Then politics within scientific disciplines becomes much like any other politics – depending upon which "school" controls the editorial board of a specific journal.

Scientific disciplines are about boundaries and about the empirical grounding of theory within a boundary. When there is no scientific consensus about disciplinary boundaries and their contents, then usually the scientific theory within such schools are not properly empirically grounded.

> Differing "schools" exist within a scientific discipline when the theories of the differing schools are not empirically grounded.

## *Illustration: Origin of European Research Universities*

In addition to scientific societies, universities play an important role in research communities – providing employment for scientists focused principally on "pure science" – on the advance of basic scientific knowledge. But not all universities provide support for research – only research universities do. Where and when did the modern research university arise? They occurred first in Europe in the 1800s, as reform in the earlier medieval universities.

### Medieval European Universities

Although there were earlier academies of learning in some ancient civilizations, universities began in Europe in the eleventh century.[3] The University of Bologna was one of the first universities, founded in 1088. The University of Paris began in 1150, growing from three earlier schools in Paris (the Palace school and those of Notre-Dame and Sainte-Genevieve). Oxford University was founded in 1167 (although teaching in Oxford existed since 1096). The University of Cambridge was founded in 1209, University of Salamanca in 1218, University of Padua in 1222, University of Valladolid in 1346, Charles University of Prague in 1348, Jagiellonian University (Poland) in 1364, University of Vienna in 1365, University of Heidelberg in 1386, etc. During the 1100s, 1200s, and 1300s, the founding of universities was a Europe-wide movement.

Medieval European universities were established as universitas (corporations) under civil or canon law, supported by fees from students. They were organized into four faculties: theology, medicine, law, and natural philosophy. The faculty of theology trained priests for the medieval Catholic Church in Europe. The faculty of Medicine trained medical practitioners. (Although the modern training of medical doctors did not begin until the late 1800s, after which the practice of medicine could be based upon the modern sciences of biology and chemistry.) The Faculty of Law trained lawyers for the administration and enforcement of law, much as modern schools of law train legal practitioners today.

It was in the Faculty of Natural Philosophy wherein would emerge the ideas leading toward modern science. The Faculty of Natural Philosophy provided what we call now a "general education." Usually, the study for a Bachelor degree required 6 years, and students studied arithmetic, geometry, astronomy, music theory, grammar, logic, and rhetoric. The courses were focused upon reading specific books, such as that of Aristotle. Medieval University 2007). After the Bachelor degree, a student could choose one of the three faculties of theology, law, or medicine for a master's degree or a doctorate degree.

University Class 1350 (http://en.wikipedia.org, Medieval university 2007)

The form of the European medieval university was important to the eventual origin of science in Europe – because they were a corporation (universitas) of scholars and not focused around a single scholar's works. Previous academies in history were devoted principally to the writings of a single scholar (such as Plato's academy in ancient Greece, or Aristotle's Lycium).

The medieval university institutionalized the principle that knowledge was pursued and communicated by communities of scholars.

The four faculties of the medieval university also focused knowledge upon both theory and practice. The faculties of natural philosophy or theology emphasized theoretical knowledge – philosophical principles or religious principles. The faculties of law and medicine emphasized practical, empirical knowledge – legal procedures or medical procedures. It was in the faculty of natural philosophy that the *connection between theory and empiricism* would be argued about (1100s–1400s) and eventually resolved. And as we saw in the Copernicus-to-Newton story, this resolution finally occurred in the 1500s – as the creation of scientific method for empirically grounding theory.

The medieval university also institutionalized the principle that knowledge was both theoretical and empirical – crossing the ancient gap between the philosopher (theoretician) and the artisan (practical empiricism).

## Prussian Reform of Universities

But the medieval university was not a research university – a university devoted to research. It was a teaching university. It was not until 1810 that the institutional transition from medieval universities toward modern universities began with the Prussian educational reform in 1810 by Wilhelm von Humboldt.[4]

Then the Prussian King, Friedrich Wilhelm III, established a new university in Berlin, donating a building that was the former palace of Prince Heinrich on the boulevard Unter den Linden. The new university was called the Friedrich Wilhelms Universitat (but now is called the Humboldt University of Berlin). It had the four classical faculties of law, medicine, philosophy, and theology. The first academic term began with 256 students and 52 faculty. But the difference between the new university and old medieval universities was that in the faculty of philosophy, there were new departments of science. The idea for science departments was that of Wilhelm von Humboldt (http:www.hu-berlin.de/uberblicken/history/huben_html, 2007).

This new organization of university knowledge upon a scientific disciplinary research base was to be the key reform for transferring the medieval university toward the form of the modern research university.

Wilhelm von Humboldt (1767–1835) was born in Potsdam near Berlin. In 1788 he entered the University of Gottingen, studying classical languages (classical philology) and then law. In 1789, the French revolution began in Paris and Wilhelm traveled to Paris (and Switzerland and the Rhineland), writing a travel journal. In 1790, he passed his university examinations in jurisprudence and entered the Prussian civil service in Berlin and married Karoline von Dacheroden. But he didn't like the civil service and a year later he resigned, spending the next years on his wife's family estates in Thuringia (near Jena). He pursued his scholarly interests in philology. Wilhelm's brother, Alexander von Humboldt, would become famous as one of the founders of the modern discipline of geology.

In 1794, Humboldt moved to Jena, where the University of Jena was becoming the center for the German movement in idealist philosophy and romanticism. He became friends with Johann Wolfgang von Goethe and Friedrich Schiller (who were to become two of Germany's most famous poets). In the fall of 1797, Wilhelm moved his family to Paris, living there until 1802 through the French Revolution. During that time, he visited the Basque country in Spain to study the Basque language and culture. In 1803, he moved to Rome to serve as a Prussian envoy to the Vatican. Meanwhile in 1804 in France, Napoleon Bonaparte became Emperor of France. A coalition of nations opposed Napoleon. In 1805, Napoleon's grand army marched into Germany and defeated the joint Austrian and Russian armies at Austerliz. Two years later, on 14 October 1806, Napoleon defeated the Prussian army at the Battles of Jena-Auerstedt.

The impact of Prussia's defeat was great and later led to major changes in the organization of government and education. In the fall of 1808, Wilhelm was appointed by the Prussian king to reform education, and he returned to Prussia. In the following two years, Wilhelm von Humboldt reformed the entire Prussian educational system – beginning in elementary to secondary education and up to university education. His idea was for free and universal education for all citizens. And at the university level, he established the University of Berlin to combine both teaching and research in the university mission.

Wilhelm von Humboldt (http:/en.wikipedia.org, Wilhelm von Humboldt 2007)

Von Humboldt reorganized the faculty of natural philosophy into departmental chairs of science – mathematics, physics, chemistry, biology. Professors were changed from private lecturers to state employees as ordinary (full professors) and extraordinary professors (associate professors). Lecturers remained paid only by student fees (privatdocent). Both professors and lecturers had to pass a "habilitation" procedure before they could teach at a university. The habilitation required a candidate to write a second thesis (after the first doctoral thesis) and defend it also before an academic committee. The habilitation thesis required research after a Ph.D. as an accumulation of research results or a research monograph on an area, based upon published work.

Research and research publications became the primary criteria for university teaching appointments in the Berlin university model. When a habilitated doctorate finally gained a professorial appointment, the professor's salary was paid by the State and research funds were also given to run a research institute (to hire research assistants and train new doctorates in a scientific discipline).

This model paid a professor for teaching but expected the professor to lead (manage) scientific research, and it became the model for all German universities in the nineteenth century. The German universities became institutions for science, involving new procedures for training, encouraging, and rewarding a successful scientific career. German professors were appointed on the basis of success, not as a teacher, but as a researcher.

By 1900, German universities became the leading research universities in the world. The pattern for the career of a German scientist in universities had become that of (1) obtaining a doctorate from a university professor, (2) employment as a research assistant in a research institute at a university, (3) writing an examination for habilitation to teach, and (4) employment as a privatdocent (in addition to sometimes remaining a research assistant). And during all this time, the new scientist was expected to publish significant scientific articles in the research literature. Only then after gaining a scientific reputation in research, would a privatdocent be considered to an appointment as an associate professor (professor extraordinarious) or to a full, chaired professor (professor ordinarious).

And from Germany this model spread throughout Europe and to America: "the structures von Humboldt created for the University of Berlin would become the model not only throughout Germany but also for the modern university in most Western countries" (http://plato.stanford.edu/entries/wilhelm-humboldt/, 2007).

## Research Universities

Scientific societies are organizations providing scientific conferences for communication of the results of new scientific research and scientific journals for publishing and archiving these results. But scientific research is not performed in scientific societies but in universities and in government and industrial laboratories. And the universities are the primary performers of basic scientific research.

The primary function of all universities is education – as undergraduate education (Bachelor's degrees) and graduate education (Master's degrees). But a subset of universities also provide doctoral education – as doctoral PhD degrees in science and engineering. These universities are called "research universities" because doctorates in science/engineering cannot be awarded unless the doctoral candidate in a thesis makes a new contribution to scientific/engineering knowledge. This requirement of a "contribution to knowledge" is unique to the doctoral education in science/education. Medical doctorates are not research degrees but practical degrees (to practice medicine). Doctorates in the humanities require a doctoral thesis, which may be a contribution to the humanities literatures but not to new knowledge of nature. In a research university, scientific research is integrated into science/engineering doctoral education. Faculty publish in the international science and engineering research literatures. In research universities, research is a significant activity, as significant as education.

### *Illustration: Discovery of Nuclear Fission*

To understand how scientific progress occurs within research universities, we will look at the historical case of the discovery of the natural "breaking apart of a nucleus" – *nuclear fission*. We recall that Rutherford established the structure of an atom as electrons orbiting a nuclear core. But what was the nature of the atomic nucleus? This basic empirical question was pursued by nuclear physicists in the first half of the twentieth century.

Henri Becquerel discovered the first instance of radioactivity – decay of some atomic nuclei. He had been studying the phenomenon of fluorescence – when he discovered a new phenomenon of nature. This he had reported on January 24, 1896 to the scientific society of the French Academy of Sciences: "One wraps a Lumière photographic plate with a bromide emulsion in two sheets of very thick black paper, such that the plate does not become clouded upon being exposed to the sun for a day. One places on the sheet of paper, on the outside, a slab of the phosphorescent substance, and one exposes the whole to the sun for several hours. When one then develops the photographic plate, one recognizes that the silhouette of the phosphorescent substance appears in black on the negative. If one places between the phosphorescent substance and the paper a piece of money or a metal screen pierced with a cut-out design, one sees the image of these objects appear on the negative.... One must conclude from these experiments that the phosphorescent substance in question emits rays which pass through the opaque paper and reduces silver salts." (http://en.wikipedia.org, Henri Becquerel 2007).

Marie Curie studied under Becquerel and advanced their discoveries in radioactivity. She married a fellow instructor, Pierre Curie, at the Sorbonne. They studied pitchblende as a radioactive material. Pitchblend is a black uranium-rich ore, containing largely uranium oxide ($UO_2$). Uranium as an oxide had been used historically to add a yellow color to ceramic glazes; and it was discovered as an element in 1789 by a German pharmacist, Martin Klaproth, precipitating a yellow compound by dissolving pitchblende in nitric acid, which he named uranium after the planet Uranus. He obtained radium from the ore.

By 1898, Pierre and Marie Curie had found radioactive elements other than uranium in pitchblende, which they named radium and plutonium. In 1903, Henri Becquerel and Pierre and Marie Curie shared the Nobel in physics for their discoveries in spontaneous radioactivity.

Henri Becquerel (1852–1908) was born in Paris and studied science at the Ecole Polytechique and engineering at the Ecole des Ponts et Chaussees. In 1892, he occupied a physics chair that the Museum National d'Histoire Naturelle, and in 1894 became chief engineer in the Department of Bridges and Highways. He shared the 1903 Nobel Prize in Physics with Pierre and Marie Curie for their discoveries in radioactivity. In the last year of his life in 1908, he was elected permanent secretary of the French Académie des Sciences.

Henri Becquerel (http://en.wikipedia.org, Henri Becquerel 2007)

Marie Curie (1867–1934) was born in Poland. She moved to Paris at the age of 24 to study science in physics and mathematics at the Sorborne. She studied under Henri Becquerel and received a Doctorate of Science, being then the first woman in France to complete a doctorate. She was to become a famous scientist, pioneering studies in radioactivity. She would receive two Nobel prizes, one in Chemistry and one in Physics. She was the first woman to win a Nobel prize and the only person to receive two Nobel prizes in different disciplines. She also became the first woman professor at the Sorborne in Paris.

Marie Curie
(http://en.wikipedia.org, Marie Curie 2007)

Pierre Curie
(http://en.wikipedia.org, Pierre Curie 2007)

We recall that during 1895–98, Ernest Rutherford investigated radioactivity and was able to distinguish between alpha, beta, and gamma rays in the radioactive phenomena of atoms. And that later, it would be found that "alpha" radiation was particles emitted from radioactive nuclei in decay that were equivalent to the nucleus of a Helium

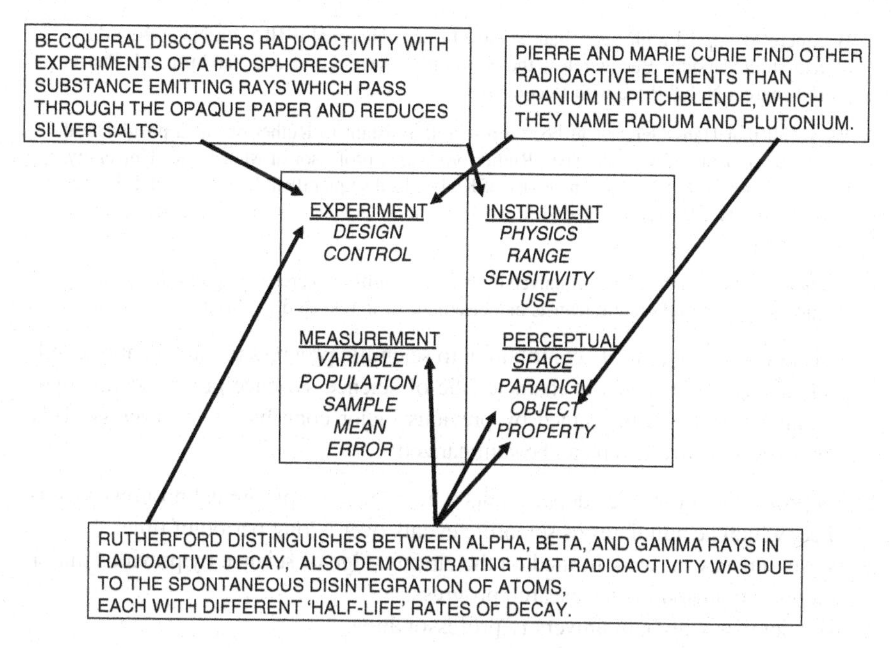


**Fig. 5.1** Empiricism

atom (two protons and two neutrons). Beta radiation was electrons emitted from radioactive nuclei in decay. Gamma rays were photons emitted from radioactive nuclei in decay. Next, Rutherford demonstrated that radioactivity was due to the spontaneous disintegration of atoms and determined that different atoms had different times of a constant rate of decay, which he called the "half-life" of a radioactive atom – for which he was rewarded the Nobel Prize in Physics in 1908.

We summarize the empiricism in these first researches on radioactivity in Fig. 5.1.

## Scientific Communities and University Professorships

But let us pause now (before we continue the story of the discovery nuclear fission from university research) to examine the institution of a research university. We have seen that – in the German reform of universities in 1800 by von Humboldt – university professorships changed from a focus upon education to a focus upon research. The research focus of professorships in research universities provided scientists with two important conditions for performing scientific research: (1) a salaried professorship provided a career for a scientist devoting a life's work to research, and (2) the graduate program for doctoral training as an activity of the professorship, enabling the training of new generations of students in scientific method and finding scientific issues to be researched.

Becquerel occupied a physics chair at the Museum National d'Histoire Naturelle, Marie Currie studied under Becquerel at the Sorborne. Marie Curie and her husband, Pierre, would share a Nobel Prize with Becquerel.

We recall that Hans Geiger had been a research assistant to Rutherford as a postdoctoral fellow at the Rutherford Lab. Then Rutherford was a professor at Manchester University. Geiger became a professor at University of Kiel and subsequently at University of Tubingen and University of Berlin. Chadwick worked under Geiger and became a professor at Liverpool University.

We have seen this is the model of university-based scientific research – progress accumulating by pieces by many different scientists in a discipline or in related disciplines.

This kind of "piece-meal" contribution to scientific progress in individual research projects as scientific events is characteristic of science. Science progresses in small communities of scientists, trained by previous major contributors to progress. This model of university research can be summarized:

1. As professors in science departments who contribute piecemeal progress in science, which accumulates as scientific events over a long period of time.
2. By scientists who were trained on the cutting edge of science by previous major scientific contributors to research progress.
3. As scientists who held university professorships.
4. To include the financial resources to hire research assistants.

## *Illustration (Continued): Discovery of Nuclear Fission*

Next, Irene Curie, the daughter of Marie and Pierre Curie, played an important role in the evolving field of nuclear science. In 1928, Irene Joliot-Curie and her husband, Frederic Joliot discovered that radioactivity could be induced in an element upon bombarding the element with alpha particles – induced radioactivity. This was an exciting discovery, that radioactivity could be induced in a chemical element by the absorption of particles.

Irene Curie (1896–1956) was born in Paris. In 1914 at the age of 18, she entered the Sorborne. But that same year, World War I began, and Irene went with her mother Marie to Britanny to help run mobile field hospitals. There she became acquainted with x-ray machines and served as a nurse radiographer. When the war ended, Irene Curie went to Paris to study at the Radium Institute (which her parents had founded). Her doctoral thesis in 1925 was research with alpha rays of polonium (an element that her parents had discovered). Just before then in 1924, she was asked to teach laboratory techniques to a young chemical engineer named Frederic Joliot. They married in 1926.

Irene Joliot-Curie (http://en.wikipedia.org, Irene Joliot-Curie 2007)

Frederic Joliot (1900–1958) had also been born in Paris and later graduated from the School of Chemistry and Physics of the City of Paris. He became an assistant to Marie Curie at the Radium Institute, and fell in love with her daughter, Irene. Joliot obtained a doctorate in science, with his thesis on the electrochemistry of radioelements.

Frederic Joliot-Curie (http://en.wikipedia.org, Frederic Joliot-Curie 2007)

In 1930 in Germany, Walther Bothe and H. Becker discovered that when alpha particles struck light elements (such as beryllium), a new kind of penetrating radiation was also emitted. At first, they thought this was gamma radiation (high energy light-wave particles) but found the new radiation was too highly penetrating for gamma radiation to be an explanation.

Walther Bothe (1891–1957) was born in Germany and studied physics under Max Planck at the University of Berlin. Bothe earned his doctorate in 1914. After World War I, he worked with Hans Geiger to study particle detection with Geiger counters, developing the first coincidence circuit. He used this technique in 1930 to discover the new form of high-energy radiation. (Much later in 1954, he (along with Max Born) would win a Nobel Prize in Physics for the invention of the coincidence circuit.)

Walther Bothe (http://en.wikipedia.org, Walther Bothe 2007)

Confirming Bothe's discovery in France in 1932, Irene and Frederick Joliot-Curie further demonstrated that if this new form of radiation fell upon paraffin (rich in hydrogen atoms), it ejected protons at high energy.

Also in 1932 in England, James Chadwick in England confirmed Joliot-Curie's experiments with the Bothe's new form of radiation. Then Chadwick suggested that the new radiation was a new kind of particle, like a proton, but electrically neutral. And he called the new particle a "neutron."

James Chadwick (1891–1974) was born in England and later studied at the University of Manchester and Cambridge University. In 1913 he went to Germany to study for a doctorate, working under Hans Geiger at the Technical University of Berlin. A year later he was jailed in Berlin as a British citizen, until released after the intercession of Geiger. (We recall that Geiger along with Marsden had performed the famous Geiger–Marsden experiment to establish the Rutherford model of the atom.) In 1935, Chadwick won the Nobel Prize in Physics for the discovery of the neutron and became a professor at Liverpool University.

James Chadwick (http://en.wikipedia.org, James Chadwick 2007)

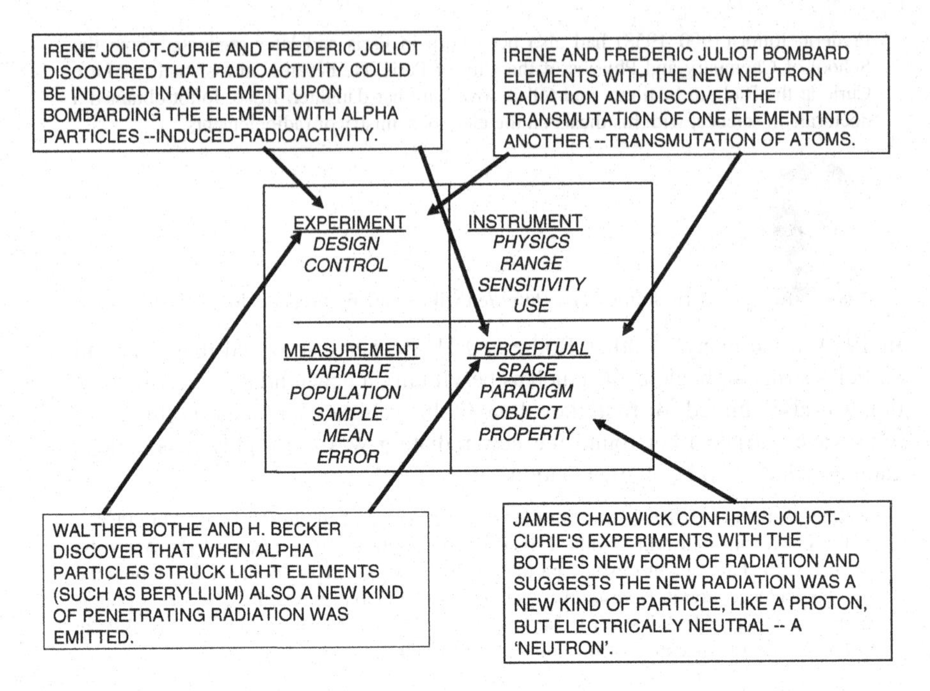


**Fig. 5.2** Empiricism

In France in 1934, Irene and Frederic bombarded elements with the new neutron radiation and discovered the transmutation of one element into another, first boron to radioactive nitrogen, then aluminum to radioactive phosphorus, and then magnesium to silicon. In 1935, they would share the Nobel Prize in Chemistry. And finally, Irene Curie would be awarded a professorship in the Faculty of Science at the University of Paris.

All these researches – by Becquerel and the Curies and Rutherford and the Curie-Joliots and Bothe and Chadwick – were necessary to build the scientific progress to lead next to the discovery of nuclear fission. And all this took place in research universities in Europe – in France and England and Germany. One can see just from this one example the incredible importance of this institution, the research university, to progress in modern science (Fig. 5.2).

## Research Teams

Earlier, we saw the importance of a research team as led by a professor, such as Rutherford's research team. Geiger and Marsden (and later Bohr) were essential members of Rutherford's research teams. Also, we also saw the idea of a "research team" in Becquerel's team, with his research assistants of Marie and Pierre Curie. Irene collaborated in a team with her husband, Frederic Joliot.

In the twentieth century, the need for a research team to study nature became an important organizational characteristic of science. The name "big science" labeled the expensive needs in science for sophisticated scientific instrumentation, organized research teams, and university laboratories. As the study of nature has gone deeper into nature's complexity, "big science" became an organizational form of university research – very necessary to some areas for scientific progress. Research universities began adding "research centers" as a complementary form to the disciplinary science departments.

In the need to create 'critical masses' of research projects, research universities have added 'research centers' to facilitate the organizing of a scientific research teams.

A research center may be needed to add together enough "pieces" of research to attain scientific progress. There are two sizes of "research projects" now in university research: (1) doctoral-sized research projects and (2) a group of research projects (critical mass of projects). The first is necessary for the university integration of education and research. The second is necessary for the advancement of scientific progress in a "critical mass" of a research effort.

The idea of a 'critical mass' of research efforts (a group of research projects) may be necessary for significant scientific progress in a university setting – requiring a university research center.

## *Illustration (Continued): Discovery of Nuclear Fission*

Next – upon learning of the Joliot-Curie researches with the neutron to induce radioactivity – Enrico Fermi in Italy started further research on the phenomenon.

Enrico Fermi (1901–1954) was born in Rome. He entered the Scuola Normale Superiore in Pisa for his undergraduate and doctorate degree. During his graduate study in 1924, Fermi traveled to see research frontiers, spending one semester at the University of Gottingen and the next at the University of Leiden, where he studied under Paul Ehrenfest. This connection to Ehrenfest was influential on Fermi, providing for his first major contribution to science. He developed a new form of statistics for nonclassical particles (quantum-mechanic particles) with half-interger spin in 1925–1926. It was called Fermi–Dirac statistics.

Enrico Fermi (http://en.wikipedia.org, Enrico Fermi 2007)

At the age of 24, Fermi became a professor in Rome at the Institute of Physics (the first position for atomic physics then in Italy). He assembled a team of assistants and students, and in 1934, Fermi's team started "using neutrons from radon-beryllium sources, in lieu of alpha particles, to activate many common elements" (Segre 1955). Members of Fermi's scientific team included Edwardo Amaldi,

Oscar D'Agostino, Franco Rasitte, and Emilio Segre. They conducted experiments with a neutron source that established that the nucleus of an element could be transformed into a different element (with a different number of protons) or a different isotope of the element with a different number of neutrons). They were particularly interested in uranium, because at the time, it was the heaviest of the naturally occurring elements, and they hoped to create new elements heavier than uranium when a uranium nucleus absorbed neutrons.

A neutron could be absorbed by an element and within the nucleus and decay to a proton – thereby transforming the element into a new element. Or if in the absorption of the neutron into a nucleus the neutron remained (did not decay into a proton), then the element was transformed into an isotope. After a neutron was absorbed, radioactive reactions were observed as decay products.

But this induced radioactivity was not much larger than the natural radioactivity of uranium. It was confusing to understand just what was happening. Segre later wrote: "In Rome, we immediately found that irradiated uranium showed a complex radioactivity with a mixture of several decay periods. We expected to find in uranium only the previously observed (radioactive) backgrounds, and so we started looking for an isotope of element 93 (the next element above uranium's proton count of 92) produced by a neutron absorption and gamma decay reaction on the 238 isotope of uranium, followed by a beta decay" (Segre 1989, p. 39).

They had hoped that the 238 weight (238 neutrons and protons) of the uranium element would absorb a neutron to transmute to a 239 weight, and they hoped a neutron would decay to a proton, by emitting an electron. Thus the element uranium (with 92 protons and weight 238) would be transmuted to a new element (of 93 protons and weight 239). Thereby, they hoped to be the first to create and discover a new element created by transmutation of another element.

This was the old dream of medieval alchemists – the transmutation of matter! But scientists are human and can make mistakes: "Here we made a mistake that may seem strange today. We anticipated that element 93 would resemble rhenium (element 75).... Products of the bombardment were indeed similar to rhenium" (Segre 1989. p. 39).

They were looking for a new element of atomic number 93 – in the products from the experiment, which had chemistry similar to the known element of rhenium. This expectation was a mistake. Because of this mistake, they did not recognize what really was happening. The decay product was not rhenium. Something very different was happening. The uranium atom was not transforming into another element; it was splitting asunder! The products of the nuclear reaction were fragments from nuclear fission. But they didn't recognize it.

The Fermi team just missed discovering the scientific phenomenon of nuclear fission (the splitting of the atom). They were later to learn their mistake came from similar chemistry. Some of the fission products were isotopes of technetium (element 43) – which also has a chemistry similar to rhenium. So nature fooled them.

Of course, scientists never anthropomorphize nature. But if they did, they would note that sometimes in research, nature appears to have cooperative moods for scientists and sometimes uncooperative moods. In this case, the confusion of similar chemistry between technetium and rhenium meant that Fermi's group failed to discover nuclear fission.

## Complexity of Nature and Research

Missing something observable in nature is possible because of the instrumental means of science. When scientists observe nature deeply, they must do so by means of scientific instruments. Only on the macro-level of natural phenomena can humans directly observe nature, with any of the human biological senses. At both smaller and larger scales of nature, scientists can only sense nature indirectly through instruments.

In the example of the discovery of nuclear fission, the different scientific teams in Italy, France, England, and Germany were all using the bombardment of neutrons upon elements as an instrumental technique.

One of the essential features of any instrumental technique is that it observes nature but only partially and obscurely.

It takes proper theoretical analysis and interpretation to understand what is seen upon observing nature through scientific instruments.

### *Illustration (Continued): Discovery of Nuclear Fission*

In 1935, after Fermi's group reported their results to the world, Otto Hahn and Lise Meitner (at the Kaiser Wilhelm Institute) began work to confirm the Fermi group's results.

And at first, they also were not able to discern the fission phenomenon: "Hahn and Meitner, working with a neutron source about as strong as the one used in Rome and Paris, started by confirming our Rome (the Fermi group's) results. This is somewhat surprising because they applied quite different chemistry. Their early papers are a mixture of error and truth, as complicated as the mixture of fission products resulting from the bombardments" (Segre 1989, p. 40).

Otto Hahn (1879–1868) was born in Germany. He went to the University of Marburg where he received his undergraduate degree and then in 1901 his doctorate in chemistry. He went to England in 1904 as an assistant to William Ramsay at University College, London.

William Ramsay (1852–1916) was a Scottish scientist who discovered the noble gases (argon, neon, krypton, neon). These are called "noble" because they show little chemical reactivity – having closed valence electron shells – neither needing more electrons for shell closure nor having valence electrons to give away for shell closure. Ramsey received the Nobel Prize in Chemistry in 1904. Working there, Hahn discovered a new isotope of thorium.

In 1905 Hahn transferred to McGill University in Canada to work with Ernest Rutherford. There Hahn discovered new radioactive elements of thorium C, radium D, and radioactinium.

Hahn went abroad from Germany to England and then Canada, learning the new chemistry of isotopes and radioactive materials (from Ramsay and Rutherford).

Otto Hahn (http://nobelprize.orgnobel_prizes/chemisry/laureates/1944)

In the summer of 1906, Hahn returned to Germany to the University of Berlin. He began collaborating with Emil Fisher at the Chemical Institute, and Fisher gave him room there for his own laboratory. In 1907, Hahn qualified to teach (habilitation thesis) at the University of Berlin. On September 28, 1907, he met a young Austrian physicist, Lise Meitner, and they began a lifelong collaboration and friendship. (Together they would discover nuclear fission – the phenomenon that Fermi and Segre just missed.)

Lise Meitner (1878–1968) was born in Vienna. In 1901, she enrolled in the University of Vienna and graduated with a doctorate in 1906. In 1907, she went to the University of Berlin. She met Hahn and began working as an assistant in his new laboratory. Ten years later in 1917, Meitner and Hahn discovered the first long-lived isotope of the element protactinium.

Lise Meitner (http://en.wikipedia.org, Lise Meitner 2007)

As we noted earlier, Fermi's team in Italy had been trying to create elements heavier than uranium. Also trying to do this was Rutherford in England and Meitner and Hahn in Germany. Others also trying were Irene Curie and Frederick Joliot in Paris. The Curie–Joliot team was bombarding neurons into the element thorium. They found a decay product that chemically resembled lanthanum. But they did not realize it was lanthanum of atomic weight 141. Had they realized that, they would have recognized it a product of nuclear fission. But they thought it might be an isotope of actinium. Both teams, Fermi's and Curie-Joliot's misdiagnosed the chemistry of the products of the nuclear reaction, and, as a result, both missed the discovery of nuclear fission.

In that spring of 1938 at the tenth International Chemistry Conference in Rome, Hahn and Joliot met and they discussed Curie's results. Hahn concluded that something was wrong. Perhaps the decay product Curie had seen was not "actinium." Upon returning to Germany, Hahn discussed the situation with Meitner, and they decided to repeat some of Curie's experiments.

But just then in that fatal spring of March 1938 for Austria, Hitler's German Army marched into Austria. Lise Meitner was an Austrian and of Jewish ancestry. Suddenly in Germany, she lost her previous protection from Nazi persecution of Jewish citizens through her Austrian citizenship. Now she was in great danger, as were all German Jewish citizens. She knew she must escape immediately from Germany to avoid arrest by Hitler's Gestapo. With the help of friends, she was able to gain permission by immigration officials to travel to the Netherlands. Boarding a train with only a suitcase of her belongings, Lise Meitner fled to Holland. In the Netherlands, she was not able to obtain an academic appointment and went on to Stockholm. There she found a post at Manne Siegbahn's laboratory. And she began to work with Niels Bohr in Denmark, who traveled regularly between Stockholm and Copenhagen.

Hahn was not Jewish and so was able to remain in Berlin. He continued the work, with the assistance of Fritz Strassmann. Hahn and Strassman concentrated on identifying the radioactive component that Curie had thought was actinium. By December 1938, Hahn and Strassmann concluded that the radioactive decay products Fermi and Curie had thought were radium, actinium, and thorium were instead barium, lanthanum, and cesium. There had been errors in chemical analysis. It was important to get the radioactive products correct. And this had required the contribution of excellent chemists (Hahn and Strassmann) to complement the efforts of the physicists in studying neutron-induced radioactivity. In these chemical fragments was the clue to nature's behavior – the splitting of the atom.

All the while, Meitner continued to correspond with Hahn. In December of 1938, she received a letter in which Hahn wrote about the results of reexamination of the chemical products: "He and Strassmann ... were coming 'again and again to the frightful conclusion that one of the products behaves not like radium but rather like barium'" (Stuewer 1985, p. 50).

Meitner continued to think about this puzzle of the nuclear fragments. She went on a skiing holiday in Sweden with some friends and her nephew, Otto Frisch. Then on the 28th of December, Meitner received a third and key letter from Hahn. She showed her nephew all the letters. In the last letter to her, Hahn had asked: "Would it be possible that uranium (of atomic weight 239) bursts into barium (with weight 138) and masurium (with weight 101 and now called technetium)?" (Stuewer 1985, p. 50)

The key feature that Hahn pointed out that was that the atomic weights added correctly: 138 nucleons from barium added with 101 nucleons from masurium totaled the 239 nucleons in uranium. Yet the puzzle continued! The atomic number did not add right! Uranium had 92 protons, and barium 56 protons, and masurium 43 protons. The sum of 56 protons and 43 protons equaled 99 protons – *instead* of the 92 protons of uranium. Hahn had written: "... so some of the neutrons would have to be transformed into protons... Is that energetically possible?" (Stuewer 1985, p. 50)

Excited about Hahn's speculations, early on the following day on December 29, the aunt and her nephew breakfasted together and went outdoors for a hike. Frisch pushed along on skis, and Meitner walked. They talked about physics. Frisch suggested that a liquid-drop model of the nucleus – which Niels Bohr had earlier proposed in the physics literature – might be used to explain Hahn and Strassmann's results. Frisch later wrote:

"I worked out the way the electric charge of the nucleus would diminish the surface tension and found that it would be down to zero around $Z=100$ (atomic number of 100 nucleons) and probably small for uranium" (Frisch, …).

Lise Meitner calculated the energies that would be available from the mass defect in such a breakup (the splitting of a liquid-drop like mass of uranium with 239 nucleons into fragments liquid-drop like sizes with of 138 and 101 nucleons): "The amount of electric repulsion of the fragments would give them about 200 MeV (million electron volts) of energy. The mass defect would indeed deliver that energy." (Stuewer 1985, p. 50)

The idea was that the uranium nucleus (like a drop of water) might split into two smaller waterdrops. When such a uranium atom split into two halves, each half would push away from the other by the repulsive positive charges of their protons. And the energy from this electrical repulsion of the halves would be about 200 MeV – which equaled the difference of the weight of the original uranium nucleus minus the two weights of the split nuclei. This was Meitner and Frisch's idea of "nuclear fission."

On the first of January, 1939, Meitner wrote to Hahn: "We have read and considered your paper very carefully. Perhaps it is indeed energetically possible that such a heavy nucleus bursts" (Stuewer 1985, p. 50). Of course, Frisch and Meitner knew that their idea of a liquid-drop model of a nucleus was only an analogy – nucleons packed together as water molecules pack into a drop of water. Frisch later wrote about this: "Bohr had given good arguments that a nucleus was much more like a liquid drop. Perhaps a drop could divide itself into two smaller drops in a more gradual manner, by first becoming elongated, then constricted, and finally being torn rather than broken in two? We knew that there were strong forces that would resist such a process, just as the surface tension of an ordinary liquid drop tends to resist its division into two smaller ones. But nuclei differed from ordinary drops in one important way: They were electrically charged. (Positive protons repel each other by an electrical charge but are held together in a nucleus with neutrons by a strong nuclear force.)"

Frisch added: "The charge of a uranium nucleus (92 protons) was indeed large enough to overcome the effect of the surface tension almost completely; so the uranium nucleus might indeed resemble a very wobbly unstable drop, ready to divide itself at the slightest provocation, such as the impact of a single neutron. After separation, the two drops would be driven apart by their mutual electric repulsion and would acquire high speed and hence a very large energy, about 200 MeV in all. Where could that energy come from? Lise Meitner calculated that the two nuclei formed by the division of the uranium nucleus together would be lighter than the original uranium nucleus by about one-fifth of the mass of a proton. Now whenever mass disappears energy is created, according to Einstein's formula $E=mc^2$; and one-fifth of a proton mass was just equivalent to 200 MeV. So here was the source for that energy; it all fitted!" (http://en.wikipedia.org, Otto Frisch 2007)

Otto Frisch (1904–1979) was born in Vienna. He graduated in 1926 from the University of Vienna with a doctorate in physics. He worked in several research laboratories before obtaining a position at the University of Hamburg, under Otto Stern. Working for Stern, Frisch did research on how impacting atoms can be scattered from other atoms in the plane of a crystal (and this is called diffraction of atomic beams). But in 1933, when Adolf Hitler came to power in Germany, Frisch knew it was time to leave. He moved to England for a time and then to Copenhagen to work with Niels Bohr, under whom Frisch specialized in nuclear physics. Frisch had moved to the research laboratories where the frontiers of science in atomic physics were being pursued.

Otto Frisch (http://en.wikipedia.org, Otto Frisch 2007)

## Scientific Competition

In this example of the discovery of nuclear fission, we can see more details about the idea of a "scientific community" – particularly in how "scientific communities" find principal employments in research universities. When research in science discovers new phenomena and opens up new paths of research, many researchers in the world take up that direction for exploration, thus creating kinds of competition for scientific discovery. In the discovery of nuclear fission, three groups, Fermi's team in Italy, Rutherford's team in England, and Hahn and Meitner in Germany were each racing to create an element heavier than uranium using neutron bombardment, whose contributions are summarized in Fig. 5.3.

Competition in the scientific community is conducted in an open manner of exchanging information. The competition is one for personal fame, but the common goal is scientific progress. Even though the Curies were in scientific competition with Hahn and Meitner, the exchange of information between Joliot and Hahn at the conference was freely given. Who was able to take advantage of the open information was up to each individual.

Open exchange and publication of information is a hallmark of the scientific community, and this is a strongly shared ethical value of scientists.

Openness of scientific exchange and publication of university research are essential to scientific progress.

Research competition divides into theoretical and experimental efforts. In Meitner and Frisch's analysis of Hahn's data, we again see an example of the importance of theoretical analysis of data in advancing scientific theory. But this theoretical analysis depended upon prior empirical analysis. In Meitner and Hahn's experiments, there was an empirical analysis of the results of the experiments in terms of the chemical products. This would establish the phenomenal "objects" involved in their relationship in the experiment. One object in nature, a neutron, smashed into a second object,

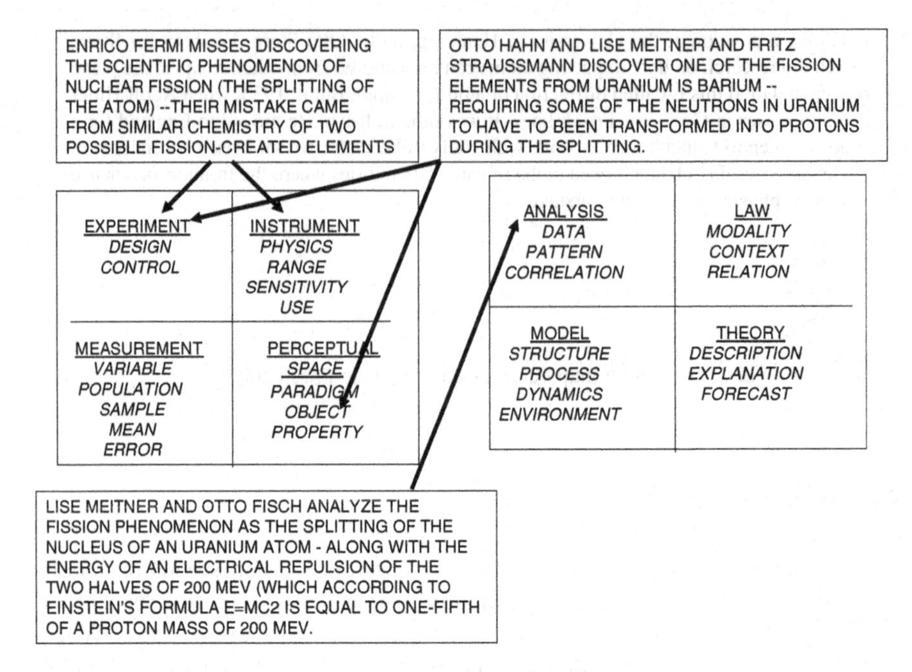


**Fig. 5.3** Empiricism and theory

a uranium atomic nucleus. From that smashup, product objects resulted, barium and masurium. But to do this in the transformation of uranium into the products, barium and masurium, some of the smaller objects in the uranium nucleus, neutrons, had to also transform into protons. This transformation was also an empirical analysis. It identified the product-objects and their relationships involved in the experiment – atoms, atomic particles, and nuclear fission. An empirical analysis in science identifies the nature of the phenomenal objects observed in an experiment.

The empirical analysis describes the observed phenomena.

Next in Meitner and Frisch's analysis of the objects' interactions was a theoretical model. The waterdrop analogy explained how the uranium atom produced two daughter atoms, barium and masurium, through splitting of the uranium nucleus – nuclear fission.

A theoretical analysis explains the observed phenomena.

We have defined science as the discovery and understanding of nature. This is why experiment (description) and theory (explanation) are the essential features of the scientific method.

When science first observes a phenomenon of nature, science discovers nature; and when science explains that phenomenon, science understands nature

A scientific representation of a natural phenomenon consists of both a description (empirical) and an explanation (theoretical) of the object in nature.

## *Illustration (Concluded): Discovery of Nuclear Fission (and Then the Atomic Bomb)*

Very excited after his collaboration with his Aunt, Frisch immediately returned to Copenhagen where he was studying under Niels Bohr. He told Bohr about the exciting news about the bursting of the uranium nucleus: "Bohr burst out: 'Oh what idiots we all have been! Oh, but this is wonderful! This is just as it must be!'" (Segre 1995, p. 42).

The next day Bohr and his wife sailed the Atlantic to attend an international physics conference in the USA. Leon Rosenfeld accompanied them and wrote of the trip: "As we were boarding the ship, Bohr told me he had just been handed a note by Frisch, containing his and Lise Meitner's conclusions; we should 'try to understand it.' We had bad weather through the whole crossing, and Bohr was rather miserable, all the time on the verge of seasickness. Nevertheless, we worked very steadfastly and before the American Coast was in sight Bohr had got a full grasp of the new process and its main implications" (Stuewer 1985, p. 52).

It was on January 26, 1939, when Bohr presented the news of Nuclear fission at the Fifth Conference on Theoretical Physics in Washington, DC, and a report from the conference stated: "Certainly the most exciting and important discussion was that concerning the disintegration of uranium of mass 239 into two particles each of whose mass is approximately half of the mother atom, with the release of 200,00,000 electron volts of energy per disintegration" (Stuewer 1985, p. 54). This was an enormous release of energy per atom!

In addition to Frisch (still in Denmark) who had started thinking about an atomic bomb, another scientist at the conference, Leo Szilard, also thought about an atomic bomb. Leo Szilard (1898–1964) was born in Budapest, then a city in the Austro-Hungarian Empire. In 1916, he enrolled in the Budapest Technical University to study engineering. In 1917, he had to leave to join the Austro-Hungarian Army. In 1919, he returned to study in Budapest. But Hungary had been split off as a nation in 1918, and the new government in Hungary was viciously anti-Semitic.

Szilard left Hungary to study in Germany at the Technische Hochshule in Berlin. He changed to physics and took classes from famous physicists, including Einstein, Planck, and von Laue. In 1923, he received a doctorate from Humboldt University of Berlin and was appointed an assistant to von Laue at the University of Berlin's Institute for Theoretical Physics in 1924. In 1927, Szilard finished his habilitation dissertation, and he became an instructor at the University of Berlin. He spent some of his time on technical inventions and obtaining patents.

But in 1933, when Hitler came to power in Germany, Szilard once again fled virulent anti-Semitism, now to England. Having learned of Chadwick's discovery of the neutron in 1932, Szilard began speculating about the idea of a nuclear chain reaction from atoms absorbing neutrons, and he tried using beryllium and indium but these did not work. Still he filed the idea for a chain reaction patent in England in 1936 and assigned the patent to the British Admiralty to ensure its secrecy. He did not want the German Nazi government to learn of the idea. In 1938 Szilard accepted an offer to do research at Columbia University in the USA. But Szilard would have to wait until 1939 at the conference in Washington, DC. to learn how to make an atomic bomb. One would use uranium.[5]

Leo Szilard (http://www.anl.gov, 2007)

In the same year that Szilard had joined the faculty of Columbia University, so too had Enrico Fermi. In 1938, when Fermi was awarded the Nobel Prize in Physics, he went to Sweden with his family to receive the prize but did not return to Italy. The family immigrated to America. His wife, Laura, was of Jewish ancestry; and Mussolini in Italy had joined Hitler in persecuting Jews.

When both Szilard and Fermi heard about the discovery of nuclear physics at the Washington conference, they returned to New York to immediately confirm the experiment. They sent neutrons into uranium and measured that more neutrons were ejected from each atomic fission of uranium. A nuclear chain reaction in uranium was possible! Later Szilard was quoted about the experiment: "We turned the switch, saw the flashes, watched for 10 min, then switched everything off and went home. That night I knew the world was headed for sorrow" (Rhodes 1986).

Also, they learned that $U_{238}$ atoms principally underwent nuclear fission by capturing neutrons slower than the speed at which neutrons emerged from a nuclear fission of a uranium atom. To make a nuclear reactor with a high-rate of nuclear fissions, they would have to surround uranium matter with something to slow down the neutrons. Neutrons bouncing off carbon atoms could do this. The slowed-down neutrons had a higher probability of being captured by a $U_{238}$ atomic nucleus. Consequently, a nuclear reactor could be constructed with a neutron-moderator (carbon) between uranium atoms.

Szilard worried that the Germans would be the first to turn the discovery into an atom bomb. Germany was the leader in physics research in the first half of the twentieth century. Szilard knew American action was urgent! He drafted a letter to be sent to the President of the USA, Franklin Delano Roosevelt. But for the President to read it, Szilard thought the letter should be signed by Albert Einstein, then the most prestigious and well known scientist in the world. The letter said: "In the course of the last 4 months it has been made probable… that it may become possible to set up a nuclear chain reaction in a large mass of uranium…. This new phenomenon would also lead to the construction of bombs …" (Szilard, Wikipedia 2007) Einstein read the letter and signed it on August 2, 1939.

Reenacted Picture of Einstein & Szilard Discussing Letter to Roosevelt (http://aip.org/history, 2007)

Einstein was Jewish and had been a professor in Germany, when Hitler came to power in January 1933. Then the Nazi government immediately passed a law removing Jews from government posts, including university professors. Nazi students burnt books publicly on May 10, 1933, including Einstein's scientific papers. Einstein had traveled to America and accepted a position at a newly founded Institute for Advanced Study at Princeton University.

On September 1, 1939, Hitler's army invaded and occupied Poland. England declared war. Szilard had given the letter to an advisor to the President, Alexander Sachs. He was able to deliver the letter to the President on October 11. After Sachs summarized the letter, Roosevelt authorized the creation of an Advisory Committee on Uranium. The Uranium Committee held its first meeting 10 days later on October 21, 1939. It was chaired by the Director of the US National Bureau of Standards, Lyman Briggs. It budgeted $6000 to Enrico Fermi to conduct neutron experiments at the University of Chicago. Fermi had moved to Chicago from Columbia and there began research on measuring the rate of capture of incident neutrons by uranium atoms.

In 1942, when Japan attacked the US naval forces in Pearl Harbor, Hawaii, the US entered World War II. Then the uranium project was enlarged to the Manhattan project, which would successfully develop the first atomic bombs.

On August 6, 1945, a US airplane dropped a uranium bomb on a southern island of Japan, on the city of Hiroshima. It directly killed an estimated 80,000 people and destroyed 67% of the city's buildings. Deaths continued after the bombing, with an estimated 60,000 more people dead of injuries or radiation poisoning. Three days later on August 9, the second bomb, a plutonium bomb, was dropped on Nagasaki, destroying the north part of the city. A latter memorial at the Nagasaki Peace Park estimated the deaths at 73,884, injured at 74,909, and diseased at 120,820.

Atomic Mushroom Cloud Over Nagaski (http://www.atomicarchieve.com, 2007)

Expanding Cloud on the Ground (http://www1.city.nagasaki.jp/na-bomb/museum, 2007)

Ruins of Mitsubishi Shipbuilding Factory (http://www1.city.nagasaki.jp/na-bomb/museum, 2007)

On August15, 1945, the Emperor of Japan announced an unconditional surrender of Japan to the USA. World War II ended, and the atomic age began.

## Summary

Communities of scientists (1) organize into scientific societies that publish scientific journals and (2) find principal employment in universities that grant Ph.D. requiring original research contributions.

1. Research universities began in Germany in 1800 when university professors were appointed on the basis of success as a researcher and not as a teacher.
2. The integration of research and education in doctoral programs in science and engineering in research universities facilitates the process of research in these universities.
3. The openness of scientific exchange and publication of university research is essential to scientific progress.
4. The distinctive feature of modern institutionalized science in the research university is the individual research project as a "doctoral-sized project."
5. Research publications proceed in units sized by the "doctoral-sized" research projects.
6. Thus, progress in science proceeds in a "piece-meal" manner expressed in these research publications
7. The professorial role is as research project manager and the role of graduate students and postdocs are as research assistants (research labor).
8. Relatively modest-sized instruments (unless a research community gains a shared government-sponsored major research facility, which becomes "big science facility").
9. Research laboratories are needed when instruments are needed to conduct the research.
10. Scientific discoveries in universities have often provided knowledge bases for the invention of new technologies.
11. Research universities play the major role in advancing science, and university research is important to technological progress in a nation.

## Notes

[1] Histories of the Royal Society include Purver and Bowen (1960) and Sprat (2003).
[2] Biographies of Bacon include Nieves (1996) and Zagorin (1999). Bacon's Novum Organum was online at: (http://www. Constitution.org/bacon/nov_org.htm. 2010).
[3] Histories of medieval universities include: Haskins (1972), Rashdall (1987), Ferruolo (1998).
[4] There are many biographies of von Humboldt, such as Sweet (1978–1980).
[5] There are many histories about the first atomic bomb, including Jungk (1956), Rhodes (1986), DeGroot (2005).

# Chapter 6
# Science and Society

## Introduction

Science interacts with society, principally through technology. We saw an example of this in the previous case of the discovery of nuclear fission and the building of the atomic bomb. The process in which scientific knowledge is transformed into technical knowledge and in which technical knowledge is designed into products and services has been called *innovation*. The societal infrastructure which institutionalizes this innovation process has been called a "national innovation system."

This system provides the organizational interactions of research in science and technology to create national technical capabilities for economic development. For science and technology (S&T) policy of a modern government, now its focus is on science and technology and innovation. In this chapter, we look at the connection between science method and science application (Fig. 6.1).

### *Illustration: Recombinant DNA Technique*

Earlier we described the history of modeling DNA[1] and the beginning of the scientific specialty of molecular biology. And, of course, this scientific discovery led to the founding of the biotechnology industry. Just how this happened depended upon one more early research in the new specialty, which was the invention of the research technique of recombining DNA, recombinant DNA. Let us briefly recount this.

After the modeling of DNA and decoding of its genetic information, several scientists began trying to cut and splice genes. In 1965, Paul Berg at Stanford planned to transfer DNA into *Escherichia coli* bacteria, using an animal virus (Svrp lambda phage). *E. coli* bacteria can live in human intestines, and the SV40 virus is a virus of monkeys which can produce tumor cells in cultures of human cells. Because of the dangerous nature of the SV40 virus, Berg decided not to proceed with the experiment, publishing a design for hybridizing bacteria in 1972. Berg organized a historic meeting on safety, the Conference on Biohazards in Cancer

F. Betz, *Managing Science*, Innovation, Technology, and Knowledge Management 9,
DOI 10.1007/978-1-4419-7488-4_6, © Springer Science+Business Media, LLC 2011

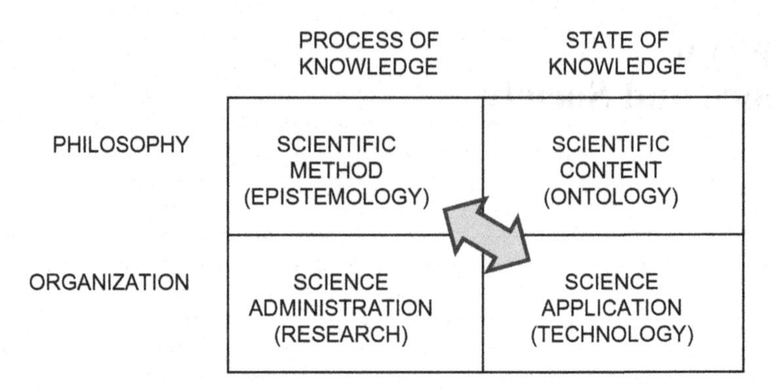


**Fig. 6.1** Philosophy and organization of science

Research in California, USA, on January 22–24, 1973 (Olby 1974). This stimulated later government action to set safety standards for biotechnology research.

Also at this time, Peter Lobban, a Stanford graduate student, had been working on a similar idea for gene splicing. Lobban was studying under Dale Kaiser of the Stanford Medical School (Kaiser had been one of the "phage school" group, which also had spawned Watson). When ideas are ripe, there often in science have been scientists hotly competing for the same scientific goal. As we noted scientists compete for the fame of being the first to discover or to understand, second place gets no recognition.

A colleague at the University of California responded to Berg's request for some EcoRI enzyme, which cleaves DNA (and leaves the "sticky ends" of the cut DNA). Berg gave the enzyme to one of his students, Janet Mertz, to study the enzyme's behavior in cutting DNA. Mertz noticed that when the EcoRI enzyme cleaved an SV40 DNA circlet, the free ends of the resulting cut and linear DNA eventually reformed by itself into a circle. Mertz asked a colleague at Stanford, to look at the action of the enzyme under an electron microscope. They learned that any two DNA molecules exposed to EcoRI could be "recombined" to form hybrid DNA molecules. Nature had arranged DNA so that once cut, it re-spliced itself automatically.

There was another professor in Stanford University's Medical Department, Stanley Cohen, who learned of Janet Mertz's results. Cohen also thought of constructing a hybrid DNA molecule from plasmids using the EcoRI enzyme. Plasmids are the circles of DNA, which float outside the nucleus in a cell and manufacture enzymes the cell needs for its metabolism (the DNA in the nucleus of the cell are principally used for reproduction). In November 1972, Cohen attended a biology conference in Hawaii. He was a colleague of Herbert Boyer, who had given the EcoRI enzyme to Berg (and Berg's student Mertz). At a dinner one evening, Cohen proposed to Boyer that they create a hybrid DNA molecule without the help of viruses. Another colleague at the dinner, Stanley Falfkow of the University of Washington at Seattle offered them a plasmid, RSF1010, to use that confers resistance to antibiotics in bacteria so that they could see whether the recombined DNA worked in the new host.

After returning from the Hawaii Conference, Boyer and Cohen began joint experiments. By the spring of 1973, Cohen and Boyer had completed three splicings of plasmid DNAs. Boyer presented the results of these experiments in June 1973 at the Gordon Research Conference on Nucleic Acids in the USA (with publication following in the Proceedings of the National Academy of Sciences, November 1973). In 1973, Cohen and Boyer applied for a basic patent on recombinant DNA techniques for genetic engineering, which was awarded to Stanford University.

After 100 years of scientific research into the nature of heredity, humanity could now begin to deliberately manipulate genetic material at a molecular level – and a new industry was born, biotechnology. Boyer and Cohen won Nobel Prizes. Also Boyer found the first new biotechnology company, Genentech, and when it subsequently went public, Boyer became a millionaire. Cohen was hired by a small firm, Cetus, to help it convert to biotechnology, and Cohen also became a millionaire. The day for biologists to move simply from "pure science" to "application" (i.e., technology) began.

The scientific experimental technique of recombining DNA provided the knowledge basis for a new industrial process of genetic engineering; and from molecular biology sprang biotechnology.

*Technology provides the societal context for use of the discoveries of science.*

Paul Berg (1926–present) was born in New York. He obtained a bachelor's degree in biochemistry from Penn State University and a PhD in biochemistry from Case Western Reserve University in 1952. He retired as a professor from Stanford University in 2000. In 1980, Berg shared a Nobel Prize in Chemistry with Walter Gilbert and Frederick Sanger for research in nucleic acids.

Herbert Boyer (1936–present) was born in Pennsylvania, USA. He obtained a bachelor's degree in biology and chemistry from Saint Vincent College and a PhD in 1963 in biochemistry from the University of Pittsburgh. He became a professor at the University of California, San Francisco. From 1976 to 1991, he served as Vice President of Genentech. In 1990, he received a National Medal of Science.

Stanley Cohen (1935–present) was born in New Jersey, USA. He obtained a bachelor's degree from Rutgers University and a PhD from the University of Pennsylvania in 1960. He became a professor at Stanford University in 1968. In 1986, he received a National Medal of Science.

## Innovation Process

This example shows the general pattern of how science and technology do interact – in an innovation process. Government research supports university science, which creates the knowledge bases for the process of technological innovation by industry. In the recombinant technique example, there was a government research agency (the National Institutes of Health) which had funded research projects performed by Berg, Boyer, and Cohen. The research for creating the recombinant technique was performed in universities. A patent of the technique was filed by Stanford University, and then the research was published in scientific journals. One of the scientists founded the first company in what would become a new biotechnology Industry, based on the technology of the recombinant DNA technique.

The innovation process for going from science to technology to utility (1) begins with the idea of nature, (2) then transforms knowledge of nature into technology, and (3) then into economic utility. If one examines any technology, one will see that some kind of nature (material, biological, or social) is being used (manipulated). The process of innovation consists of how knowledge-of-nature (science) can be connected to technology (manipulation-of-nature) which then can be connected to use-of-nature (economy), as sketched in Fig. 6.2.

1. *Research*
   In technological innovation, one begins with nature and knowledge about nature – what it is (discovery) and how it operates (explanation) – is gained by *science* through the act of *research. Scientists* are the principle kinds of people, who as researchers gain knowledge of nature.
2. *Invent*
   Scientific knowledge of nature is used as a *knowledge base* by technologists to create new technologies (manipulations of nature) through the act of *invention*. Technologists are usually scientists or engineers or other technical personnel.
3. *Commercialize*
   Technical knowledge is *embedded* within a product/service/software through the act of *design*. In a business, technological knowledge is used by engineers to develop and design new high-tech products or services or processes. *Commercialization* is the act of connecting (embodying) technology into the products/services/processes. In the product/service development procedures of a business, both technical personnel and business personnel act together in innovation teams.
4. *Market*
   A business competes by selling high-tech products/services in a market place, earning income – which become profits when the sales prices exceed the costs of producing products/services.

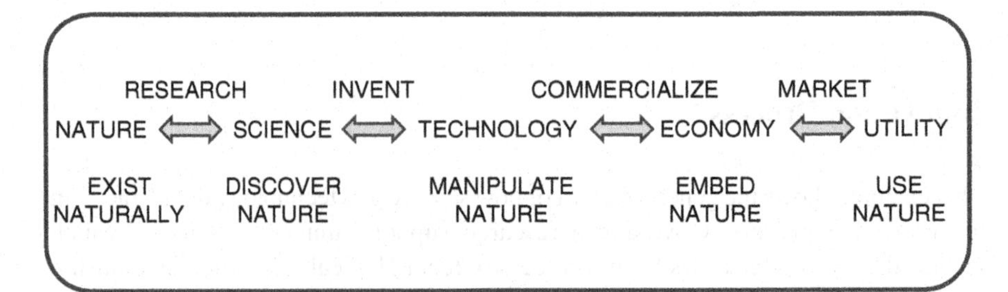


**Fig. 6.2** Innovation process

For clarity, it is helpful to define all the key terms in innovation upon a common semantic basis of "nature," as in the following.

1. *Nature is the totality of the essential qualities of the observable phenomena of the universe.*
   In the communities of scientists and engineers, the term "nature" is commonly used to indicate essential qualities of things that can be observed in the entire universe.
2. *Science is the discovery and explanation of nature.*
   The derivation of the term "science" comes from the Latin term "scientia," meaning "knowledge." However, the modern concept of scientific research has come to indicate a specific approach toward knowledge, which results in discovery and explanations of nature.
3. *Technology is the knowledge of the manipulation of nature for human purpose.*
   The technical side of the idea of technological innovation – invention – derives, of course, from the idea of technology. The historical derivation of the term "technical" comes from the Greek word, "technikos," meaning "of art, skillful, practical." The portion of the word "ology" indicates a "knowledge of" or a "systematic treatment of." Thus, the derivation of the term "technology" is literally "knowledge of the skillful and practical." This meaning of "technology" is a common definition of the term – but too vague for expressing exactly the interactions between science, technology and economy. The "knowledge of the skillful and practical" is a knowledge of "manipulation" of the natural world. Technology is a useful knowledge – a knowledge of a functional capability. In all technologies always there is nature being manipulated.
4. *Scientific technology is a technology invented upon a science base of knowledge which explains why the technology works.*
   Not all technology in the world has been invented upon a base of scientific knowledge. In fact, until science began in the world in the 1600s, all previous technologies – fire, stone weapons, agriculture, boats, writing, boats, bronze, iron, guns – were all invented before science. Consequently, technical knowledge of these understood how to make the technologies work but not why the technologies worked. What science does for technology is explain why technologies work. After science, all the important technologies in the world have been invented upon a knowledge base of science.
5. *Engineering is the design of economic artifacts embodying technology.*
   Technologies are implemented in products and services by designing the technical knowledge into the operation of the products/services, and engineers do this design. Engineering designs enable businesses to use nature in adding economic value through its activities. What engineers design in the commercialization phase of technological innovation are new products or services or processes that embody the technical principles of a new technology.
6. *Economy is the social process of the human use of nature as utility.*
   The products/services provide utility to customers who purchase them. Through products/services, the concept of "utility" provides the functional relationship of

a technology to human purpose. Thus, economic utility is created by a product or service sold in a market, and which provides a functional relationship for its customer. For example, xerography products provided the functional relationship of copying (duplicating) the contents of printed papers, which was useful to the customer. Since in a society, its technology connects nature to its economy, we will use a meaning of the term "economy" which indicates this. The common usage of the term "economy" is to indicate the administration or management or operations of the resources and productivity of a household, business, nation, or society. But we will note the term "economy" in the use of nature as utility.

7. *Management is the form of leadership in economic processes.*
   Business organizations provide the social forms for economic activities. The leadership in an economic organization is provided by the management staff of the business.
8. *High-tech products/services/processes are commercial artifacts which operate on the principles of a new technology.*
   The monetary value of science and technology emerge eventually as the prices of high-tech products and services. This is the basic interest of business in science and technology.

## *Illustration: Origin of the Biotechnology Industry*

Soon after the news of recombinant DNA, an entrepreneur, Robert A. Swanson, heard of the new DNA technique and saw the potential for raising venture capital to start a genetic engineering firm. The story is that Swanson walked into Boyer's office and introduced himself. He proposed that they start a new firm. They each put up $500 in 1976 and started Genentech. Early financing in Genentech was secured from venture funds and industrial sources. Lubrizol purchased 24% of Genentech in 1979. Monsanto bought about 5%.

Robert A. Swanson (1947–1999) was born in the USA and attended the Massachusetts Institute of Technology. In 1970, he obtained both a bachelor's degree in chemistry and a Masters of Business Administration from MIT. Mr. Swanson served as CEO of Genentech from its founding in 1976 through 1990, and from 1990–96 was chairman of its Board of Directors. (http://en.wikipedia.org. Robert Swanson 1999)

Founded earlier in 1971, Cetus provided a commercial service for fast screening of microorganism; but in 1976, Cetus changed its business to designing gene-engineered biological products. Cetus retained Stanley Cohen as one of its 33 scientific consultants. Cohen became head of Cetus Palo Alto (Business Week 1984). Further investment in Cetus came from companies interested in the new technology. A consortium of Japanese companies owned 1.59% of Cetus. Standard Oil of Indiana purchased 28% of their stock. Standard Oil of California bought 22.4%. National Distillers & Chemical purchased 14.2%. Corporate investors wanted to learn the new technology.

Both Genentech and Cetus offered stock-options to their key scientists. Genentech and Cetus were the first of the biotechnology firms to go public. Genentech realized net proceeds of $36 million. At the end of fiscal year 1981, it had $30 million cash but required about a million yearly for its R&D activities. In its public offering, Cetus raised $115 million at $23 a share. Of this, $27 million was intended for production and distribution of Cetus-developed product processes, $25 million for self-funded R&D, $24 million for research administrative facilities, $19 million for additional machinery and equipment, and $12 million for financing of new-venture subsidiaries.

For new firms, it is important that early products create income. In 1982, Genentech's product interests were in health care, industrial catalysis, and agriculture. In 1982, early products included genetically engineered human insulin, human growth hormone, leukocyte and fibroblast interferon, immune interferon, bovine and porcine growth hormones foot-and-mouth vaccine. Genentech's human insulin project was a joint venture with Eli Lilly, aimed at a world market of $300 million in animal insulin. Genentech's human growth hormone project was a venture with KabiGen (a Swedish pharmaceutical manufacturer), a world market of $100 million yearly. The leukocyte and fibroblast interferon was a joint venture with Hoffmann-La Roche, and the immune interferon with Daiichi Seiyaku and Toray Industries. The bovine and porcine growth hormones were a joint venture with Monsanto, and the foot-and-mouth vaccine, with International Minerals and Chemicals.

In comparison in 1982, Cetus was interested primarily in products in health care, chemicals, food, agribusiness, and energy. Their commercial projects included high-purity fructose, veterinary products, and human interferon. The high-purity fructose project was a joint venture with Standard Oil of California. In 1983, Cetus introduced its first genetically engineered product, a vaccine to prevent scours, a toxic diarrhea in newborn pigs.

But because of the complexity of biological nature, the path to riches for the new biotechnology industry had not gone smoothly: "Early expectations, in hindsight considered naive, were that drugs based on natural proteins would be easier and faster to develop.... However, ... Biology was more complex than anticipated" (Thayer 1996, p. 17).

For example, one of the first natural proteins, alpha-interferon, took 10 years to be useful in antiviral therapy. When interferon was first produced, there had not been enough available to really understand its biological functions. The production of alpha-interferon in quantity through biotechnology techniques allowed the real studies and experiments to learn how to therapeutically begin to use it. This kind of combination of developing the technologies to produce therapeutic proteins in quantity and to use them therapeutically took a long time and many developmental dollars.

Cetus spent millions of dollars and bet everything on interleukin-2 as an anti-cancer drug and failed to obtain the US Federal Drug Administration's (FDA) approval to market for this purpose. Subsequently, in 1992, Chiron acquired Cetus. Even in 1995, interleukin-2, which was eventually approved by the FDA, Chiron only earned 4% of its revenues of $1.1 billion in 1995. About this, George B. Rathmann commented: "... the pain of trying to get interleukin-2 through the clinic just about bankrupted Cetus and never has generated significant sales" (Thayer 1996, p. 17).

The innovation process for biotechnology industry in the USA included (1) developing a product, (2) developing a production process, (3) testing the product for therapeutic purposes, (4) proving to the US's FDA that the product was useful and safe, and (5) marketing the product. In fact, the recombinant DNA techniques was only a small part of the technology needed by the biotechnology and the smallest part of its innovation expenditures. The testing part of the innovation process to gain FDA approval took the longest time (typically 7 years) and the greatest cost. Because of this long and expensive FDA process in the USA, extensive partnering occurred between US biotech firms and the larger, established pharmaceutical firms. For example in 1995, pharmaceutical companies spent $3.5 billion to acquire biotechnology companies and $1.6 billion on R&D licensing agreements (Abelson 1996). Also pharmaceutical firms spent more than $700 million to obtain access to data banks on the human genome that was being developed by nine biotechnology firms.

The US Government role in supporting science was essential to the US Biotechnology industry: "The government has a very big role to play (in helping) to decrease the costs. Support of basic research through NIH (National Institutes of Health) is very important to continue the flow of technology platforms on which new breakthrough developments can be based" (Henri Termeer, chairman and CEO of Genzyme and chairman of the US Biotechnology Industry Organization) (Thayer 1996, p. 19).

The scientific importance of understanding the molecular nature of biology was an essential methodology to develop new drugs. Yet the making of money from the technology of recombinant DNA was harder and took longer than expected because biological nature turned out to be more complicated than anticipated. Yet all the expense going into the biotechnology industry has turned out to be worth it for molecular biology continues to be the future of the pharmaceutical industry, as George Rathmann summarized: "It doesn't have to follow that science automatically translates into great practical results, but so far the hallmark of biotechnology is very good science and that now is stronger and broader than ever…. The power of the science is ample justification that there should be good things ahead for biotechnology" (Thayer 1996, p. 18).

## Science Bases for Technology

Historically, the knowledge interaction between science and technology has been science providing "knowledge bases" upon which technologies are developed.

1. Scientists pursue research that asks very basic and universal questions about what things exist and how things work. (In the case of genetic engineering, the science base was guided by the questions: What is life? How does life reproduce itself?)
2. To answer such questions, scientists require new instrumentation to discover and study things. (In the case of genetic research, the microscope, chemical analysis techniques, cell culture techniques, x-ray diffraction techniques, and the electron

microscope were some of the important instruments required to discover and observe the gene and its functions.)

3. These studies are carried out by different disciplinary groups specializing in different instrumental and theoretical techniques: biologists, chemists, and physicists. (Even among the biologists, specialists in gene heredity research differ from specialists in viral or bacterial research.) Accordingly, science is pursued in disciplinary specialties, each seeing only one aspect of the existing thing. Nature is always broader than any one discipline or disciplinary specialty.
4. Major advances in science occur when sufficient parts of the puzzling object have been discovered and observed and someone imagines how to put it all together in a model (as Watson and Crick modeled the DNA molecule). A scientific model is conceptually powerful because it often shows both the structure and the dynamics of a process implied by the structure.
5. Scientific progress takes much time, patience, continuity, and expense. Instruments need to be invented and developed. Phenomena need to be discovered and studied. Phenomenal processes are complex, subtle, multileveled, and microscopic in mechanistic detail. (In the case of gene research, the instruments of the microscope and electron diffraction were critical, along with other instruments and techniques. Phenomena such as the cell structure and processes required discovery. The replication process was complex and subtle, requiring determination of a helix structure and deciphering of nature's coding.)
6. From an economic perspective, science can be viewed as a form of societal investment in possibilities of future technologies. Since time for scientific discovery is lengthy and science is complicated, science must be sponsored by government as a social "overhead" cost for a technically advanced society. Without the overhead of basic knowledge creation, technological innovation eventually stagnates for lack of new phenomenal knowledge for its inventive ideas.
7. Once science has created a new phenomenal knowledge base, inventions for a new technology may be made by either scientists or by technologists (e.g., scientists invented the recombinant DNA techniques). These radical technological inventions start a new industry.
8. When the new technology is pervasive across several industries (as genetic engineering is across medicine, agriculture, forestry, marine biology, materials, etc.), the technological revolution may fuel a new economic expansion. The long-waves of economy history are grounded in scientific advances that create basic new industrial technologies.

## National Innovation System

Organizationally, how is the innovation process institutionalized within a nation? This infrastructure has been called a "National Innovation System," consisting of three institutional R&D sectors – universities, industries, and governments. The innovation processes in the system are the procedures by means of which the

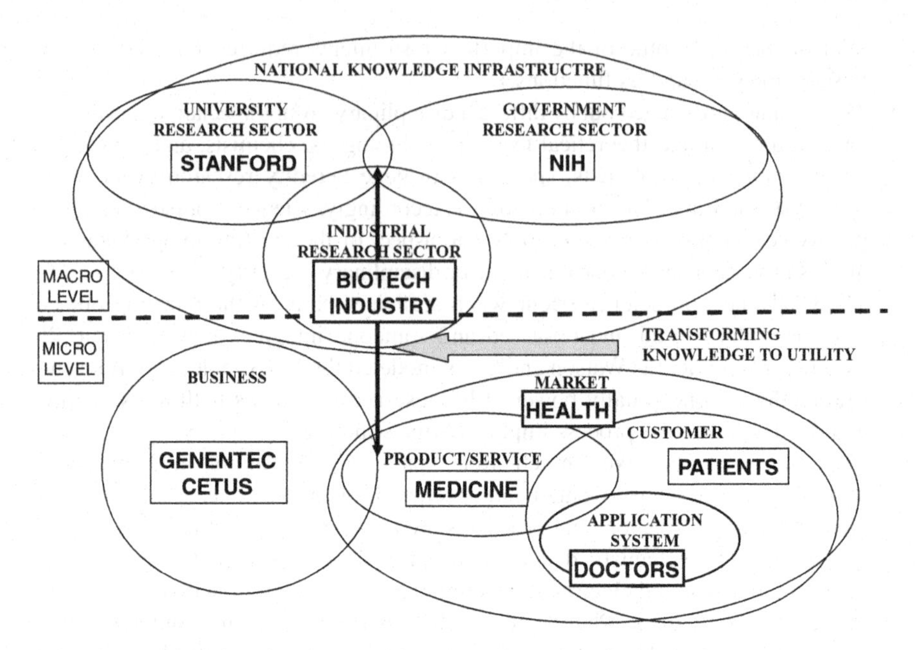


**Fig. 6.3** National innovation system institutionalizing the transformation of knowledge to value

three sectors interact for technological innovation. We can sketch this as a set of institutional areas (Venn diagrams) organized for the transformation of knowledge from nature to utility, as shown in Fig. 6.3.

The first group of sectors (university, industry, and government research-sectors) constitutes a macro-level of innovation research in a nation. Research-on-nature as science is the focus of university research. Support-of-science is the primary responsibility of government research funding; government research-performance in government laboratories focuses upon technology research. Industrial research in different companies focuses primarily upon technology research. The second group of high-tech businesses and markets constitute a micro-level of innovation in a nation.

Both levels are essential in an innovation system. The macro-level of research creates research progress in generic knowledge (science and technology) which provide knowledge bases for new high-tech products/services. It is the micro-level of research in particular businesses that translate generic knowledge into the design and production of high-tech products/services for sale to specific markets. The transformation by research of nature to utility occurs in the research transmission from macro- to micro-levels of innovation.

## *Illustration: US R&D Infrastructure*

We have emphasized how science has enabled in the invention of new technologies – the activity of applying science. Science has provided the knowledge bases for many new scientific technologies. Thus from an economic perspective, science can be viewed as a form of societal investment in possibilities of future technologies. And since time for scientific discovery is lengthy and science is complicated, science must be sponsored and performed as a kind of overhead function in society. Without the overhead of basic knowledge creation, technological innovation eventually stagnates for lack of new phenomenal knowledge for its inventive ideas. Once science has created a new phenomenal knowledge base, inventions for a new technology may be made by either scientists or by technologists (e.g., scientists invented the recombinant DNA techniques). Such radical technological inventions start a new technology and even a new industry (e.g., biotechnology).

In a national R&D infrastructure, the institutional sponsors of R&D are principally industry and the federal government (with some support by state governments and private philanthropic foundations). The performers of R&D are industrial research laboratories, governmental research laboratories, and universities. Industry is the principal producer of technological progress. Universities are the principal producers of scientific progress. (Government laboratories participate in varying degrees by country in performing some technological and scientific progress.) Governments have become the major sponsor of scientific research and a major sponsor of technological development in selected areas (such as military technology). One can see where the processes of innovation are performed in organizational sectors in Fig. 6.4.

In the second half of the twentieth century in the USA, the R&D expenditures in gross domestic product (GDP) ran to about 2.5%. Other nations R&D/GDP ranged from 1% to 3%. In the USA in 2004, about 70% of R&D was sponsored by

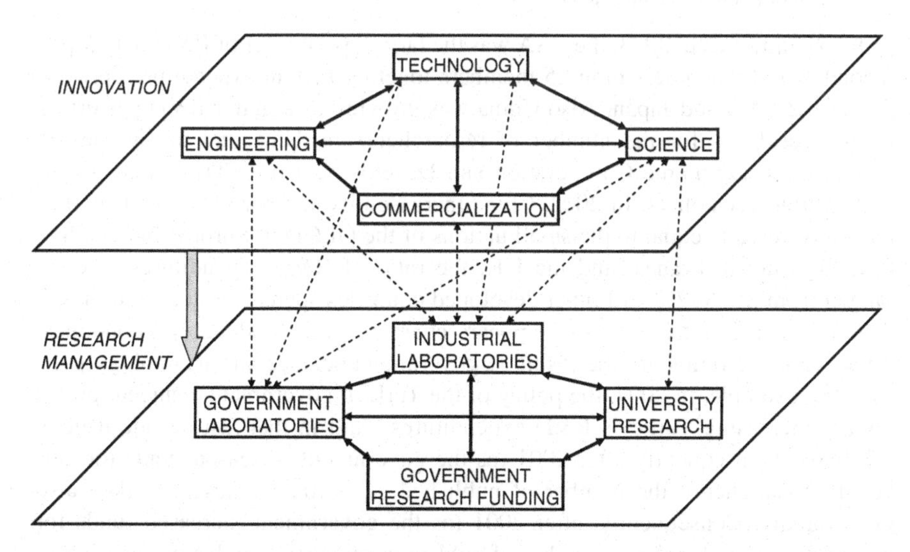


**Fig. 6.4** Innovation processes performed in institutions

industry and 30% by government. In 2004, about 70% of US R&D was performed by industry, 14% by universities, and 12% by governmental laboratories (NSB 2007). Source of the R&D funds were about 71% from industry, 24% Federal government, and 5% university.

Industrial support of R&D was primarily in a few industries: computers, information, chemicals, transportation, biotechnology, and professional services. This means that at any time in industry, only a few industries are investing in technology research. Technology change in products does not last forever. High-tech industries remain high-tech only as long as technology progresses. But technology progress ultimately depends upon nature and science. R&D investment alone cannot provide technology progress. As a basis for new technology progress, one also needs new science and new nature to be studied and manipulated.

In the USA, a major impact of the Second World War on the US R&D infrastructure was to provide a massive increase in federal funding for university research. Previously, some academic research had always been connected to industry prior to the war (particularly in engineering), but then the federal funds altered the balance. For example, N. Rosenberg and R. Nelson noted: "One consequence (of World War II) was a shifting of emphasis of university research from the needs of local civilian industry to problems associated with health and defense." The changes in US R&D infrastructure created a kind of division of labor between industry and universities: "R&D to improve existing products and processes became almost exclusively the province of industry, in fields where firms had strong R&D capabilities.... What university research most often does today is to stimulate and enhance the power of R&D done in industry, as contrasted with providing a substitute for it" (Rosenberg and Nelson 1994, p. 340).

> University research now advances the science base of knowledge upon which industrial research develops technological progress.

In expenditures in 2003, the USA was the largest performer of R&D with Japan second. The Economic Union 25 members together had an expenditure between that of the USA and Japan. Also China was growing as a major R&D performer in the world. As for the number of researchers, the USA and the nations of Organization Economic Co-operation and Development (OECD) had each over one million researchers. In 2002, China increased its number of researchers dramatically, already equal to the small nations of the OECD in Europe. In the USA, Germany, Japan, France, and the UK, the ratio of R&D expenditures to GDP ranged from 3% to 2%. All other developed countries spend less at a ratio of less than 1.5%. This range of 2–3% appears characteristic of advanced economies. These kinds of ratios are the result of government science and technology policies. For example, in 2007, the policy of the Turkish government changed toward rapidly increasing Turkey's R&D expenditures. And in that last decade it grew 0.3–0.8%. Consequently after 2001 (as the government increased funds for university researchers), the number of published scientific articles in Turkey also grew rapidly. Consequently, after 2001 (as the government increased funds for university researchers), the number of published scientific articles in Turkey also grew rapidly.

## *Illustration: Innovation of the Internet*

We turn from the example of innovation in biotechnology to an example of innovation in information technology. At the end of the twentieth century, research in science and technology created economic development by the innovation of the Internet.[2] In the twenty-first century, the Internet continued to technically fuel economic development around the world. The Internet is an idea, a technology, but it is implemented as an interconnected set of products and services provided by businesses as sketched in Fig. 6.5.

The Internet is constructed of many, many units which continually are connecting into or out of the network at different times – either as businesses directly connecting to the Internet or as home-based customers connecting to the Internet through connection services. The operations of this functional system enable users (as businesses or as consumers) to log onto the internet – through their respective personal computers or web-servers – and thereby communicate from computer to computer. The technological innovation of the Internet was commercialized by a set of businesses:

1. Sale of personal computers (e.g., Dell), containing a microprocessor (e.g., Intel CPU), an operating system(e.g., Microsoft Windows), and a modem
2. An Internet service provider (e.g., AOL, Vodaphone, Comcast, etc.)
3. A server and router (e.g., Cisco, Dell, IBM, etc.)
4. A local-area network or wide-area network in a business (e.g., Cisco, Erickson, etc.)
5. An Internet backbone communications system (e.g., AT&T, Sprint, Vodaphone, etc.)
6. Internet search services (e.g., Google, Yahoo, etc.)

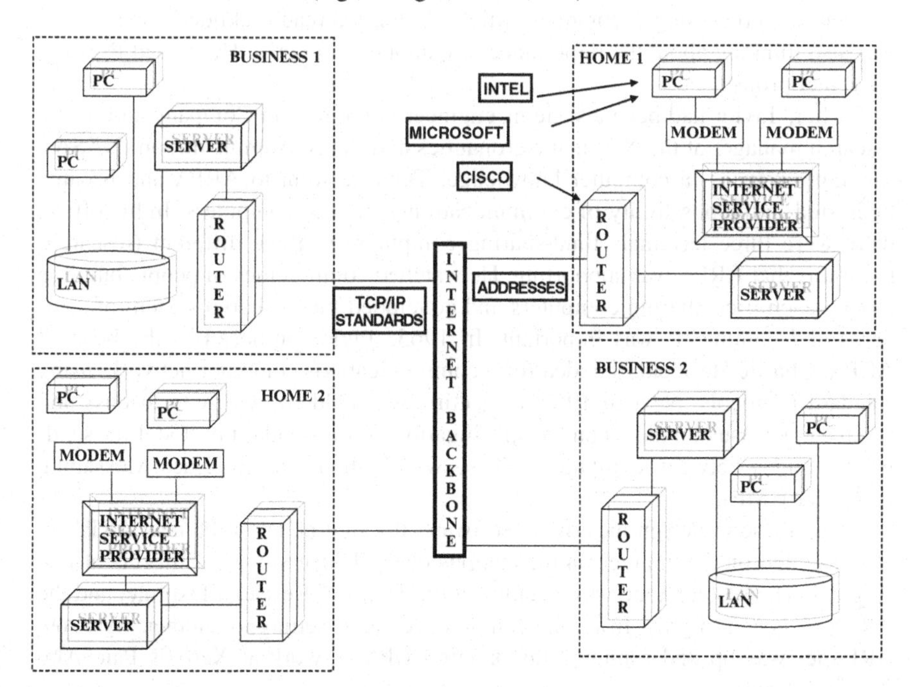


**Fig. 6.5** Architecture of internet

The invention of the Internet technology can be traced to an earlier computer network then called ARPAnet. ARPAnet's origin, in turn, can be traced to Dr. J.C.R. Licklider. Licklider served in 1962 in the government agency, Advanced Research Projects Agency, funding military research projects for the US Department of Defense. He headed research in ARPA on how to use computers for military command and control (Hauben 1993). Licklider began funding projects on networking computers. He wrote a series of memos on his thoughts about networking computers, which were to influence the computer science research community.

About the same time, a key idea in computer networking derived from research of Leonard Kleinrock. Kleinrock had the idea of sending information in packaged groups, packet switching. He published the first paper on packet switching in 1962 and a second in 1964. Packet switching enabled computers to send messages swiftly in bursts of information – without tying up communication lines very long and thus vastly increasing communication capacities of network lines.

In 1965, Lawrence Roberts at the Massachusetts Institute of Technology (MIT) connected a computer at MIT to one in California through a telephone line. In 1966, Roberts submitted a proposal to ARPA to develop a computer network for a military need (defense) for protection of US Military communications under a nuclear attack. This was called the Advanced Research Projects Administration Network, or ARPAnet (and was to develop, eventually, into the Internet).

Robert W. Taylor had replaced Licklider as program officer of ARPA's Information Processing Techniques Office. Taylor had read Licklider's memos and was also thinking about the importance of computer networks. He funded Robert's ARPAnet project.

Earlier, Taylor had been a systems engineer at the Martin Company and next a research manager at the National Aeronautics and Space Administration (NASA), funding advances in computer knowledge. Then he went to ARPA and became interested in the possibility of communications between computers. In his office, there were three terminals time-sharing computers in three different (research) programs that ARPA was supporting. He watched communities of people build up around each time-sharing computers and concluded that a shared system of standards and commands was important. In 1965, Taylor proposed to the head of ARPA, Charlie Herzfeld, the idea for a communications computer network, using standard protocols. Next in 1967, a meeting was held by ARPA to discuss and reach a consensus on the technical specifications for a standard protocol for sending messages between computers. These were called the "Interface Messaging Processor (IMP)."

Using these to design messaging software, the first node on the new ARPAnet was installed on a computer on the campus of the University of California at Los Angeles. The second node was installed at the Stanford Research Institute, and the ARPAnet began to grow from one computer research setting to another. By 1969, ARPAnet was up and running. Taylor left ARPA to work at Xerox's Palo Alto Research Center.

J.C.R. Licklider

Leonard Kleinrock

Robert W. Taylor
(http://en.wikipedia.org. Internet 2009)

As ARPAnet grew, there was the need for control of the system. It was decided to control it through another protocol, called network control protocol (NCP); and this was begun in December 1970 by a private committee of researchers called the Network Working Group.

The ARPAnet grew as an overall interconnected independent multiple sets of smaller networks. In 1972, a new program officer at ARPA, Robert Kahn, proposed an advance of the protocols for communication as an "open architecture" accessible to anyone. These were formulated as the Transmission Control Protocol/Internet Protocol (TCP/IP). They became the open standards, upon which later the world's Internet would be based.

While the ARPAnet was being expanded in the 1970s, other computer networks were being constructed by other government agencies and universities. In 1981, the National Science Foundation (NSF) established a supercomputer centers program, which needed to have researchers throughout the USA able to connect to the five NSF-funded supercomputer centers (in order for researcher's computers to use these supercomputers). NSF and ARPA began sharing communication between the networks, and the possibility of a truly national Internet became envisioned. In 1988, a committee of the National Research Council was formed to explore the idea of an open, commercialized Internet. They sponsored a series of public conferences at Harvard's Kennedy School of Government on the "Commercialization and privatization of the internet."

In April 1995, NSF stopped supporting its own NSFnet "backbone" of leased communication lines; and the Internet was privatized. The Internet grew to connect over 50,000 networks all over the world. On October 24, 1995, the Federal Network Council defined the Internet as:

- Logically linked together by a globally unique address space based on the Internet Protocol (IP)
- Able to support communications using the Transmission Control Protocol/Internet Protocol (TCP/IP) standards

One can see in this case that the innovation of the Internet occurred at a macro-level of a nation – motivated by researchers seeking ways to have computers communication with each other – a new kind of functional capability in computation. The invention of the computer networks required the creation of several technical ideas, and together these constitute the *technology* of the Internet:

1. *Computer-to-Computer Communications*. Computers would be electronically connected to each other.
2. *Packet-Switching*. Computer messages should be transmitted in brief, small bursts of electronic digital signals – rather than a continuous connection used in the preceding human voice telephone system.
3. *Standards*. Formatting of the digital messages between computers needed to be standard in format to send message packets, and these open standards became the Internet's (TCP/IP) standards.
4. *Routing*. A universal "address" repository would provide addresses so computers could know where to send messages to one another.
5. *HTML*. Web pages would be written in a language which allowed computers to link to other sites
6. *World Wide Web (www)*. World Wide Web registration of directory of Web sites allows sites to be connected through the Internet.
7. *Browser*. Software on computers allowed users to link to the World Wide Web (www) and find sites.
8. *Search Engine*. Software allowed users to search for relevant sites and link to them.
9. *Web Page Publication*. Software facilitated the preparation and publication of sites on the Internet.

A technology consists of the technical ideas which together enable a functional transformation. (The functional transformation of the Internet technology provides communication between and through computers).

All these technical ideas together enabled the new Internet technology. Next commercialization of the new technology occurred when NSF transferred network management from the government to private companies. Thus, the innovation of the Internet did occur in a common pattern of technological innovation – first the *invention* of new technical ideas (as ARPAnet) and second the *commercialization* of new products and services embodying these new ideas (in the privatization of the Internet).

Technological innovation consists of both the invention and commercialization of a new technology.

## Economic Long-waves

The Internet impacted the US economy initially with a dramatic growth in markets. As the twentieth century ended, there was a financial boom from 1995 which then burst in 2000, historically known as the "dot.com" financial bubble of the US stock market. In the year 2000, the US NASDAQ stock market peaked at a then historic high of 6,000 (Fig. 6.6).

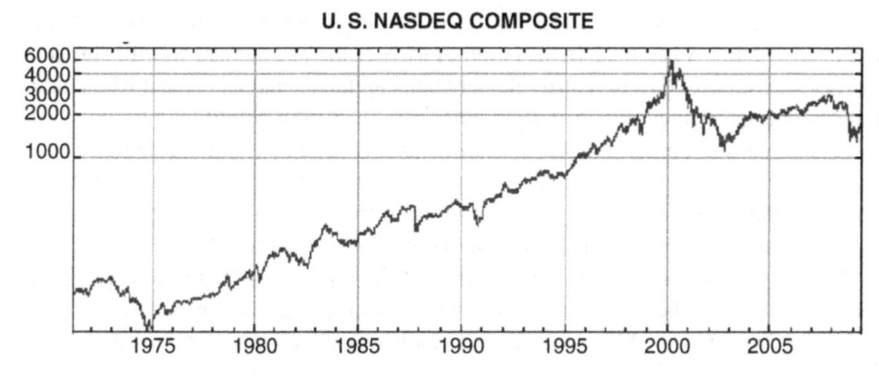


**Fig. 6.6** US stock market dot.com financial bubble

Joseph Nocera and Tim Carvell nicely summarized that time: "The Internet decade has seen the unscrupulous rewarded, the dimwitted suckered, the ill-qualified enriched at a pace greater than at any other time in history. The Internet has been a gift to charlatans, hypemeisters, and merchants of vapor… and despite all that, it still changes everything" (Nocera and Carvell 2000, p. 137). At the time of the commercialization of the Internet in 1995, there also occurred, coincidentally, a US government deregulation of the US telecommunications business. It was in both the deregulation and the new Internet that many entrepreneurs saw new business opportunities. They launched both new telecommunication businesses and many new businesses in electronic commerce, "dot.com businesses."

Historically, the Internet stock bubble was just one of the many examples of *excessive financial enthusiasm* about technological innovation. (An earlier example in the late 1800s in the USA was a financial bubble in railroad stocks.) There are important economic lessons one can learn about radical innovation, in this pattern:

1. Scientific research provides the knowledge base for the invention of new basic technologies.
2. New basic technologies create new economic opportunities.
3. The path to wealth in economic opportunities is often financially rocky.
4. However rocky the entrepreneurial path, economic development from technological innovation does become a permanent feature of societal structures – for example, despite that stock bubble, the Internet and electronic commerce was in the world to stay.
5. Technological innovation and economic development is neither simple nor inevitable.

Technological innovations, such as the Internet, have historically produced "booms and busts" in economies. The first economist to empirically document this pattern was the Russian economist, Nicoli Kondratiev. In the 1920s, he studied the economic development in the English economy of the late 1800s and early 1900s.

Kondratiev did his pioneering studies in Russia under the communist dictators Lenin and Stalin. A key tenet in the communist doctrine of Marxism was that modern history was supposed to be a struggle between capitalists and labor; and when capitalists dominated, they would starve labor. This starvation would lead to the inevitable decline and collapse of capitalism. Marxists assumed that capitalists would never allow labor to have sufficient wages to create a viable internal market which could sustain capitalism over the long term. Kondratiev measured the long-term economic activity of the capitalist country of England to test Marx's prediction. But he found (instead of Marx's prediction of the inevitable decline of capitalism) there actually had been recurrent cycles of economic expansion and contraction in England's economy. Moreover, the net result was increasing economic activity in England, not decreasing.

Kondratiev asked the question: How was capitalism in England periodically renewed and overall expanding? Kondratiev's answer was "technology." He plotted a correlation between times of basic innovation to times of economic expansion. Overall, the capitalistic economies were expanding rather than contracting because of periodic innovations of new technologies.

This economic idea is now called a "long economic cycle" or a "Kondratiev wave." The idea is that long periods of economic expansion in modern economies are stimulated by the invention of a new basic technology – such as the airplane or radio or computers or biotechnology. New businesses begin to economically exploit the new technology, and new markets emerge to purchase the new-technology (or high-technology) products and services. This expansion of businesses and markets drives the economy over a long time during the rising part of a Kondratiev wave.

Nicoli Kondratiev (1892–1938) was born near Moscow. He studied at the University of St. Petersburg. After the abdication of the Czar Alexander, a government was formed under Alexander Kerensky, and in 1917, Kondratieff was appointed as Minister of Supply. After the Bolshevik revolution in 1919, Kondratieff was appointed to a teaching position at the Agricultural Academy. Then in 1920, he founded the Institute of Conjuncture for the study of business cycles. During 1923–25, he worked on the Soviet 5-year plan for agriculture. In 1924, he published his book detailing his theory of major economic cycles. He then traveled to Europe and the United States. After the death of Lenin in 1924 and the rise of Stalin in 1928, Kondratieff was removed from his institute directorship. In 1930, he was sentenced to eight years in prison. In a prison at Suzdal, near Moscow, he continued his research and prepared new books. But then in 1938 in one of Stalin's "great purge" trials, Kondratiev was sentenced to an additional ten years but executed by a firing squad that same day. Kondratiev was imprisoned and executed because he had violated a cardinal rule of dictatorships - instead of reiterating party dogma, he told the truth. (In dictatorships, you get shot when you tell the truth.)

Nicoli Kondratiev (http://en.wikipedia.org, Nicoli Kondratiev 2007)

While Kondratiev was imprisoned in Russia in the 1930s, another economist in Austria, Joseph Schumpeter, had read and shared Kondratiev's idea – that innovation was important to economic growth: "... the work of Schumpeter ... put emphasis on innovation and on the subsequent burst of entrepreneurial investment activity as the engine in the upswing of the long cycle, a la Kondratiev ..." (Ray 1980, pp. 79–80). In his book, *Business Cycles*, Schumpeter thought of the idea of economic "innovation" could be of several kinds:

1. The introduction of a new good (product or service) to a consumer (or new quality in a good).
2. The introduction of a new method of production of a good.
3. The opening of a new market to the sales of a good.
4. The obtaining of a new source of supply (inventing or finding or developing new material or energy sources).
5. The implementation of a new form of organization in a business which provides a competitive advantage to the business (Schumpeter 1939).

What technological progress can do for these kinds of innovation is (1) provide a new good or improve the quality of a good (e.g., telephone, radio, airplanes, automobiles, plastics, computers, etc.), (2) improve the quality of production and/or lower the cost (e.g., new semiconductor chip production processes), (3) access a new market by providing a new functional capability to the market (e.g., e-commerce by means of the Internet), (4) locate new sources of resources or provide improved means for obtaining and processing the resources (e.g., seismic techniques for discovering oil deposits or horizontal drilling techniques). Thus Schumpeter saw the idea of "economic innovation" as any means which (1) creates new markets or market expansion, and/or (2) provides a competitive advantage to a business serving these markets.

This is the fundamental idea for connecting science to society, science to economy – technological innovation. In particular, innovation based upon new scientific technologies has been a major force in modern economic developments.

Later, Schumpeter expanded on the impacts of radical innovation as an economic process using the term "creative destruction." He argued that radical technological innovations destroyed existing industries and created new industries. This was a process of economic renewal, destroying the old to create the new (Schumpeter 1942).

Joseph Schumpeter (1883–1950) was born in Moravia (now part of the Czech Republic). He studied law at the University of Vienna, obtaining his PhD in 1906. In 1909, he became a professor of economics at the University of Czernowitz and then moved in 1911 to the University of Graz. In 1919, he served as the Austrian Minister of Finance. In 1920, he became head of the Biederman Bank until it collapsed in 1924 (due to the post-war inflation in Germany and Austria). From 1925 through 1932, he was a professor at the University of Bonn. But in 1932 with the rise of the Nazis, Schumpeter fled to the U.S., where he became a professor at Harvard. Schumpeter fled the Nazi dictatorship not because of his ideas but of his ethnicity. (Dictators hate truth and also often have very weird ideas. Historically, science thrives better in open societies than in prejudiced dictatorships.)

Joseph Schumpeter (http://en.wikipedia.org, Joseph Schumpeter 2007)

In 1990, Robert Ayers brought more up-to-date data about Kondratiev's earlier empirical correlation – between European industrial expansion and contraction and the occurrence of new-technology-based industries (Ayers 1990).

---

New-Technology Industries and Early Economic Expansion in Industrialization

1770–1800 Economic expansion

The beginning of the industrial revolution in England and next in Europe was based on the new technologies of steam-engine power, coke-fueled steel production, and textile machinery factories.

1830–1850 Economic expansion

A second acceleration of the European industrial revolution was based upon the innovation of the new technologies of railroads, steamships, telegraph, and coal-gas lighting.

1870–1895 Economic expansion

Contributions to a third wave of economic expansion were made by innovations of new technologies in electrical light and power, telephone, and chemical dyes and petroleum production.

1895–1930 Economic expansion

A fourth wave of economic expansion in both Europe and North America was based on innovations of new technologies in automobiles, airplanes, radio, and plastic materials.

---

1770–1800 – The invention of the steam-powered engine required the science base of the physics of gases and liquids. This new scientific discipline of physics provided the knowledge base for Thomas Newcomen and James Watt's inventions of the steam engine. Coal-fired steel required knowledge of chemical elements from the new science base of chemistry, which was to be developed in the middle of the 1700s. The new disciplines of physics and chemistry were necessary for the technological bases of the industries of the first wave of economic expansion from the beginning of the industrial revolution in England.

1830–1850 – The second economic expansion to which the telegraph contributed was based on the new discoveries in electricity and magnetism in the late 1700s and early 1800s.

1870–1895 – In the third long economic expansion, it was again the new physics of electricity and magnetism that provided the science bases for inventions of electrical light and power and the telephone. Also advances in the new discipline of chemistry provided the science base for the invention of the chemical dyes. Artificial dyes were in great economic demand because of the expansion of the new textile industry. Producing these dyes added to the production of gun powder, that industry expanded into the modern chemical industry.

1895–1930 – The fourth long economic expansion was fueled by the invention of the automobile, which depended upon the earlier invention of the internal combustion engine. This internal combustion engine invention required knowledge bases from chemistry and physics. Radio was another new invention based upon the advancing science in physics of electricity and magnetism. Chemical plastics were invented from further experimentation in the advancing scientific discipline of chemistry.

The contractions of economic activity also followed after the expansions of economic activities due to the innovation of new technologies. Why should economies first expand and then later contract after technological innovation? Historically in the late 1800s and early 1900s, why didn't the English or other European economies (which Kondratiev followed) continue to grow smoothly as the new technologies created new industries? Kondratiev had also observed the successive economic contractions in England. The answer is in the interaction between technology and finance.

---

Economic Cycles in England from 1792 to 1913

The English economy expanded from 1792 to 1825 but then contracted from 1825 to 1847. Kondratieff argued that temporary excess production capacity in iron production, steam engine production, and textile production increased economic competition and drove prices down for an economic recession.

After the second economic expansion in England from 1847 to 1873, there followed another economic contraction from 1873 to 1893 due to excessive investments in the new industries of railroads, steamships, and telegraph industries.

The third economic expansion from 1893 to 1913 was interrupted by the First World War in Europe. Then the economic expansion was renewed in North America and Japan, only to end in a worldwide depression – due to (1) a collapse in the stock market in the USA and (2) excessive inflation in Germany. The global depression did not end in the different several countries until military production in their economies restarted their industries – for weapons production.

---

This long-term interaction between technology and finance can be partly described in a "model" of the Kondratiev long wave (Betz 1987). At first, technological innovation attracts capital investment which produces returns on investment as new high-tech products and services are sold into an expanding market. But this business expansion attracts competition. Further financial investment launches competing businesses in the new market for high-technology products and services. And even while the new high-technology-based industry markets grow, industrial production can grow even faster. This leads to a temporary excess production-capacity, which results in price competition and some business failures. Business expansion can be followed by business contraction when financial investment creates temporary excess supply to demand. The stages in this Kondratiev long wave consist of:

1. Science discovers phenomena which can provide for a new manipulation by technological invention of nature.
2. New basic technology provides business opportunities for new industries.
3. A new high-tech industry provides rapid market expansion and economic growth.
4. As the new industry continues to improve the technology, products are improved and prices decline and the market grows toward large volume.
5. Competitors enter the growing large market, investing in more production capacity.

6. As the technology begins to mature, production capacity begins to exceed market demand, triggering price cuts.
7. Excess capacity and lower prices cut margins and increase business failures and raise unemployment.
8. Turmoil in financial markets may turn recession into depression.
9. New science and new basic technologies may provide the basis for a new economic expansion.

History is not deterministic. There is no historical inevitability in Kondratiev's long-wave pattern. It can begin (1) only after scientists discover new nature and (2) after technologists invent new basic technologies. Then basic technological innovation can provide business opportunities for economic expansion. But there is no guarantee that new science and new technologies will always be invented.

It is the innovative interaction between science and technology and economy which provides the basis for long-term economic expansion. Yet also within the competitive financial processes of industry, there is a second pattern – called a short-term economic cycle. In the normal business cycle, after periods of economic expansion, there has often developed excess production capability, which lowers prices as any new technology-based industry matures. The important point about the long-wave pattern is that one should expect eventual over-production, even in a new high-tech industry as technology matures and many competitors enter the new market. This always will cut profit margins, even in a relatively new industry. High-tech industry will never be a high-profit-margin industry for long, competition will intensify after the technology begins maturing.

Neither technological progress nor economic development is smooth and continuous – but both involve discontinuous (technology) and cyclic (economic) processes.

## Performance of National Innovation Systems

In OECD countries, there are many studies trying to evaluate the performance of national innovation systems. The Organization for Economic Development Co-operation and Development periodically reviews the innovation policies of different countries. For example, in 2006, OECD reviewed the innovation policy for Switzerland. It noted that strengths of Switzerland in innovation were:

1. "Strong industrial research and innovation. Switzerland has a strong and varied industrial research base.
2. A high-quality research-oriented university sector and a well-developed research infrastructure.
3. A strong services sector. This sector, which includes a highly developed financial industry, plays an increasing role in the Swiss economy and innovation system.
4. Orientation towards high quality. A pervasive orientation toward high-quality products and services throughout the Swiss economy contributes to high standards, performance and reputation" (OECD 2006, p. 8).

If this example is indicative of the kinds of strengths in an innovation system, a nation needs to have in the modern globally competitive world. There are three aspects of the research sector system are important:

1. A national innovation system should have in its industrial sectors strong research capabilities.
2. A national innovation system should have a high-quality research-oriented university sector.
3. A national innovation system should have at least one strong internationally competitive industrial or service sector.
4. A national innovation system should have a culture of valuing high-quality of performance.
5. A national innovation system should have government policies that strongly fund appropriate R&D activities in universities and selected mission areas.
6. A way to identify cutting-edge science to jump over current technology should be a national science and technology policy priority.
7. Science and technology policy must balance research for technology improvement in current industries and research to establish new internationally competitive industries in new technologies.

In evaluating the performance of a national innovation system, it is essential to distinguish between two concepts about a national economy – productivity and innovation. Productivity is the quantity of goods and services produced per capita in a nation. Innovation is the quantity of new high-tech goods and services products introduced into the markets of a nation. All businesses and universities in a country contribute to the nation's productivity but not all necessarily contribute to innovation.

Only high-tech businesses and research universities contribute to innovation. High-tech businesses invent and commercialize new technology products. Research universities train new doctorates in science and engineering and contribute to scientific literature.

This distinction is important for evaluating the contribution of human resources to innovation in a nation. All the universities in the nation contribute to productivity – because a college-educated proportion of a population is per-capita more productive than a non-college-educated proportion. But all the universities do not equally contribute to innovation. In innovation, only research universities contribute to innovation through their production of PhD scientists and engineers and Master degree graduates in engineering. Doctoral scientists and engineers perform the research necessary for innovation. And master's level engineering graduates design the high-tech products for innovation. Thus general education contributes to economic productivity in a nation but not necessarily to innovation. Still the standard of general education is crucial to current economic productivity.

Doctorates in science or engineering or computer science manage research in the stages of innovation; while master's graduates in science or engineering or computer science design the new high-tech products or services for innovation.

Since the research universities in the university sector are the primary producers of scientific progress in a nation, one should pay special attention to the funding and performance of the research universities in a national innovation system.

For an internationally competitive science base in a nation, its national innovation system should have a set of research universities operating at international standards of scientific progress.

In summary, for scientific technology to be developed and commercialized in a nation in order for the nation to be globally competitive in high-tech products and/or services, the respective specializations of the different research sectors must all be strong and also interconnected:

1. The nation should have a national science base in research universities, which are contributing significantly to international progress in science and engineering research and are producing research doctorates in science and engineering. This is the "science base" for the national innovation system.
2. The nation should have an internationally competitive set of industries, exporting products and/or services competitive in the global market.
3. The nation should have government agencies which provide research funding to universities and industries for progress in science and technology – at standards of international frontiers of science and technology.

## Summary

1. Three institutional sectors of research are important to a nation's capability for technological innovation: university, industry, and government.
2. The logic of technological innovation requires successive stages of research focus – from science to technology to engineering to commercialization.
3. To be globally economically competitive, a modern nation needs to have its innovation system operating at international standards in science and technology research.

## Notes

[1] More recent histories of DNA include (Morange 1998).

[2] There are many histories of the Internet, such as Hafner (1998), Sherman (2003), Banks (2008); and some of the history of the early Internet can be found on sites, such as (http://www.nsf.gov/about/history/nsf0500/internet/internet.htm) or (http://www.livinginternet.com).

# Chapter 7
# Paradigms and Perceptual Spaces

## Introduction

We have looked at the scientific method and research techniques, communities of scientists, application of science, and organization of innovation. Now, we turn to the idea of the content of science and how content differs among the scientific communities.

For a field of science, the content of science: (1) defines the current state of science and (2) provides future research challenges. The scientific contents of a discipline are published as scientific papers and as science textbooks. Scientific papers record the latest research in a discipline and are published in scientific journals. Science text books are used in science courses to train the next generation of researchers. Science is divided into disciplines, and disciplinary scientists know in depth only portions of science (usually a specialty area in the discipline) and the basics of the discipline.

But scientific content is vast! This creates a problem as very few scientists (probably none) know all of science – particularly in knowing the recent scientific publications in all scientific specialties. How can scientists communicate across disciplines? How can scientists from different specialties or disciplines cooperate in research teams – when the research requires knowledge and skills from different specialties or disciplines and research techniques? And much cutting-edge research is interdisciplinary and/or multidisciplinary – requiring research teams. The answer, making interdisciplinary and multidisciplinary research cooperation and teams possible, lies in the idea of a *scientific paradigm*. Now, we look at this meta-level idea of a scientific paradigm. How many paradigms are used in science? And which disciplines use which paradigms?

As indicated in Fig. 7.1, we now look at the interaction between scientific method and scientific content. We will see how scientific paradigms provide a meta-logic (intellectual framework) for the research techniques of scientific method.

F. Betz, *Managing Science*, Innovation, Technology, and Knowledge Management 9,
DOI 10.1007/978-1-4419-7488-4_7, 

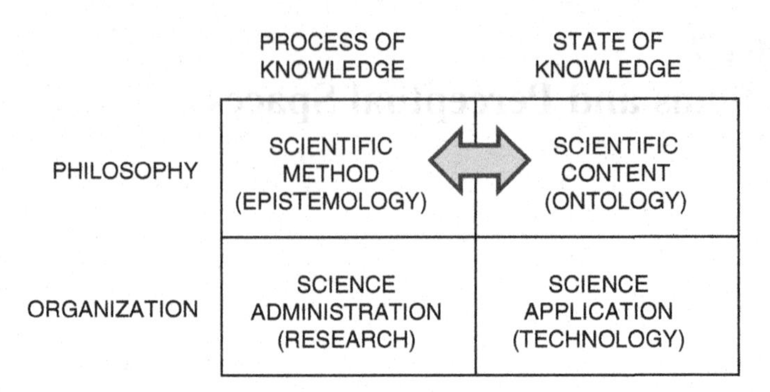


**Fig. 7.1** Philosophy and organization of science

## *Illustration: Center in Neuromorphic Systems Engineering*

Understanding the concept of a scientific paradigm is important because it enables multidisciplinary research, research requiring the efforts of several different disciplinary researchers. And the frontiers of scientific and technological research are mostly multidisciplinary. One could see this vividly, at the end of the twentieth century, in one of the most innovative research centers of the time, the Center for Neuromorphic Systems Engineering (CNSE) at the California Institute for Technology (CalTech) in Pasadena, California, USA. The Center's research program crossed two disciplines, biology and electrical engineering. It had asked the research issue of: Could a new field of electronics be developed, learning from how biological organisms sensed and thought?

The Center posted its research strategy as: "Vision. Olfaction. Hearing. Touch. Learning. Decision making. Pattern recognition. These are all things that even simple biological organisms perform far better and more efficiently than the fastest digital computers. The scientists and engineers at CNSE are working to translate our understanding of biologic systems into a new class of electronic devices that imitate the ways animals sense and make sense of the world" (http:/www/cnse.caltech.edu, 2007).

That multidisciplinary research vision between biology and electrical engineering began in the early 1990s. The Center had been proposed by Carver Mead for funding by the US National Science Foundation in its Engineering Research Center (NSF/ERC) Program. Mead's research idea was to make electronic sensing and control systems that mimicked how nature performed certain functions.

To understand this research vision at the time of the origin of the Center, an NSF site review team in 1993 had traveled to Cal Tech to evaluate their research proposal for a new multidisciplinary center between biology and engineering. Then, Carver Mead who was a member of the Cal Tech faculty team had proposed the center to NSF and explained their vision. Mead had drawn a schematic of the "mind" (Fig. 7.2).

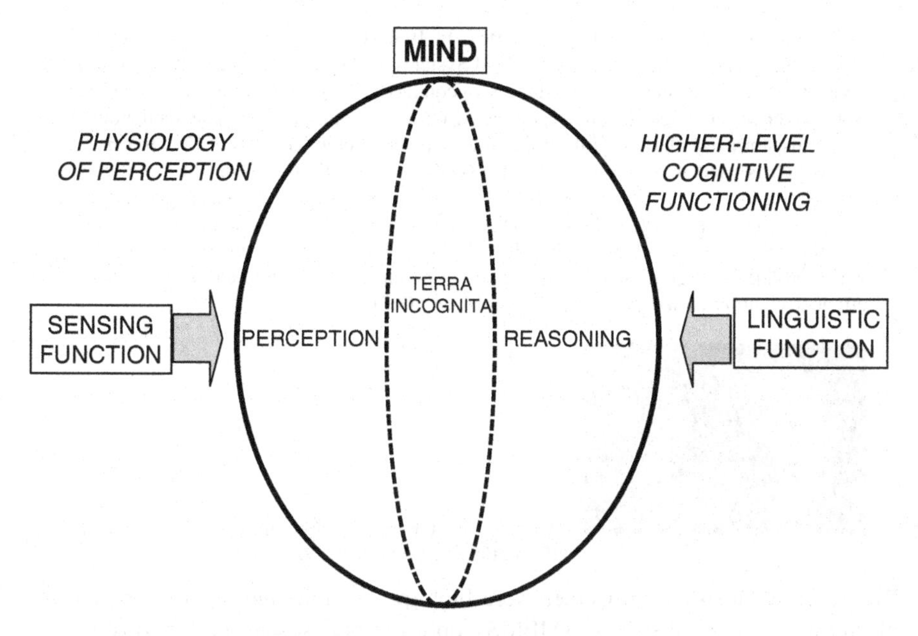


**Fig. 7.2** Schematic of the mind from perception to language

Carver Mead argued that the mental processing of the mind as having two sources of inputs from either side of a mental universe, sensing and language. A mind had two different kinds of inputs: (1) senses from the external physical world and (2) linguistic activity from the internal mental world of learned language capabilities. He proposed that (1) the study of the physiology of the sensing function from physical inputs to the mind was an entirely different kind of research from (2) the study of linguistic function which produced reasoning in the mind. The research strategy of the proposed center was to develop electronic ideas from (1) the physiology of sensing and not from (2) higher cognitive functioning. At the presentation, Mead deliberately was not at the time concerned for the mental connection of sense physiology to linguistic cognition. He simply indicated this connection as a biological "terra incognita" (unknown land) of how perception was transformed into reasoning. Exactly how, Mead asserted, present science did not then know.

The focus of the Center was to be upon the physiological of sensing side of the mind divide – to the left of Mead's scheme of terra incognita of the brain. Electrical engineers would focus on this side of sensing, as biological research into neural processing of sensory data (early 1990s) was then a rapidly advancing research field.

Carver Mead was born in 1934 in California, USA. In 1956, Mead obtained a bachelors degree from Caltech and a PhD degree in 1960 in electrical engineering. He began as an Instructor in Electrical Engineering in Caltech in 1958 and became an assistant professor in 1959 advancing to full professor in 1967. In 1980, Carver Mead with Lynn Conway wrote a pioneering text, Introduction to VLSI Systems, which has been used to train generations of electrical engineers in designing semiconductor integrated circuits, at a transistor density then of very large system integration. Also Mead invented an important configuration of a transistor particularly useful for carving in a silicon chip, GaAs MESFET transistor, and which is used in wireless electronics (Galium Arsenide transistors in a particular configuration). He then turned to studying how animal brains compute and created the idea of an electronics approach to draw lessons from biological neural system processes, which he called Neuromorphic Engineering. In 1999, he founded a new company, Foveon Inc, which produced a digital camera with image sensors in silicon to capture three pixel colors efficiently. (http://en.wikipedia.org, Carver Mead 2007)

Carver Mead Photo by Jon Brennis (http://web.mit.edu/invent/a-winners/a-mead.html, 2007)

The research areas in the Center were biology, learning and algorithms, micro-electronic-mechanical systems (MEMS), optics, robotics, sensors. We look at two projects in these areas: (1) how a computer can interface with a brain and (2) how a computer chip technology for processing visual images could be improved.

An example of a project in 2007 in the biology group of the Center was one performed by Rajan Bhattacharyya, and Richard Andersen: "Technological developments in the past decade have accelerated the pace of research in brain computer interfaces. Multiple research groups across the country are pursuing this area of research as a possible solution to spinal cord injury. (We) …specialize in studying brain areas in the parietal cortex, which is associated with vision and motor planning, and in particular the Parietal Reach Region (PRR) which encodes the plan for the next intended reach movement ...") This project was doing biological experiments in physiology of the brain for vision and motion control.

An example of a research project on the engineering side of the Center was in the Sensor group, performed by Ania Mitros under the supervision of Christof Koch: "Feature extraction is a first step for many existing computer vision algorithms. This computation is also often one of the most time- and resource-intensive steps because the same local computation must be performed at each pixel. To head towards a real-time, small-size, energy-efficient implementation... a (new) feature extraction algorithm was implemented in silicon… each feature detector worked splendidly (from the algorithm) but transistor mismatch killed the performance of the silicon array. This project improves the silicon performance by changing the transistors in the array" (http:/www/cnse.caltech.edu, 2007). The goal of this project was not biological but technological – to improve the technology of semiconductor integrated circuit (IC) chips that process visual images.

These examples are technical and difficult to understand to those of us outside the disciplines of biology and electrical engineering. But what is interesting is that both projects were crossing between biology and electronics. And what we can understand is that in biologically trying to understand how biological organisms sense and think, the multidisciplinary team was trying to invent new electronics to artificially sense and think effectively as had some organisms so evolved. The first research-project example was biological research – in the physiology of sensing and control in the brain. The second research-project example was electronic research – in improving signal processing speed in semiconductor chips for computerized image processing. What was striking in the research of the Center was bringing very diverse disciplines together – biology and electronics. This multidisciplinary research center cut across biology and engineering and involved both biology and electrical engineering faculty at Cal Tech. In this Center, one can see how the two different fundamental perspectives of biology and electrical engineering were combined in a multidisciplinary research program.

## Scientific Paradigm

Now we look at the idea of how a fundamental perspective in a science or engineering discipline derives from an intellectual framework in which each discipline observes nature and formulates scientific theory; and this is now called a scientific paradigm. It was Thomas Kuhn who introduced the term into the philosophy of science (Kuhn 1996).

A "scientific paradigm" is an "intellectual framework behind scientific theories." A scientific paradigm does not describe the "details of research" at the cutting edge of disciplinary specialties. Instead, a paradigm describes the meta-theory, the framework in which the research details (experimental formulation and theory) are constructed. A paradigm is an intellectual framework within which the scientists observe, describe, and explain nature. A paradigm is a "meta-logic" to the "logic" in a theory – a kind of "meta-theory" to the theories in a scientific discipline.

As examples of this in science, Kuhn used the two paradigm shifts in physics in the beginning of the twentieth century: (1) from Newtonian physics to special relativity and (2) from classical mechanics to quantum mechanics. Both shifts, he argued were accepted within the physics community as "generational changes," with younger scientists more easily making the intellectual change than many older scientists.

Kuhn's book had a major impact upon sociologists because it introduced the idea of group consensus as a methodological issue in science. Kuhn argued that scientific consensus in a community was not always easily nor smoothly attained. Instead, consensus depends upon how big an intellectual leap was being conceptually proposed as "progress in science." Kuhn argued that science does not always progress by a steady accumulation of knowledge but sometimes makes large conceptual leaps in the forms of a paradigm shift.

Thomas Kuhn (1922-1996) was born in Ohio, USA. In 1943, he received a bachelor's degree from Harvard University and then a PhD in 1949 in physics. From 1948 to 1956, he taught a history of science course at Harvard. IN 1957, he went to the University of California at Berkeley to join there both the philosophy department and history department. In 1964, Kuhn moved to Princeton University and then to MIT in 1991. In 1962, Kuhn had published his seminal book in the sociology of science, The Structure of Scientific Revolutions.

Thomas Kuhn (http://en.wikipedia.org, Thomas Samuel Kuhn, 2007)

## *Illustration: Kant's Critique of Pure Reason*

An illustration of how paradigms provide an intellectual framework for scientific theory (a meta-logic to the logic in theory), we will refer to a major work in traditional philosophy, Immanuel Kant's *Critique of Pure Reason*. In his book, Kant proposed that a mind must have two kinds of a priori capabilities before a sensory experience can be perceived. By the term "a priori," Kant meant certain capabilities must be built into the mind in order for the mind to process any experience.

Kant was interested in the philosophical question, what kinds of a priori capabilities must exist before any mind can reason and experience mental images in sensing nature. He identified two general kinds of a priori capabilities, which he called respectively transcendental aesthetics and transcendental logic. We can use a modern analogy of a computer to understand what Kant meant by his terms of "transcendental aesthetic" and "transcendental logic" (Fig. 7.3).

For data input to the computer, there must exist two prior capabilities in the computer before the input of the data: (1) a *format* recognized in the computer for structuring (formatting) the data and (2) a *stored program* in the computer as instructions for processing the formatted data.

The modern term "data format" is equivalent to Kant's term of "transcendental aesthetics," and the modern term "stored program" is equivalent to Kant's term of "transcendental logic."

For example, in physiology of the brain, people must be born with physical eyes, optical nerves, and optical processing portions of the brain. These physical mechanisms must exist in a body for vision experience to be possible. These mechanisms must be "a priori" – existing before – any visual experience in the brain. People born blind in the eyes can never see, as people whose eyes are severely damaged lose the capability of vision. All sensory experience by the human mind does require prior existing mechanistic capabilities of the body – "a priori" capabilities.

Both the "data format" and a "stored program" must be put into (exist) in a computer prior to (*a priori*) any information processing by the computer. In Kant's terminology, both transcendental aesthetics (data format) and transcendental logics (stored program instructions) must exist "a-priori" – before the experience of

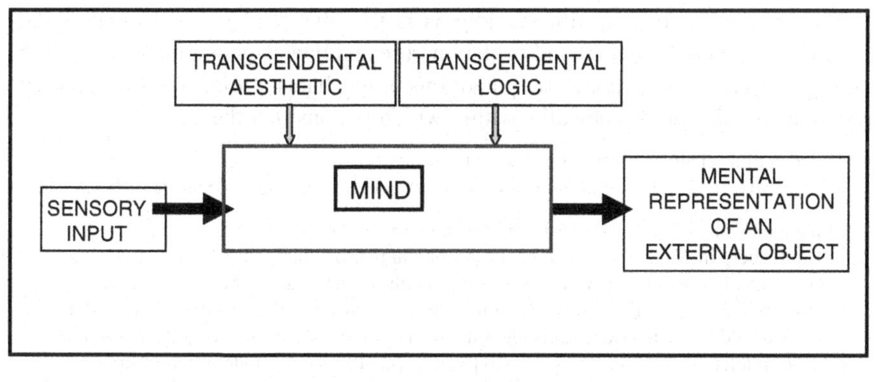


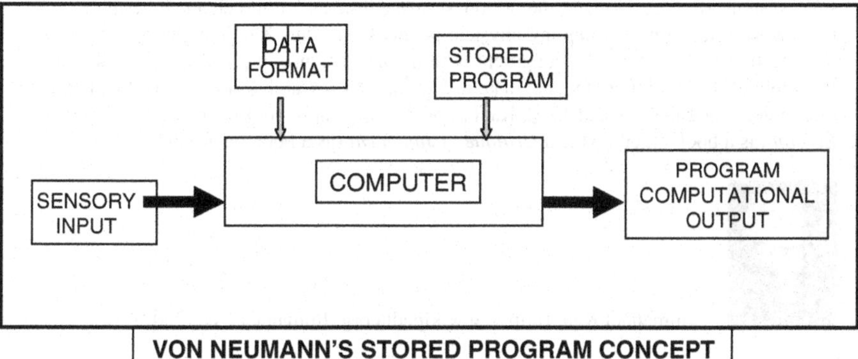


**Fig. 7.3** Kant's critique of pure reason and the computer

computation by the computer. In Kant's model of pure reason, one can see how a computer can be viewed as a kind of "reasoning mind" – a mind capable of processing input information from an environment into mental images. The computation is a form of reason, pure reason. And the source of the data input is an object external to the mind of the computer.

For the human mind displaying pure reason, Kant said that the *transcendental aesthetic* of the mind is the capability of formatting sense perceptions prior to any sensory experience of the world. The transcendental logic of the mind is the capability of formatting synthesis of perceptions into mental representations of objects as the source of sensory experience.

Today, we call these kinds of prior capabilities of the mind to perceive and represent the world as "mechanisms" of the brain. The mechanisms of the brain must provide two general capabilities for the brain to function: (1) capability of formatting sensory perceptions – *transcendental aesthetic*, and (2) capability of formatting mental reconstructions of the sensory perceptions as objects (things outside the mind) – *transcendental logic*.

> The scientific meta-level of a Scientific Paradigm is equivalent to Kant's idea of a Transcendental Logic in the operation of pure reason – observing nature.

The importance of scientific paradigms is that they underlie a discipline and provide how scientists depict and explain nature and how they communicate within the discipline. Scientists use either paradigms of mechanism, function, logic, and/or system as intellectual frameworks within which to construct theory.

The idea of a scientific paradigm is the "intellectual framework" in which scientific theory explains nature, and the paradigm is a logical meta-regression above scientific theory.

Immanuel Kant (1724-1804) was born in Konigsberg in Prussia and entered the University of Konigsberg in 1740. There he read philosophical works – including the new mathematical physics of Newton which then was being taught in the natural philosophy faculties of German universities in the 1700s. Newton's idea of a space/time descriptive framework for physics would influence Kant in his ideas in his major philosophical work, *Critique of Pure Reason*, published in 1781. From Newton's perceptual space of modeling the Copernican solar system, Kant generalized the notion as space/time as a transcendental aesthetic – a framework prior to measuring any physical phenomena. Thus Kant's philosophical work was the first philosophy to be congruent with the new science of physics – providing the first "model" of mind (pure reason) matching to the new research techniques of Newtonian mechanics. Later in 1788 and 1790, Kant published two more books, *Critique of Practical Reason* (as a book on ethics) and *Critique of Judgment* (as a book on aesthetics).

Immanuel Kant (http://en.wikipedia.org, Immanuel Kant 2007)

A footnote about logic. This idea of a "logical meta-regression" is a key idea to scientific content and method. A scientific paradigm is a logic meta-regression to a scientific theory. The concept of a "meta-regression" arose in traditional logic from understanding the linguistic relationship of a grammar to a language. Any language consists of terms (nouns, verbs, adverbs, adjectives, connectives) and a grammatical structure (sentences, paragraphs, arguments). The grammatical structure of a sentence is the composition of a sentence from (1) a noun as a subject and (2) a prepositional clause as an object and (3) a verb as a relationship between subject and object. Grammar provides a meta-language for the language. Grammar is a one-level logical regression above language.

Similarly, in science, theory just doesn't arise purely from the abstractions of experiment. Theory is formulated within a priori intellectual frameworks (such as space/time or matter/energy). For example, Newton's second law of motion, $F = ma$, required the prior assumption of (1) a space–time framework and (2) a Cartesian geometry to describe the locations of a particle at different times (particle motion).

## Modern Scientific Paradigms

Kuhn did not elaborate upon the different kinds of paradigms that are used in science. (In the previous example of the Cal Tech Center, both biological and engineering researches were pursued; and both areas operated within different scientific paradigms

**Fig. 7.4** Scientific paradigms: meta-frameworks for scientific expression (description and explanation)

| | WORLD | SELF |
|---|---|---|
| MATTER (SUBSTANCE) | MECHANISM | FUNCTION |
| MIND (IDEA) | SYSTEM | LOGIC |

appropriate to the biological or physical disciplines.) We next examine what are the paradigms in modern science. In a previous book, the author proposed that several paradigms are used in science (Betz 2003). One constructs a taxonomy of scientific paradigms based upon the two philosophical dichotomies about nature: (1) *matter and mind* and (2) *self and world*, as shown in Fig. 7.4.

## Matter and Mind

Consider the basic logical dichotomy of matter and mind. Everything in the world is philosophically composed of matter or ideas (mind). In this dichotomy of mind/ideas and matter/substance, the physical idea of "matter" includes the idea of energy, since in modern physics quantities of "matter" can be converted to quantities of energy, according to the formula $E=mc^2$. But what is not matter, nonmaterial, is an idea.

> All things in nature must exist either as phenomena of matter or as phenomena of ideas. "Matter" is a property of physical substance; while ideas are properties of a "Mind."

Minds think in ideas. And minds are a part of nature, as is matter. Humans have both a brain (material) and a mind (ideas). Thus, everything sentient in nature can be thought of in intellectual frameworks that are either material or idea, matter or mind. Intellectual frameworks (in Kuhn's and Kant's perspectives) provide the stuff of paradigms – the meta-logic to logic of theory. So matter or mind can divide the intellectual frameworks (or meta-logics or paradigms) of science.

A second set of ideas about the intellectual frameworks is the dichotomy of self and world. This dichotomy is an individual-centric view of everything.

> All things in nature exist as belonging to a world or to a self in that world.

From the perspective of an individual person, all totality of the world can be seen in reference to the self. This is an important philosophical dichotomy because consciousness or the world (awareness of the world) is purely and solely a property of an individual sentient being. All consciousness in the world belongs only to individuals. There is no such thing, in science, as a general consciousness in the world. (Although in religion, a general consciousness of the world is assumed to be "God"; but this kind of idea is not dealt with in science.) In science, only individual minds are conscious, whether that mind is in a person or an animal. This is the striking fact about the scientific idea of mind! Only individual minds display consciousness. Without individual minds in the world, there is no awareness of the world, no consciousness about things of the world. The dichotomy of self and world basically covers everything in the universe about the perception of the universe.

## *Matter and World: Scientific Paradigm of Mechanism*

Next consider an intellectual framework in which everything in nature is seen as both material and of the world – this is to say, that everything in the world is seen as matter/energy. The scientific paradigm which views nature primarily in this perspective has been called mechanism – a view of the world as consisting of mechanisms.

The atom, molecules, DNA molecules, photons, electromagnetic radiation, planetary systems – all these are mechanistic ways of perceiving nature. We can recall that historically, mechanism was the first paradigm created by scientists (principally by Copernicus, Brahe, Kepler, Descartes, Galileo, and Newton in the 1600s). Then in the 1700s, Newtonian mechanics became the focus of scientific research as a mechanistic paradigm. In 1905, Einstein altered the paradigm with special relativity for higher speeds of phenomena. Einstein also challenged the Newtonian paradigm for gravitational phenomena with an alternate paradigm of general relativity. From 1900 to 1930, Planck, Einstein, Bohr, Heisenberg, and others replaced the particle formulation of atoms in a Newtonian paradigm with a wave description as quantum mechanics. Dramatic paradigm shifts in physics did occur in the 1900s. We will call a physical relationship $R_p$ of material things in the world ($M_1$ and $W_2$) as a relationship described in the paradigm of mechanism as $M_1R_pW_2$. For example, in the Copernican solar system, the Earth orbits the Sun attracted together by a gravitational force ($R_p$).

In the scientific paradigm of Mechanism, physical things in the world ($M_1$, $W_2$) relate to each other through physical forces $R_p$, such that $M_1R_pW_2$.

## *Matter and Self: Scientific Paradigm of Function*

Next consider an intellectual framework in which everything in nature is seen as both material and from the perspective of self. The scientific paradigm which views nature in this intellectual framework is called function.

We saw in reviewing the history of DNA research that this view of function became important in the new biology discipline of the 1800s, particularly when Charles Darwin introduced his theory of evolution in 1859. In that theory, the idea of function was central to the evolution of a species as some structural feature gave the species a functional advantage to survival in a specific environment. In biology, the paradigm of mechanism was insufficient to completely describe living beings.

Another scientific paradigm of function was needed be added to fully represent the living phenomena.

The function of a mechanism can be defined as the relevance of a mechanism to an organism as a self – as an individual perspective on the world. The paradigm of function is the idea of the relevance of a material feature of the world or self to an organism. Both mechanism and function provide necessary intellectual frameworks to describe and explain the living beings in nature. We will call a functional relationship $R_F$ between a material thing ($M_1$) and a self ($S_1$) as a relationship described in the paradigm of function. For example, the DNA molecule ($M_1$) provides functional information ($R_f$) for chemical construction of the material proteins of an organic self ($S_1$).

In the scientific paradigm of Function, physical things ($M_1$) relate to a self ($S_1$) in a functional relationship $R_F$, such that $M_1R_FS_1$.

## *Mind and Self: Scientific Paradigm of Logic*

Next, let us consider the taxonomic category of what is both mind and self. In the history of philosophy, one of the most distinctive features about the human mind and its consciousness of self is rationality, thinking as deductive or inductive inference.

For example, one needs only to read the many classical works in philosophy, such as Plato or Descartes or Kant to see that the idea of rationality has been the philosophical focus about human intelligence. And the paradigm of rationality is logic. We remember how in Plato, the Socratic "dialectic" was the logical method of probing at the essential meaning of ideas. Or we remember in Aristotle how the syllogism was the logical form of inference: (All men are mortal. Socrates is a mortal. Therefore Socrates is mortal.) Even going down to the beginning of computer science, we can see how Boolean algebra became the logical foundation of computational techniques. So "logic" is a good name for the intellectual perspective which views all of nature as things that are both mental and individual – mind and self. We will call a functional relationship $R_L$ between a mental idea ($I_1$) and a self ($S_1$) as a inferential relationship ($R_L$), as described in the paradigm of Logic. For example, the syllogistic form ($I_1$) structures deductive inference ($R_L$) for thinking by a self ($S_1$).

In the scientific paradigm of Logic, linguistic forms ($I_1$) relate to a self ($S_1$) in an inferential relationship $R_L$, so that $I_1R_LS_1$.

## *Mind and World: Scientific Paradigm of System*

Finally consider the taxonomic category of what is both world and mind, an object of the world expressed as an idea. The term for this has been called a "system," or a system model. We saw the classical example of this as Newton's gravitational model of Copernicus's system theory about the totality of the sun and its planets, solar system. The idea of a worldly object described as a "system" is a way to describe the idea of the totality of a dynamic object. The description of an object of science as a system has (1) internal components, (2) connected as a process, and (3) interacting as a whole object with an environment. The totality of the sun and planets together as a system (1) is composed of distinct components (sun and earth and other planets) and has (2) the relation (connectivity) between the sun and planets by a force of gravity and with (3) the sun and its circling planets existing in an environment of space.

The concept of system is central to physics, such as the system of an atom or of a molecule. The concept of system is central to biology, such as the functional systems of the body (respiratory system, circulatory system, reproductive system, digestive system, etc.) The concept of system is central to computer science, such as a computational system. The concept of system is essential to technology, such as radar system, an airplane system, an automobile system, a building system, etc.

We will describe the modeling ($R_S$) of a worldly object ($W_1$) as a system idea ($I_1$) by the expression: $W_1R_SI_1$.

> In the scientific paradigm of System, the totality of an object of nature ($W_1$) can be modeled ($R_S$) as the idea of a system ($I_1$), so that $W_1R_SI_1$.

## Paradigms and Scientific Disciplines

As earlier noted, science is divided into disciplines. Different disciplines use different paradigms. Thus scientists in disciplines see the world of nature differently from each other – using different paradigms. Within a discipline, a paradigm mentally provides a scientist with a priori frameworks to observe nature – nature as mechanisms, nature as biological functions, nature as mathematical logic, nature as natural systems. A disciplinary scientist will see in nature how the paradigm formats sensual data and how the paradigm logically structures the form of a natural object.

We next look at all the science disciplines. The disciplines of modern science have been organized as: (1) departments in a modern university and (2) disciplinary scientific societies. One can classify all the science disciplines into those of the physical sciences, biological science, mathematical sciences, and social sciences, as shown in Fig. 7.5. This classification groups the disciplinary fields of nature into: (1) inanimate (without life), (2) animate (living), (3) cognitive (thinking), and (4) societal (human populations). From this, we can see that all human knowledge is organized with relevance to human existence. We humans see in the universe what is relevant to us – as inanimate, animate, cognitive, and social nature. (This is not too unreasonable, since modern knowledge (science) is, after all, a human activity.) The classification also lists the kinds of scientific paradigms which each disciplinary area uses in describing its perspective on nature – meta-frameworks of scientific description and explanation.

| INANIMATE | ANIMATE | COGNITIVE | SOCIETAL |
|---|---|---|---|
| PHYSICAL SCIENCES | BIOLOGICAL SCIENCES | MATHEMATICAL SCIENCES | SOCIAL SCIENCES |
| PHYSICS<br>APPLIED PHYSICS<br>CHEMISTRY<br>ASTRONOMY<br>EARTH SCIENCES | MOLECULAR BIOLOGY<br>CELL BIOLOGY<br>SYSTEMIC BIOLOGY<br>POPULATION BIOLOGY<br>ECOLOGY<br>SOCIO-BIOLOGY | MATHEMATICS<br>COMPUTER SCIENCE | ECONOMICS<br>SOCIOLOGY<br>ANTHROPOLOGY<br>PSYCHOLOGY<br>POLITICAL SCIENCE<br>MANAGEMENT SCIENCE |
| MECHANISM<br>SYSTEM | MECHANISM<br>SYSTEM<br>FUNCTION | LOGIC<br>SYSTEM | LOGIC<br>SYSTEM<br>FUNCTION |

**Fig. 7.5** Disciplines of science

## *Inanimate: Physical Sciences*

Physics describes the nonliving objects in nature as matter existing in a framework of space and time and moving over time through space. Interactions between material objects occur due to forces that alter the energy of the material object. At molecular scales, chemistry is a physical discipline that elaborates upon atomic and molecular interactions as chemical interactions. At the planetary scale, environmental sciences is a physical discipline that elaborates upon the physical systems in planetary processes. At a celestial scale, astronomy is a discipline that elaborates upon the physical systems in stellar processes.

> The disciplines of the physical sciences (physics, chemistry, earth sciences, astronomy) differ from each other by specialization on spatial scale – all using scientific paradigms of mechanism and system.

## *Animate: Biological Sciences*

The biological science is also arranged partly by spatial scale, studying the molecular level of life to the cell level to the organism level to the populations of organisms – molecular biology, cellular biology, physiology, ecology. Modern biology is unified by the principles of gene theory, cell theory, homeostasis, and evolution:

Gene theory – A living organism's traits are encoded in DNA.
Cell theory – All living organisms are composed of cells.
Homeostasis – Physiological processes enable an organism to sustain living chemical processes by means of taking in energy from an environment.
Evolution – Genetic mutations enable functional change in generations of a species, providing variation in a species with increased chance of survival in a specific environment.

The specialties of biological science provide a description and explanation of living forms based upon carbon-based chemistry – using scientific paradigms of mechanism, function, and system.

## *Cognitive: Mathematics and Computer Sciences*

Mathematics is the logic of quantitative inference – relating quantitative statements to quantitative statements. Quantity is a property of similar things, a set of things that do not significantly differ and so can be counted as repeats of the same thing. Formally, in mathematics, a set of things can be defined as a group of things which share a similar property, this property defining the condition for anything to belong to the set. (For example, the set of all apples in an apple orchard in season consist of all the fruit on the apple trees or fallen from the trees in the orchard.) The idea of a quantity of things in nature complements the idea of a quality of a thing in nature. Quantity is the number of things sharing the same quality. Traditionally, mathematics began with numbers, counting similar things and expanded into algebra, quantitative expressions among variables. Also, traditionally, mathematics began also with abstraction of spatial forms and their similarities, expanding into the topic of geometry. Modern mathematics still deals with numbers and structures of numbers as groups, algebras, vector spaces, etc. Modern mathematics still deals with forms as geometry, trigonometry, differential geometry, topology, fractal geometry. Modern mathematics also deals with change (as calculus, dynamical systems, chaos theory) and with mathematical logic (set theory) and statistics.

Computer science is a recently new scientific discipline providing a science base for the development of computer and information technology – which was invented in the second half of the twentieth century. It focuses upon the theoretical foundations for dealing with computation and information in computer systems. Its theories include foci upon mathematical foundations of computation, computational theory, algorithms and data structures, programming languages and compilers, computational procedures and architectures, artificial intelligence, and computer graphics.

Mathematics and computer science focus upon linguistic expression of quantity – providing quantitative and calculation languages and logics – for expressing quantities and performing inference and calculations about quantities.

### *Societal: Social Sciences and Management*

The social sciences all focus upon societal phenomenon in the human species but are divided into different perspectives of what they see in a society – perspectives of economics, sociology, anthropology, political science, psychology, and management science.

Economics looks at exchange of utility in societal interactions
- Economic systems

Sociology looks at social interactions in industrial cultures
- Social systems

Anthropology looks at cultural patterns in pre-industrial cultures
- Cultural systems

Political science looks at governmental patterns in industrial societies
- Political systems

Psychology looks at individual behaviors in societies
- Individual psychology

Management science looks at the decisions and control of organizations
- Decision science

The social sciences are the least integrated disciplines of science, compared to the physical, biological, and mathematical sciences. There is no overall "social science," nor any integrated departments of social science in research universities nor any integrated journals of social science. This lack of integration reflects both the absence of a common theoretical framework for observing society and shared empirically grounded theory.

> The challenge of whether or not there can be scientifically constructed an integrated "social science" versus a disintegrated set of "social sciences" is one of the great methodological issues of modern science.

## Perceptual Spaces

In addition to the a priori role of scientific paradigms in theory (as kinds of transcendental logics), there is also a similar a priori role for perception in science (as kinds of transcendental aesthetics) – which we have called a *perceptual space*. Earlier, we noted that a "perceptual space" is one of the research techniques for scientific empiricism. To understand this concept of a "perceptual space," we return again to Kant's philosophy. Therein, we will see that the idea of a perceptual space as a research technique in science is equivalent to Kant's idea of a "transcendental aesthetics" in pure reason (Fig. 7.6).

As scientific paradigms provide a kind of "transcendental logic" for scientific theory, so then do perceptual spaces provide a kind of "transcendental aesthetics" for scientific observation.

> Perceptual spaces enable both the description and measurement of properties of a natural object.

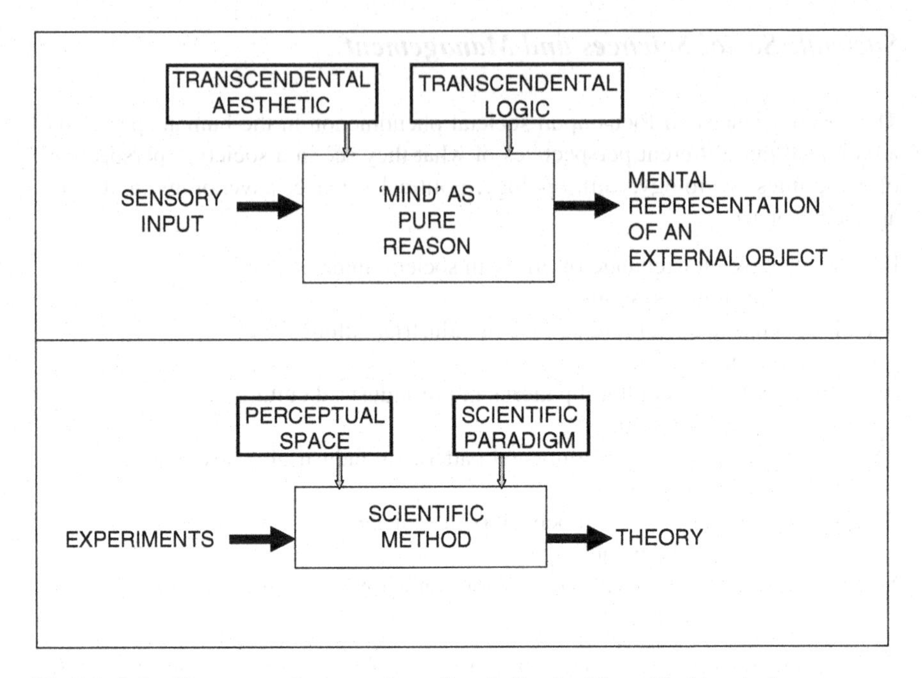


**Fig. 7.6** Scientific perceptual-space and paradigm in Kant's philosophical terminology

Measurement determines the quantity that can be experienced about the quality of an object in nature. The quality can be of a property of the object or the object itself. The measured quantity of a property determines how much of a property an object displays. The measured quantity of an object determines how many of these objects exist in a field of nature.

> To construct a measurement space, one needs to define the field of perception of an object as the dimensions and directions of perception.

## *Perceptual Space for Observing Physical Nature*

We next look at two kinds of perceptual spaces: for the physical sciences and for the social sciences. For the physical sciences, we have seen that Newton used Descartes' analytical geometry in which to describe Copernicus's sun-centric model in a perceivable space of physical space and physical time. Space/time provide the observational frameworks for perceiving (describing) physical existence. Space is the methodological concept of how material objects can coexist in nature at the same time. Time is the methodological concept of how material objects can occur at different positions in space as a sequence of temporal events. Newton had formulated his calculus for describing instantaneous motion in space, within the

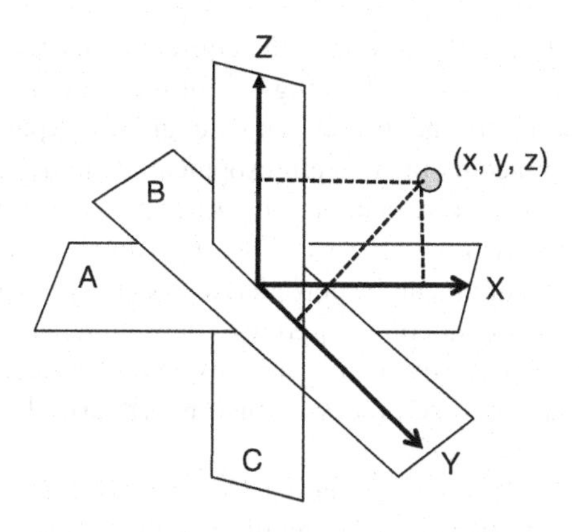


**Fig. 7.7** Three-dimensional geometric physical space

mathematical framework of Descartes' Analytical Geometry. This is illustrated in Fig. 7.7, wherein the three geometrical planes (A, B, C) are positioned mutually perpendicular.

This provides a geometrical description of a three-dimensional space. Next, three mutually perpendicular arrows (X, Y, Z) are positioned respectively on each plane, so that their intersection (0) describes the mid-point of intersecting planes. The three arrows (X, Y, Z) and origin (0) provide a reference frame. Each position of an object in space can be mathematically described by three position numbers $(x, y, z)$. This set of numbers $(x, y, z)$ is mathematically called a vector (v) where $v=(x, y, z)$. A space described by such vectors is called a vector space. In Newtonian mechanics, we live in a three-dimensional vector space (length, width, and depth). In relativistic physics, we live in a four-dimensional vector space (three dimensions of space and a fourth dimension of time).

> All physical things (matter) in a material universe are individuated from each other in a spatial and temporal framework.
>
> Two physical objects in material nature are said to be different because they can exist at different points of space at the same time.
>
> Any object phenomenally observed in a given space can be mathematically described as to its position by a set of spatial coordinate numbers (x, y, z) upon a reference frame (X, Y, Z) of the space.
>
> Description of position is the first step in any mechanistic representation of physical things in nature.

## *Illustration: Jung's Personality Types*

As an illustration of a perceptual space in the social sciences, one can look at Myer-Briggs' measurement space of cognitive styles – derived from earlier work of Carl Jung.

In 1913, Carl Jung described a typology of cognitive processes at the Munich Psychological Congress and published this in 1921 as a book called *Psychological*

*Types*. He described the conscious operations of the mind as occurring in different processes, of thinking (T), of feeling (F), of intuition (I), and of sensation (S). One way to understand Jung's cognition types is to place them in a taxonomy, determined by the two philosophical dichotomies (mental–body and analysis–synthesis) – which the author had earlier presented (Betz 2001). Thinking can be considered a cognitive process, which is primarily mental in focus and analytical in approach. Intuition can be considered a cognitive process, which is primarily mental in focus and synthetic in approach. Sensation can be considered a cognitive process, which is primarily bodily in focus and analytical in approach. Feeling can be considered a cognitive process, which is primarily bodily in focus and synthetic in approach, as in Fig. 7.8.

In 1941, Katharine Cook Briggs and Isabel Briggs Myers introduced sets of questions to help identify which of the cognitive functions (T, I, S, F) are preferred in frequency of use by a given individual's cognition, called the Myers–Briggs type indicator. To construct a measurement space of Myers-Briggs cognitive personality types, one can construct a two-dimensional geometric space with the analysis-synthesis dichotomy as the dimensions and the mind–body dichotomy as directions (Fig. 7.9).

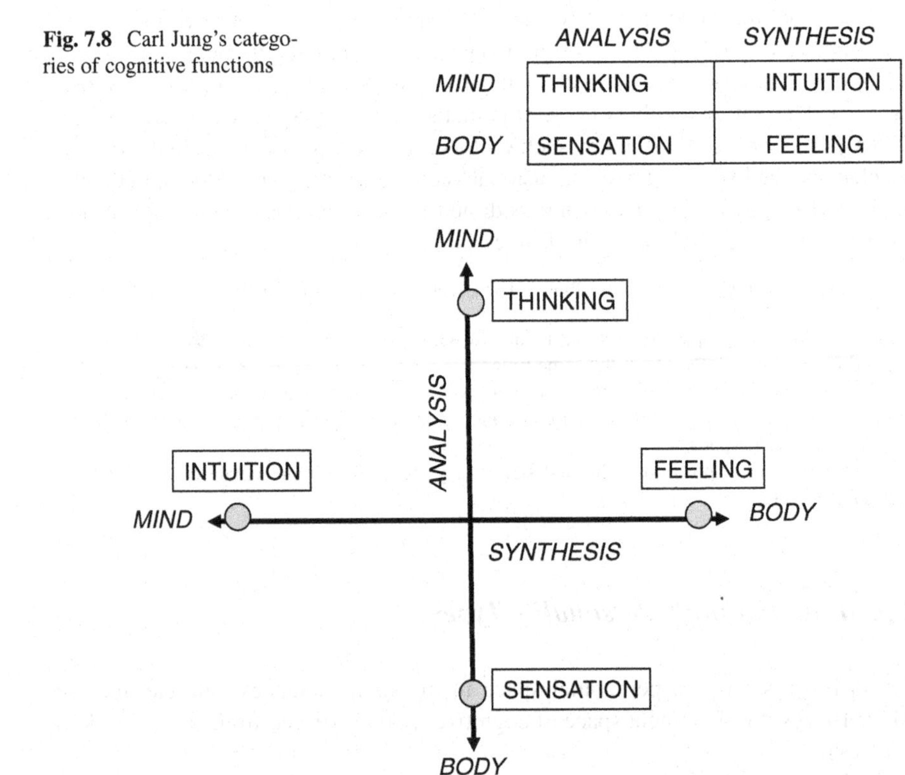


**Fig. 7.8** Carl Jung's categories of cognitive functions

**Fig. 7.9** Two-dimensional psychological/cognitive perceptual space

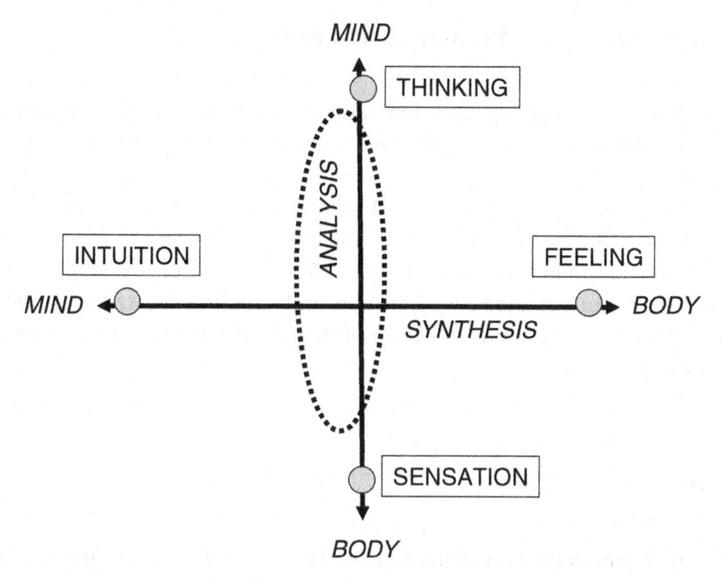


**Fig. 7.10** Two-dimensional psychological/cognitive perception space for Meyar brings measurements

One can see that in the measurement space, the cognitive functions are points on each dimension (analysis or synthesis) of the space. Accordingly, what Meyer–Briggs questions do in measuring a personality's preference for cognition is to plot frequency of use of the different cognitive functions. For example, suppose a person was measured by a Meyer–Briggs' instrument to display the following frequencies of preference in use of cognitive style: (1) mostly "thinking," (2) next frequent "sensate," and (3) lastly equally "intuition" and "feeling" equally. Then, this person's cognitive style could be plotted in a Jungian perceptual space, as shown in Fig. 7.10.

Measurement spaces are meta-logic basic dichotomies used as measurement in which to define and measure properties of an observed object – physical object or psychological object or societal object.

Carl Jung (1875–1961) was born in Thrugau, Switzerland. He attended the University of Basel, graduating in medicine in 1900. He practiced psychiatric medicine and in 1906 sent copy of his book on Studies in Word Association to Sigmund Freud, after which they became friends for the next 6 years.

Carl Jung (http://en.wikipedia.org, Carl Jung 2007)

## Perceptual Spaces in the Social Sciences

The perceptual space depends upon the nature being observed. So one will expect that perceptual/measurement spaces will differ between the physical and social sciences. In the present history of the social sciences, the social sciences do not share a common perceptual space – as do the physical sciences in their transcendental aesthetic of space and time. Perceptual spaces in the social sciences must be constructed within the Functional paradigm that a social science discipline is using (and how it is using it). Thus, we can now only look at the research idea of a perceptual space in the social sciences – not in general – but in particular examples of social objects.

## Summary

1. A scientific paradigm is a meta-logical framework for describing the natural context of a field of nature.
2. A scientific paradigm provides the intellectual framework for formulating theory and experiments in a scientific discipline.
3. The modern scientific paradigms are mechanism, function, logic, and system.
4. Different scientific disciplines use different paradigms.
5. Using Kant's philosophical terminology – a s*cientific paradigm* is equivalent to a "transcendental logic" and a *perceptual space* is equivalent to a "transcendental aesthetic" – in the pure reason of science – in *scientific method.*

# Chapter 8
# Paradigms of Mechanism and Function

## Introduction

We have constructed a taxonomy consisting of four paradigms of science: Mechanism, Function, Logic, and System. Now we will examine in detail the paradigms of Mechanism and of Function. In biology, for example, we saw in the case of the discovery of DNA and its modeling that both paradigms of Mechanism (chemistry of DNA) and Function (function of DNA in genetics) were used. The basic research issue of "what is life" required both mechanistic and functional explanations. The paradigm of mechanism perceives the world in terms of space, time, matter, energy, and force. But to fully describe and explain life, science needs additional concepts to those of Mechanism: concepts such as *purpose, intention, will,* and *reason.* These are central concepts in the paradigm of Function.

## *Illustration: Ontology of Physics and Chemistry*

In the modern paradigm of mechanism, the ontology of the modern world also began in the seventh century. The first step toward the modern discipline of chemistry was made by Robert Boyle (1627–1691), occurring at the same time of Newton's formulation of mechanics. Robert Boyle began to establish chemistry on a modern basis from the preceding medieval alchemy.[1] Alchemy had been principally concerned with the transmutation of elements – elements which today we recognize as molecules. Boyle was born in Ireland; and at the age of 8, he was sent to Eton College. At 11 years of age, he traveled abroad with a French tutor and lived in Italy. In 1642, the same year in which Galileo died, Boyle had gone to Florence to study Galileo's works. In 1645, he returned to England, inheriting a manor in Dorset, along with estates in Ireland. He devoted himself to the new movement in science, joining a group that met frequently in London at Gresham College and sometimes in Oxford.

In 1657, he read of Otto von Guericke's new air pump. Guericke (1602–1686) lived in Germany and invented the vacuum pump in 1650 (consisting of a piston and cylinder with one-way flaps to pull air out of a vessel to which it was connected).

F. Betz, *Managing Science*, Innovation, Technology, and Knowledge Management 9,
DOI 10.1007/978-1-4419-7488-4_8, © Springer Science+Business Media, LLC 2011

Guericke demonstrated the considerable force (pressure) of the atmosphere by evacuating a metal sphere cut in two and showed how hard it was to pull the half-spheres apart due to the external pressure on the spheres by the atmosphere (a pressure of about 14 lb/in.$^2$ at sea level).

Boyle constructed a pump for himself with the assistance of Robert Hooke. He studied the relation of volume to pressure in a gas and expressed the relation as a law in which the volume of the gas varies inversely to the pressure – Boyle's Law. Although Boyle began as an alchemist, his experimental studies led him to distinguish between mixtures and compounds. He supposed that elements were made of particles, setting chemistry on the path to seeing molecules as particles composed of atoms. Thus chemistry, which studies the substances of nature, was on the path toward a physics of nature composed of particles, molecules, and atoms.

Antoine Lavoisier (1743–1794) is also considered a "father of modern chemistry."[2] He first stated the law of conservation of mass in chemical reactions and co-discovered both oxygen and hydrogen. But Lavoisier is a scientist of the eighteenth century, and so chemistry really begins a half century later than Newton's physics and Boyle's early chemistry in the late 1600s. So chemistry too arose as the substance of molecules at the same time as mechanism became a paradigm of science, through the Newtonian synthesis of mechanical physics. Chemistry was perceived in space and time filled with particles in motion and interacting with one another through forces and is the basic descriptive framework for the paradigm of mechanism. Chemistry was no longer the alchemy of a world composed of air, earth, fire, and water.

The substance (ontology) of the natural world is either as material particles or energetic waves, both moving in space and time.

Robert Boyle (http://en.wikipedia, Boyle 2009)

Portrait of Monsieur Lavoisier and his Wife, by Jacques-Louis David. (http://en.wikipedia, Lavoisier 2009)

## Scales of Nature as Mechanisms

The major changes in the paradigm of mechanism that occurred in the twentieth century were special and general relativity and quantum mechanics. None of these changes in mechanism invalidated the older Newtonian paradigm of mechanics but made a special case of this in physics – empirically true at slow speeds and medium scale of the universe. Special relativity is a more general formulation of mechanics at high speeds, and general relativity is a more general formulation of mechanics at stellar scale, and quantum mechanics is a more general formulation of mechanics at an atomic scale. The idea of the *scale of space* affects both the kinematical experience of the physical world (motion) and also the dynamical experience of the physical world.

Newtonian mechanics provided a sophisticated and adequate intellectual framework for the physical experience of nature (experiments) at a macro-level – but not at a micro-level (atomic scale), nor even at a cosmic-level (galactic scale), and not for motion near the speed of light (light speed).

An atom is about 10 Å in width, and one Angstrom is $10^{-10}$ m. An atom is thus about one-billionth of a meter in width. Since we live in a meter-sized world (about 3 English feet), an atom is about one-billionth smaller than are we. This vast change in spatial scale carried with it a vast change in physics – Newtonian mechanics down to Quantum mechanics.

Also, one could imagine jumping up to galactic scale of space. The universe is about 15 billion years old ($15 \times 10^9$). Light travels at a speed of 1 billion kilometers per hour ($1 \times 10^9$). There are $24 \times 365 = 8{,}760$ h in a year. Light travels $8.76 \times 10^{12}$ km a year. There are 1,000 m ($10^3$) in a kilometer. So light travels rapidly and far: $8.76 \times 10^{15}$ m in a year.

Multiplying this distance times the age of the universe at $15 \times 10^9$ years, one can estimate the size of our expanding universe as about $15 \times 10^9 \times 8.76 \times 10^{15} = 131.4 \times 10^{24}$ m in radius. This is $1.3 \times 10^{27}$ m wide. We stand about 2 m in height. So the universe is about $\frac{1}{2} \times 10^{27}$ bigger than us. And at the galactic scale of space, physics is described by Albert Einstein's theory of General relativity.

So jumping across the entire scale of the universe, from the very tiny atomic scale of nature, $10^{-10}$ m, to the very gigantic cosmic scale of nature, $10^{27}$ m, the range one would jump is $10^{37}$. So we should not be too surprised that the physics of nature can differ over this vast range of scale. Newtonian mechanics is accurate at medium scale and slow speeds (earth size). Special relativity is accurate at high speeds (speed of light). Quantum mechanics is accurate at small scale (atomic scale). General relativity is accurate at large scale (stellar and galactic scales).

*Physics changes as the scales of observation of nature change, and the kinematics and dynamics of nature differ in the different speeds and different scales of nature.*

## *Illustration: Einstein's Special Relativity*

We recall that a perceptual space in Newtonian mechanics used a space/time a priori framework (transcendental aesthetic) to provide an observational framework for perceiving (describing) physical existence. Space is the methodological concept of how material objects can coexist in nature at the same time. Time is the methodological concept of how material objects can occur at different positions in space as a sequence of temporal events. Newton had formulated his calculus for describing *instantaneous motion* in space, within the mathematical framework of Descartes' analytical geometry. Next, as an illustration of a paradigm change in Mechanism, we look at the case of Einstein changing the perceptual space from Newtonian physics to Special Relativity.

Albert Einstein formulated special relativity: a space/time framework in which the speed of light is a fundamental empirical constant of the physical universe. Albert Einstein (1879–1955) was born in Wurttemberg, Germany. Einstein graduated with a teaching diploma from the Swiss Federal Institute in Zurich, Switzerland in 1901. He looked for a teaching position, but on not finding one, he took a job as an assistant patent examiner in the Swiss Federal Office for Intellectual Property in 1903. Then in 1905, the physics journal, Annalen der Physik, published four key papers by the young Einstein:

1. Special relativity, which postulated that the speed of light was a constant in the universe with the same value as seen by an observer and implied that the mass of an object increased as the velocity neared the speed of light.
2. Equivalence of matter and energy, which postulated that mass could be converted into energy at the quantity $E = mc^2$.
3. The photoelectric effect, which demonstrated that light interacted with electrons in discrete energy packets.
4. Brownian motion, which explained the random paths of particles suspended in a liquid as direct evidence of molecules.

Finally in 1908, Einstein obtained a doctorate and stayed on to teach at the University of Bern as a privatdozent. In 1909, he published a paper on the quantization of light as photons. In 1911, he became an assistant professor at the University of Zurich and, soon after, a full professor at the Charles University of Prague. In 1912, he returned to Switzerland as a professor at the Swiss Federal Institute. In 1914, he moved to the University of Berlin. In 1915, he published a paper of general relativity, viewing gravity as a distortion in the space–time framework due to matter. In 1919, an eclipse of the sun showed the deflection of light (photons have a zero mass) by the sun's gravity, as predicted by Einstein's theory of general relativity. In 1921, Einstein received the Nobel Prize in Physics. In 1924, Einstein in collaboration with Satyendra Nath Bose developed a statistical model of a gas (Bose–Einstein statistics). In 1932, Einstein moved to the Princeton University in the United States to avoid remaining in Germany with the rising power of the Nazi party.[3]

Albert Einstein in 1905 (http://en.wikipedia.org, Albert Einstein 2007)

# Scientific Paradigm of Mechanism

The physical scientific disciplines of physics and chemistry represent the world in the paradigm of natural mechanisms. This paradigmatic concept of mechanism is complex as it contains within it several ideas:

1. A kinetic framework for motion
2. A dynamic framework for explanation
3. Causal prediction

## *Kinematics in Mechanism*

Einstein's theory of special relativity specifies that the invariant length of an object observed by two different observers is the squares of the sum of the special coordinates (x, y, z) minus that of the temporal coordinate t. This invariant length is expressed as $x^2+y^2+z^2-(ct)^2$. What this means is that in Newton's physical space, two different observers see the same spatial length of an object $(x^2+y^2+z^2)$. But in Einstein's physical space, two different observers see the same length of an object as $(x^2+y^2+z^2-c^2t^2)$. This comparison is shown in Fig. 8.1.

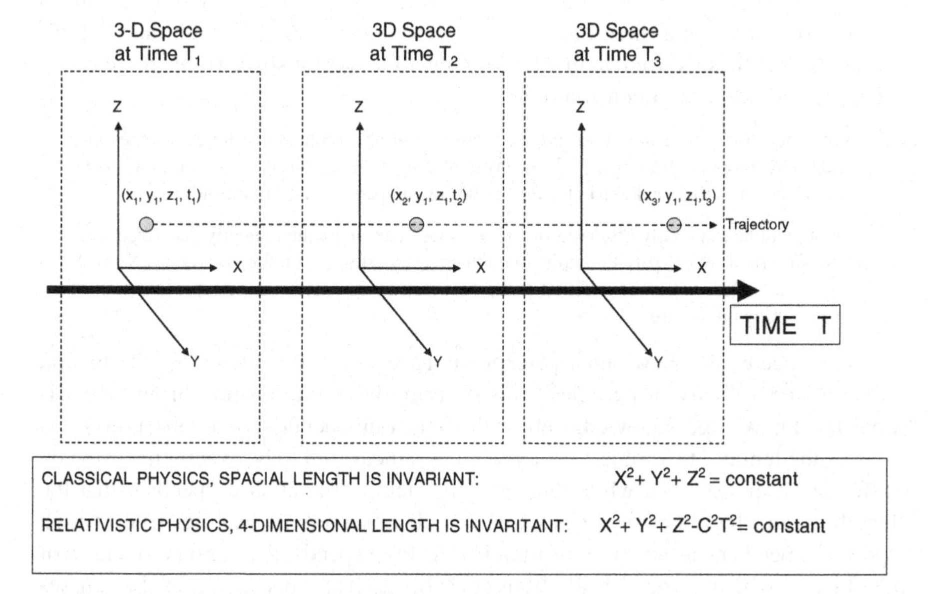


**Fig. 8.1** Four-dimensional space–time framework

This provides a geometrical description of a three-dimensional space. Next, three mutually perpendicular arrows (X, Y, Z) are positioned respectively on each plane, so that their intersection (0) describes the midpoint of intersecting planes. The three arrows (X, Y, Z) and origin (0) provide a *reference frame*. Each position of an object in space can be mathematically described by three position numbers (x, y, z). (This set of numbers (x, y, z) is mathematically also called a vector ($v$), where $v = (x, y, z)$, and a space described by such vectors is called a vector space; (later in the chapter on measurement, we will review this concept of a vector).

In contemporary physics, the description of motion in a space–time framework is called the "kinematics" of the mechanistic description.

Whose perceptual space is real and accurate in the description of physical nature? Einstein's perceptual space is accurate of nature – as verified by many experiments and also in the design of equipment, such as proton accelerators. And it has physical consequences. No physical object can travel faster than the speed of light in a vacuum. Traveling nearly at the speed of light, one observer (relative to another observer traveling with the object) will see the mass of the object as increasingly massive and the clock of the observer with the object as slowing down. And one does observe exactly these effects while experimenting with speeding protons and radioactive atoms (their masses increase and the half-lives of atoms decrease).

Special Relativity is an empirically grounded theory – theory verified by physical experiments.

In a mechanistic paradigm, the physical world that science depicts is a natural world represented in a three-dimensional vector space (length, width, and depth). In relativistic physics, we live in a four dimensional vector space (three dimensions of space and a fourth dimension of time).

All physical things (matter) in a material universe are individuated from each other in a spatial and temporal framework. Two physical objects in material nature are said to be different because they can exist at different points of space at the same time.

Any object phenomenally observed in a given space can be mathematically described as to its position by a set of spatial coordinate numbers (x, y, z) upon a reference frame (X, Y, Z) of the space. Description of position is the first step in any mechanistic representation of physical things in nature.

At slow speeds, the Newtonian perceptual space objectively describes the natural position and velocity of particles. This description of the position of an object is *objective* knowledge (knowledge about the object independent of all observers) – as long as the length of the object ($x^2 + y^2 + z^2$) was measured to be exactly the same by different observers. And while this was empirically true at slow speeds – that the length of an object ($x^2 + y^2 + z^2$) is true for all observers, it was not true at high speeds – speeds near the speed of light (c). At high speeds, the conserved length of an object – which is true for all observers (objectivity) – is the sum of the squares of the spatial dimensions and time dimension ($x^2 + y^2 + z^2 + c^2t^2$).

This means that at slow speeds, if two observers (physicists) compare their meter sticks for measuring lengths, their meter sticks will remain relatively the same length, even if the two observers move at different speeds with reference to each other. However, at high speeds, the other observer's meter stick will appear to each observer as shortening in the direction of their relative velocity. (And their relative clocks for measuring time will appear to each that the other's clock slows down.) Thus, the validity of theory is constrained to the range of time and space for which the theory is empirically verified. Newtonian mechanics is true at slow speeds. Special relativity is true at speeds near the speed of light. At slow speeds, Einstein's special relativity reduces to Newtonian kinematics – so there are no mathematical contradictions in the physics at slow and high speeds.

## *Dynamics in Mechanism*

Next in the mechanistic paradigm, one can impose forces over the points of the space and one then obtains a representation of the changes in motion – the "dynamics" of the mechanistic representation. We recall that Newton formulated Galileo's physical laws for the dynamical laws in the mechanism – the three universal laws of motion:

1. The law of inertia – The motion of a body is constant unless acted upon by an external force.
2. The law of force – The effect of an external force upon a body is to change its acceleration, proportional to the body's mass: $F = ma = m\,\mathrm{dv/dt}$.
3. The law of action-reaction. For every action (force) upon a body, there is an equal and opposite reaction (reactive force).

Then Newton invented differential calculus to quantitatively express the law of force as:

($F = m\,\mathrm{dv/dt}$). And we recall that in modern calculus, there are standard ways in calculus to solve many differential equations – that is to find the kinds of algebraic solutions that fit a differential equation. Newton formulated the gravitational force $F_g$ and substituted it into the law of force equation:

$$F_g = m\mathrm{dv/dt}$$

$$gMm/r^2 = m\mathrm{dv/dt}$$

(where $g$ is the gravitational constant as 32 ft per $s^2$ and $M$ and $m$ are the two masses that are attracted by gravity.)

Also, we recall that Newton solved this differential equation (using his new mathematical methods of calculus) to find that the solutions to this equation are (in analytical geometry form) the quantitative formulae that describe either an ellipse or a parabola. Thus, the first model expressed in the new Newtonian paradigm of mechanism was a physical model of Copernicus's solar system (Fig. 8.2).

Modern physics gets complicated because physics can be expressed in two different modes, as particles or as fields: (1) in a particle form, a particle has a

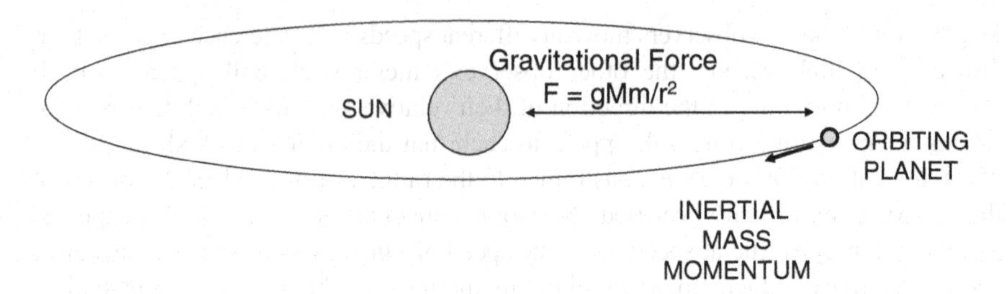


**Fig. 8.2** Newton's gravitational solar model

force field around the space of the particle, while (2) in a field form, a force field exists that can create or annihilate particles. In Newtonian physics, a particle formulation is used (e.g., the Copernican solar system model). In quantum chromo-electrodynamics physics, a field formulation is used (e.g., high-energy particle collisions).

## *Prediction in Mechanism*

In a mechanistic model, the dynamics of force and energy connect states of a physical system in a *causal sequence*. A cause-and-effect relationship between two events A and B is said to exist if the occurrence of a prior state A is both necessary and sufficient for the occurrence of a subsequent B. Then A is said to cause B as an effect.

> Prediction of natural events is possible when a physical theory provides an explanation of physical events as causally related.

Prediction of future states of a physical system is possible when all the mechanistic states of the system can be calculated from a model of the forces in the system. Prediction is very powerful for both science and technology. If a scientist can model the physical phenomenon of the physical morphology of the technology, then the technologies can predict how and when future states of the physical morphology occur under technological manipulation. For example, exact models have been constructed of the electron transport in transistor structures (physical morphology of the transistor). Electrical engineers use these models when they design transistorized electrical circuits to predict precisely what will happen to an electrical signal as it is processed by the circuit. As another example, when the US Space agency planned the trips to the moon, they used precise predictions as calculations of flight paths to get the astronauts there and back.

> Because of the invariance of physical laws as independent of any particular observer, the dynamics of physical laws can explain nature in cause–effect relations – causality.
>
> But social laws in the perceptual spaces of the social science are never wholly independent of observers, and therefore causality does not empirically appear as explanation in the social sciences. (And we will discuss this later.)

In summary, the five key ideas making up the paradigm of physical mechanism are spatial description, temporal kinematics, force and energy dynamics, scales of explanation, and causal prediction. This paradigm of Mechanism makes modern physical theory possible. Physical theory allows all physical morphologies of any technology to be represented as mechanisms and enable manipulations of nature by the technology to be predictable.

## *Illustration: Darwin's Theory of the Evolution of Species*

Mechanism is a powerful scientific paradigm used not only in physics and chemistry but also in biology. But in biology, Mechanism is insufficient to explain all of a biological phenomenon. Description and explanation of biological phenomenon requires a second and complementary paradigm, that of the *scientific paradigm of Function*. And next we will turn to this paradigm of Function. To illustrate it, we first review Charles Darwin's theory of evolution. This is a key theory in biology that connects mechanism to function in the variation of species.

Charles Darwin (1809–1882) was born in Shrewsbury, England.[4] In 1827, he entered Christ's College in the University of Cambridge to study theology, graduating in 1831. But he was interested in natural history. As a naturalist, he joined a second exploratory sea expedition on the HMS Beagle, captained by Robert FitzRoy. FitzRoy gave Darwin a book on geology by Charles Lyell (Principles of Geology) so that Darwin could understand landforms as resulting from geological processes over time. The Beagle sailed from England across the Atlantic, down the coast of South America, west around Cape Horn, up the coast of South America and then west to the Galapagos Islands. HMS Beagle returned to England via the Indian Ocean, around Cape of Good Hope in Africa, and north in the Atlantic.

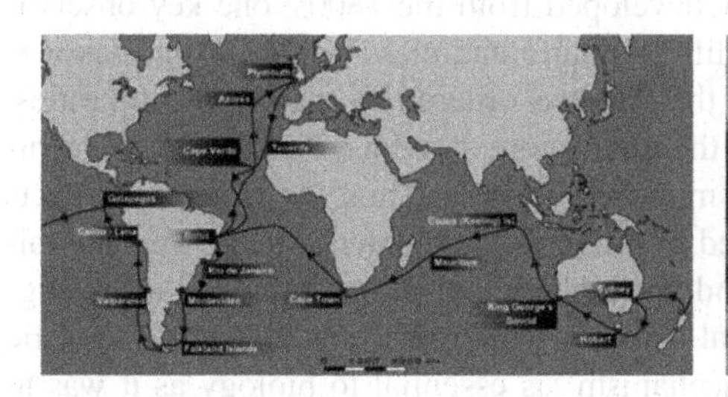

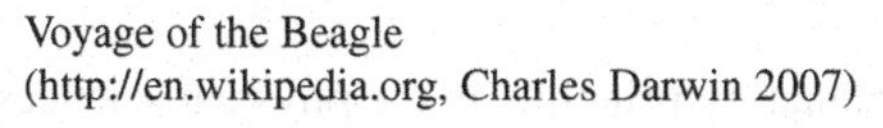

Voyage of the Beagle
(http://en.wikipedia.org, Charles Darwin 2007)

Beagle in Tierra del Fuego
(http://en.wikipedia.org, HMS Beagle 2007)

Darwin wrote a diary and also wrote geological letters; copies of these were widely read. When Darwin returned to England in 1836, he was acknowledged by scientific circles, particularly for fossils he had collected on the journey. In 1837, Darwin was elected to the Council of the Geographical Society and then moved to London to continue his scientific work.

In 1859, Darwin published his book, *On the Origin of Species*, which laid out his theory of natural selection in the evolution of species. He argued that in evolution, *mechanistic structures* played a role in the natural evolution of a species according to the *function* the structures of a species played for survival in an environment. Over time, species could evolve as the mechanistic structures of a species could alter through mutations and success in the function of survival.

Charles Darwin (http://en.wikipedia.org, Charles Darwin 2007)

Charles Darwin's theory of the species provides a *functional explanation* of the complexity of mechanisms in the different species of life. Mutations in the genes of organisms create different organic mechanisms. Organisms with functional mechanisms propagate as surviving species – when the mutated mechanism provides a species a functional advantage for survival in a specific environment.

In evolutionary theory, function is the value of a biological mechanism to survival of the organism in its environment.

The instinctive behavior of individuals of a species is selected in the process of the species adapting to an environment through reproduction of individuals with functionally successful genes.

## Scientific Paradigm of Function

As the modern discipline of biology developed from the 1800s, one key observational technique was in seeing wildlife in nature and classifying life into taxonomies. The idea of the taxonomy was that life was ordered by structural similarities. Also, we recall that the invention of the microscope provided biology a key instrument, with which to observe life at a microscopic scale. Thus, the structural aspects of life were observed to be composed of microscopic elements, cells. This combination of the ideas of "structure" and "cellular organization" provided a paradigmatic strategy for the new biological science, which one now calls "mechanistic reductionism." The paradigm of "mechanism" is essential to biology as it was to physics and chemistry.

Just as spatial scale is used as an explanatory principle in the physical sciences, so is scale used in the life sciences. At the microscopic scale, the smallest complete unit of life is the cell alongside bacteria. Below them, at the molecular level, are structures and processes of organic chemistry. Bacteria, cells, and viruses are the smallest units of life. They are constructed from information encoded in molecules that mechanically replicate themselves: DNA and RNA.

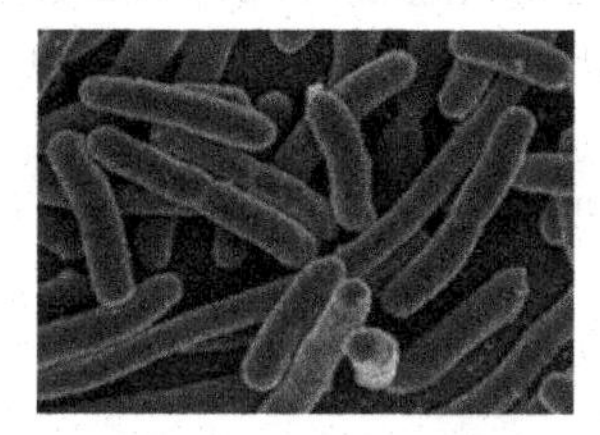

*E. coli* Bacteria, Magnified 25,000 times

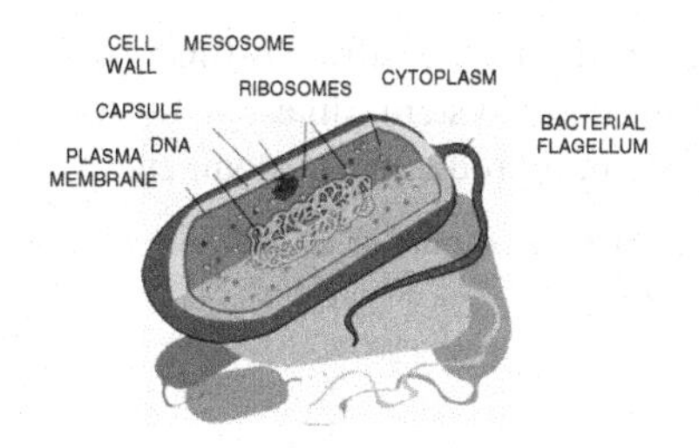


Diagram of the Cellular Structure of a Bacterium (http://en.wikipedia.org, Bacteria 2007)

But the Mechanism paradigm is insufficient to discriminate animate living beings from inanimate nonliving things. The second paradigmatic idea needed for biology is the idea of Function. One must not only elucidate the mechanisms in biology, but also interpret the *functional benefit* of the mechanisms to an organism. The paradigm ideas of both *mechanism and function* are needed to fully describe and explain (represent) the phenomenon of life, or *animate matter*.

For example, we recall that the DNA molecule is structurally in the form of a double helix of identical chains of polymers constructed of different sequences of four amino acids. This is a description of Mechanism. But the four amino acids provide information for the construction of proteins by the RNA in groups of three (the genetic code); and this is a description of Function.

This is the basic difference in the *scientific explanations* between inanimate and animate matter, the need to explain life not only in the paradigm of Mechanism but also the paradigm of Function. Inanimate material can only undergo physical processes. But animate material not only experiences physical processes but can also convey information about biological function. To explain animate physical processes, we need to describe both the *physical processes* and also the *relevance* of these processes to the animate being. Physical processes are described in Mechanism; whereas relevance is described in Function. The functional relevance of physical processes is experienced by an animate being as valuable.

Biological function is a description of the relevance of a physical process to an animate being.

Biological purpose is a description of the relevance of an action of an animate being.

Biological value is a description of the experience of the relevance of a physical process or of an animate action by an animate being.

Function, purpose, and value are basic ideas in the paradigm of Function, which describes relevance of existence to an animate being. And this is why the information aspect of DNA is important – the coding for the construction of proteins by amino acids – as triplet sequences of bases in DNA. This coding conveys information about the function of the DNA mechanism – its value to a living being in constructing useful proteins. In multi-celled organisms, genetic information provides cell specialization to create complex mechanisms for different functions in an organism. For example, at the macroscopic level of the human organism, mechanisms are partitioned into the different functional systems of the human body – skeletal system, muscular system, digestive system, circulatory system, immune systems, neural system, reproductive system, etc.

Each mechanistic system of an organism provides a different functional capability in the organism.

## Function and Behavior

The use, by an organism, of a functional capability of its mechanisms to facilitate its survival, can be called the "behavior" of the organism.

Biological behavior is a description of the actions of an animate being in terms of its purposes.

This view of behavior was emphasized by John Dewey, who was one of the philosophers of the American school of pragmatism of the early twentieth century. Dewey emphasized how both biological and cultural functions are connected to mechanisms of life in a basic relationship of *energy between life and environment*: "Whatever else organic life is or is not, it is a process of activity that involves an environment... An organism does not live in an environment; it lives by means of an environment.... For life involves expenditure of energy and energy expended can be replenished only as the activities performed succeed in making return drafts upon the environment – the only source of restoration of energy." (Dewey 1938, p. 25)

The seeking out of materials and energy for living is the biological ground of animate "purpose."

Dewey was emphasizing that the explanatory idea of *function* in biology derives from a description of life not merely as mechanism but also as biological *action* – action by living organisms. Living matter alters its environment through behavior – behavior is purposeful activity of life. Inanimate matter makes up (composes) an environment in a purely physical and unintentioned manner. Yet animate matter acts within and upon an environment in an intentioned manner, through active behavior. *Action* is an animate organism's purposeful engagement of its environment. And animate beings act with a purpose, and intention is the anticipation of the outcome of an action.

Biological intention is a description of the anticipated relevance of an action by an animate being.

Intention and action are properties of animate matter.

John Dewey (1859–1952) was born in Burlington, Vermont. In 1879, he obtained a bachelor's degree from the University of Vermont and in 1884, a PhD from Johns Hopkins University.[5] He then became a professor of philosophy at the University of Chicago. He published essays about the pragmatic school of philosophy and founded the University of Chicago Laboratory Schools to develop educational pedagogy. In 1899, he published his first book on education, *The School and Society*. In 1904, he moved to Columbia University in both the philosophy department and Teachers College. Dewey is one of the three central figures of the philosophical school of American pragmatism, along with Charles Sanders Pierce and William James. In his reform of Turkish education in the 1930s, Ataturk invited him to Turkey as a consultant.

John Dewey (http://en.wikipedia.org, John Dewey 2007)

## Function and Will

Concepts such as function, will, intention are concepts that science needs to use (in addition to the concept of mechanism) to fully describe life. And this idea of biological intention also connects the modern scientific paradigm of Function to the older traditions in philosophy – about the idea of will. For example, do even simple one-celled animals such as amoeba display (at least) a simple kind of will? According to Dewey's view about behavior, they do. The behavioral action of an amoeba is to move its pseudopodium out to encompass organic matter to obtain its food. Even an amoeba can be ascribed an "intention" about its movements – an intention to obtain food.

> Biological will is a description of the determination of an animate being to achieve a purpose.

The behavior of all animate matter can be explained as showing will – a *determination* to achieve a purpose. The action of a plant is to grow leaves and orient them toward the sunlight to absorb the energy of light, while also growing roots into the soil to absorb the materials of water and nitrogen for growth. A plant orienting its leaves toward the direction of sunlight can be ascribed an "intention" to obtain light energy. The action of a herbivorous gazelle is to graze on the green grasses of the plains, and the action of a carnivorous lion pack is to seize and eat a gazelle. Intentions of the herbivore in grazing and of the carnivore in hunting are both to obtain food. Humans show also intention in their behaviors, such as the actions of the laborer going to work and punching in the computerized time clock, or that of the white collar worker signing off biweekly on the computerized time sheet or that of the executive exercising stock options. Biological actions such as intentions to gain food or energy or money are all alike – the noble lion, the graceful gazelle, the peaceful plant, the lowly amoeba, the struggling worker, the wealthy executive – all alike in seeking food and energy.

Action is the basic description of matter as animate – action necessary for survival, prosperity, and reproduction. In action, the concept of *function* requires an *actor* (one who acts). Any action by an actor occurring in the present is intended to attain a future value from the action.

> Willful action by a biological actor is performed in the present for future anticipated value to the actor.
>
> The means of an action is the way the action is carried out by an actor and the ends of an action is the actor's intended outcome of the action.

## *Illustration: Human Brain*

The brain is a central mechanism of the body in the neural systems of higher biological organisms for both instinctive and reasoned behavior. The mechanism of the brain is marvelous and complex. The brain consists of a high density of neural cells connected in synaptic structures. These structures are organized into complex sets of folded planes that are interconnected three dimensionally. There are $10^{12}$ (a million billion) of the neurons that communicate with each other. The total weight of the brain is about 1,300 g.

Thought processes occur as the effect of electrical and chemical changes among these communicating neurons. Moreover, different regions of the brains process different kinds of sensory information and operate different mental activities. Also the brain is but one component of a larger neural system of the body, with its connection to the rest of the neural system. The neural system operates both with chemical and electrical signals.

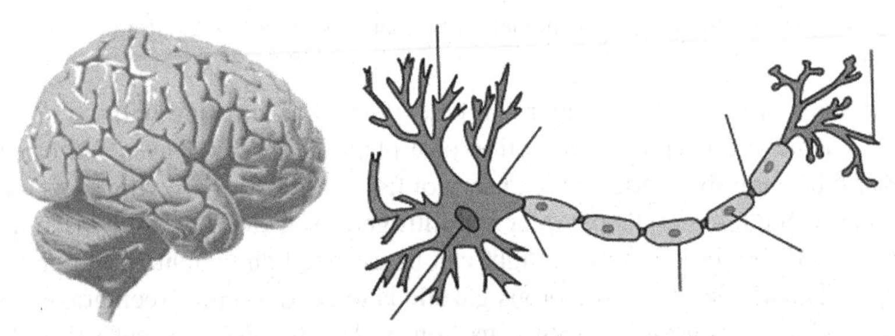

Human Brain Diagram of a Neuron Structure (http://en.wikipedia.org, Brain 2007)

Different spatial areas in the brain are specialized for different cognitive functions. The lower involuntary functions of the brain center in the medulla, which controls breathing, swallowing, digestion, and cardiovascular coordination. The hypothalamus area is involved in expression of many basic instincts. The cerebellum plays an important role in coordinating motor and voluntary movement.

The processing of visual information takes up a large part of the brain. Many of the higher functions of the brain, such as moral judgment, is processed in the frontal lobes of the brain. Locomotion is coordinated between "balance" located in the brain and "motor synapses" located in the spinal cord. Control of the heart itself is primarily located in the structure of the heart. Control of sleeping and waking is coordinated by hormonal activity in the brain.

The brain is divided into two hemispheres, with each hemisphere coordinating the kinesthesis of opposite sides of the body. Also, the two hemispheres of the brain in humans have evolved different reasoning specializations. The left side of the human brain has specialized in verbal reasoning and the right side in spatial reasoning.

This description of the "brain" is in the paradigm of Mechanism, such as the physiology of the neural system – physical processes. In contrast, a description the mental activity of the neural system requires the paradigm of Function – "mind."

*The brain is a mechanism, and the mind is the function of the brain mechanism.*

## Reason and Information Processing

In the last half of the twentieth century, scientific research made significant progress in understanding the mechanisms of the brain and also in understanding mind as cognitive activity. Psychologists study this problem, as do biologists and sociologists. Economists use a model of the human mind as being economically rational.

But computer science has contributed new ideas about the mind as an information-processing activity. And this concept of information has become central to any model of the mind. *Information input* to the mind begins with the sensual mechanisms of the neural system of the body, of which the brain is a central processing unit. *Information output* is in the form of decisions to action. A decision to act is processed in the brain through *mental reasoning*. Let us compare this information input and output information model to that of Immanuel Kant's description of "reason." Figure 8.3 illustrates that the idea of *data format* is equivalent to Kant's idea of *transcendental aesthetic*; and the idea of a *stored program* is equivalent to Kant's idea of *transcendental logic*.

With this concept of reasoning as an information-processing activity, we may proceed to use modern psychological theories to construct a functional model of the mind – as opposed to a physiological model of the brain. The paradigm of Function provides a very different description of mind than does the paradigm of Mechanism – brain versus mind. "Brain" is a description of intelligence in the paradigm of Mechanism; whereas "mind" is a description of intelligence in the paradigm of Function.

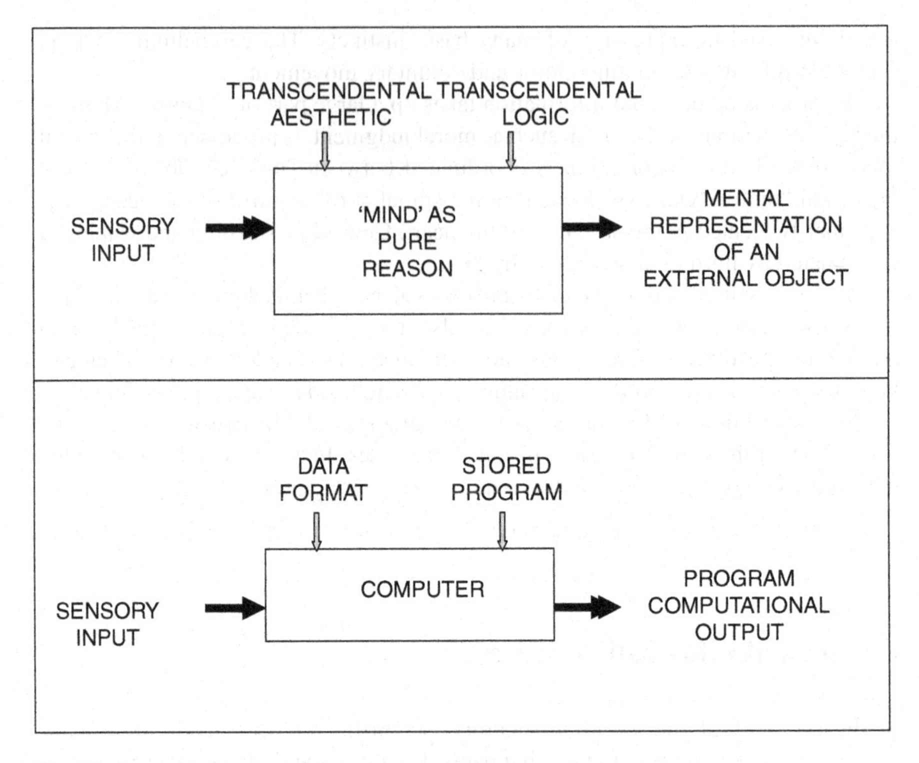


**Fig. 8.3** Reason in human and in computer minds

## Functional Model of Intelligence

One can construct a functional model of intelligence, mind, by using basic functional concepts described by the different schools of psychology, including those of Wendt, Nietzsche, Adler, Freud, James, Jung, and Campbell. The point of such a psychological model of the mind is to appreciate the difference between a functional model of the mind (how cognitive processes operate) and a mechanical model of the mind, (such as the spatial location of functions). Mechanical models show where cognitive functions occur spatially in the brain, but not how the cognitive processes create mental reasoning.

### *Stimulus–Response Model: Perception of the World and Action in the World*

One can start with the simplest functional model in modern psychology, a stimulus–response (S-R) model. Stimulus–response partitions the mental world into "world" and “mind.” The mind is stimulated by (perceives) objects in the world and responds by taking action on nature in the world. Now the connections of stimulus (perception) to response (action) consist of the mental processes of the mind. What are these? The next one can add theories from the different schools of psychology – to construct some of the component operations of the S-R mind.

The S-R model of the mind traces back to the founder of modern psychology, Wilhelm Wundt. He established an experimental approach to the study of psychology. William James later emphasized that the S-R model need not be a direct physiological connection between stimulus and response, but that the response can be mediated by learning. Later, Carl Jung introduced the idea that some personality types were primarily sensitive to external stimulus – inputs from the world (extroverts) and some to internal stimulus – inputs from the self (introverts). Physiology aims to describe the operation of the nervous system as a mechanism (paradigm of Mechanism); whereas psychology aims to describe the functioning of the nervous system (paradigm of Function).

Wilhelm Wundt (1832–1930) was born in a town near Mannheim, Germany. He graduated in medicine from the University of Heidelberg in 1856. He joined the staff of Heidelberg as a research assistant. There he taught the first course in scientific psychology, stressing the use of experimental methods. This began the specialty of physiology, and in 1874, he published the *Principles of Physiological Psychology*.

Wilhelm Wundt (http://en.wikipedia.org, Wilhelm Wundt 2007)

The psychological school of behaviorism followed upon Wundt’s methodology of measuring physical stimuli and behavioral reactions. The most famous behavioral experiment was performed by Ivan Pavlov who trained dogs to expect food at certain signals.

Other scientists studying psychology at that time emphasized that the stimulus–response linkage can be moderated by learning. William James described the stimulus–response model of the mind but emphasized that physiologically the stimulus could not be directly hardwired to a response but the connection must be interruptible for the response to be learned.

Ivan Pavlov (1849–1936) was born in Ryazan, Russia. In 1879, he attained a doctorate in natural science from the University of St. Petersburg. He studied the gastric function of dogs and noticed that dogs tended to salivate before food was actually delivered. He called the training of dogs to anticipate food (salivate) upon a signal as “conditional reflexes.”

Ivan Pavlov (http://en.wikipedia.org, Ivan Pavlov 2007)

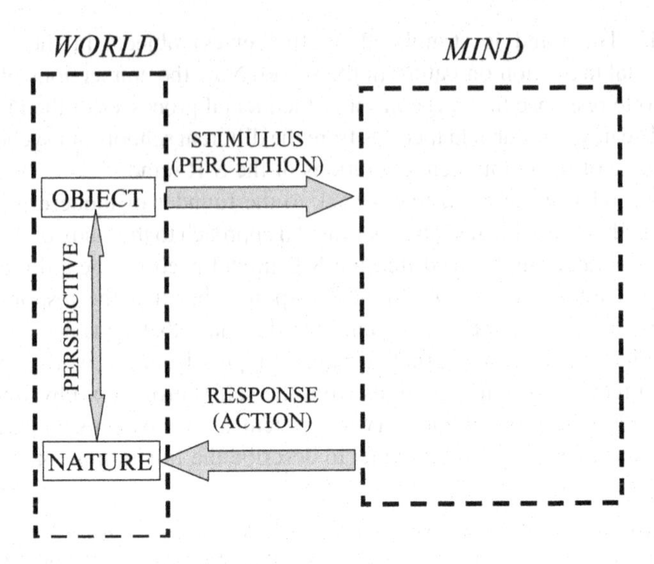


**Fig. 8.4** SR mind model: self and world

William James (1842–1910) was born in New York City in the USA. In 1869, he attained a medical degree from Harvard University. He was appointed an instructor in psychology at Harvard in 1873. In 1890, he published *Principles of Psychology*.

William James (http://en.wikipedia.org, William James 2007)

A stimulus–response model of mind would appear as in Fig. 8.4. The WORLD is both a source of stimuli to the MIND and is operated upon by responses from the mind in the form of actions upon the world. Objects in the world are the sources of stimuli to the mind in the perceptions of the mind. Actions upon nature in the world are the responses of reasoning and decision making (information processing) in the mind.

All perceptions of objects in nature by a mind are made in the particular perspective of the mind viewing and acting upon the world.

## *Pure Reason: A Priori Capabilities of Mind*

Between perception and action, there must be cognitive processes in the mind that use perceptions as information inputs and actions as decision outputs. This cognitive process should begin in reason, as we have learned from Immanuel Kant. This is a second philosophical/psychological theory we can use to construct a functional model of mind. The first step of the cognitive process of reasoning is to create a

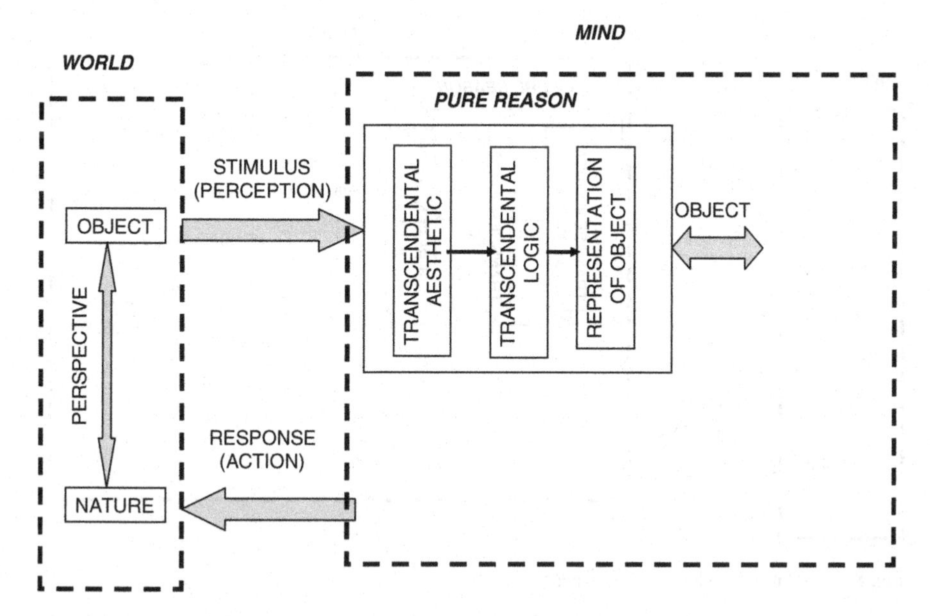


**Fig. 8.5** SR mind model: pure reason

mental representation of an object, which we add into the S-R model. Figure 8.5 diagrams how perceptions of objects in the world are formatted in sensual data in the mind (transcendental aesthetic) and the synthesized by cognitive logic (transcendental logic) into mental representations of external objects in the world. These "ideas" of objects are passed onto a second cognitive process – consciousness.

## *Consciousness: Cognitive Processes*

Next we can take Carl Jung's taxonomy of cognitive functions (thinking, intuition, sensation, feeling), which we earlier reviewed in Fig. 8.6, and one can add to the S-R model of mind, this process of consciousness by placing Jung's categories after "pure reason" so as to cognitively operate upon ideas of objects, as shown in the next diagram.

## *Sub-conscious: Cognitive Functions beneath the Conscious Level of the Mind*

However, we also know from modern psychology that not all cognitive activity of the mind is conscious to the mind. Sigmund Freud showed that subconscious mental

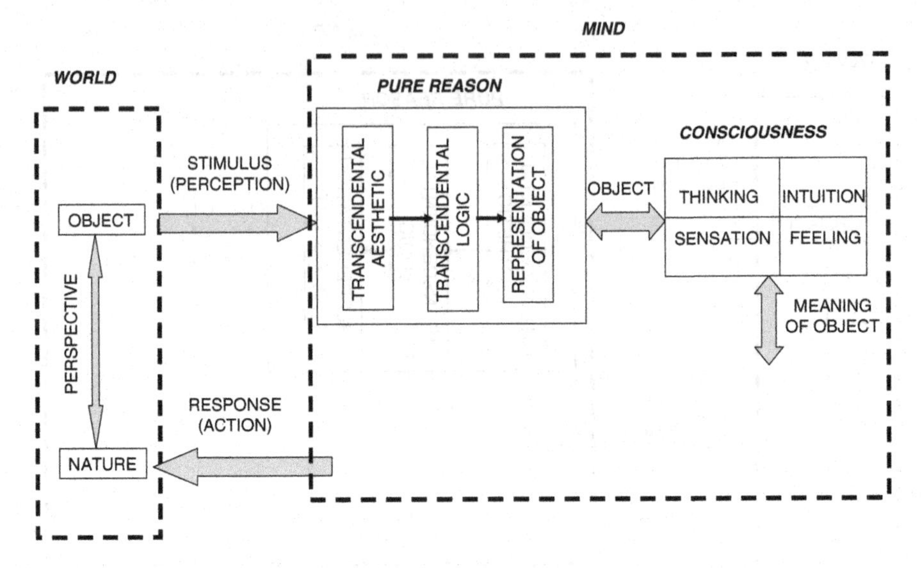


**Fig. 8.6** SR mind model: consciousness

processing does occur – such processes not aware to the consciousness of the mind while occurring. Freud introduced the terms "unconscious mind" and "repression" to indicate a cognitive activity beneath the conscious level and sometimes not accessible to the consciousness due to "repression" of subconscious cognitive ideas. He identified aspects of the mind as divided into "id," "ego," and "superego." The "id" is an aspect of self that is driven by primitive drives. The "ego" is an aspect of self that deals with drives from the id; and the "superego" is an aspect of self that imposes moral judgments upon the self.

Carl Jung's addition to Freud's ideas about the unconsciousness of a personality was that there could be a "collective" aspect to any individual's unconsciousness (e.g., a cultural aspect). Jung called these collective aspects as kinds of "psychological archetypes." Joseph Campbell expanded upon Jung's idea of archetypes as kinds of mythic roles in the self.

Sigmund Freud (1856–1939) was born in Moravia, then part of the Austrian-Hungarian Empire and now part of the Czech Republic. He attended the University of Vienna to study medicine, graduating in 1881. In 1886, he opened his own medical practice, specializing in neurology. In 1900, he published *The Interpretation of Dreams*, in which he argued the logic in dreams could depict unconscious and repressed desires. http://en.wikipedia.org, Sigmund Freud 2007

Sigmund Freud (http://en.wikipedia.org, Sigmund Freud 2007)

Joseph Campbell (1904–1987) was born in White Plains, New York, U.S.A. He attended Dartmouth College, graduating with a bachelor's degree in English literature in 1925 and a master's degree in medieval literature in 1927. In 1934, Campbell became a professor at Sarah Lawrence College. Campbell then published a book, *The Hero with a Thousand Faces*, in which he elaborated upon Jung's idea of a collective unconsciousness as a kind of mythic self.

Joseph Campbell (http://en.wikipedia.org, Joseph Campbell 2007)

Historically, Freud drew some of his ideas from the philosopher, Friedrich Nietzsche, particularly that of the "id." Forty years before Freud, Nietzsche had emphasized that beneath and central to all personality is a strong instinct for a "drive-for-power" – a "willen-nach-kraft" – a will-to-power. This idea of an underlying drive was probably the stimulus for Freud's later idea of a raw force in the human psyche that he called the "id" and certainly a precursor.

The English term of "will-to-power" is a literal translation of Nietzsche's German term of "willen-nach-kraft." But in terms of what Nietzsche meant by this idea, perhaps a better English term for his concept would be the term "pride." Certainly, social organizations that strive to shape the will of individuals to that of the group appeal to an individual's pride. For example, military organizations train soldiers to do their military duty, even at the cost of sacrificing their lives, as a soldierly pride, pride in being a soldier. Scientists take pride in the correct use of scientific method to observe the secrets of nature. Pride is an essential feature in shaping individual will toward the goal of a social group, pride in the social role. For example, the pride of a medical doctor is expressed in their Hippocratic oath to first do no harm. Thus, one interpretation of Nietzsche's German term of "willen-nach-kraft" is not simply "will-to-power" but also the *human pride*.

Later in the psychoanalytic literature, Alfred Adler emphasized that Nietzsche's idea about "power" is central to the unconscious aspects of personality. Adler introduced the idea of "inferiority complex" to describe problems of low self-esteem, problems in an individual's pride.

Friedrich Nietzsche (1844–1900) was born near Leipzig, Germany. In 1864, he attended the University of Bonn and then went to the University of Leipzig, studying philology (the origin of words). In 1869, Nietzsche received an appointment to the University of Basel as professor of classical philology. His first book was on aesthetics, entitled *The Birth of Tragedy out of the Spirit of Music*. Due to ill health, he resigned from Basel in 1879. For the next 10 years he traveled in Switzerland in the summers and lived in Turin, Italy, in the winters. He wrote several books, focused around a key idea that "will-to-power" was the central drive of the self.

Friedrich Nietzsche (http://en.wikipedia.org, Friedrich Nietzsche 2007)

Alfred Adler (1870–1937) was born in Austria. He trained as a medical doctor. He helped found the school of individual psychology. He collaborated with Sigmund Freud as part of the psychoanalytic movement. In 1912, he broke with Freud and in his future work emphasized the importance of social relationships in psychology.

Alfred Adler (http://en.wikipedia.org, Alfred Adler 2007)

Thus for Nietzsche's and Adler's ideas of the role of "power" as an overriding concept in unconscious cognition, one can indicate a self of "pride" – the role an individual's sense of "pride" in unconscious thinking – absorbed in protecting or increasing one's power. In Fig. 8.7, we can next add a cognitive state of unconscious processes as operating beneath the conscious cognitive processes in an S-R model of mind. The black arrows in the diagram indicate that the subconscious cognitive processes operate in parallel to and beneath cognitive processes in consciousness. And to incorporate all the ideas about the cognitive unconsciousness processes of Freud, Jung, and Campbell, I have used the philosophical idea of "self" to indicate that unconscious processes are all aspects of "self" – human identity in an individual.

The subconscious aspect of self operates underneath the conscious aspect of self and in parallel.

For Freud's ideas of repression, one could indicate the processes of a critical aspect of self – criticizing and censoring conscious thought – a critical self. For Jung's and Campbell's idea of collective and mythic unconscious personalities, one could indicate a process of personality type, which is storytelling a mythmaking – making up stories of the real personality of one's self in a role of an archetype – a mythic self. For Nietzsche's and Adler's ideas of the role of "power" as an overriding concept in unconscious cognition, one can indicate a self of pride. And for Feud's idea of an ego, as influenced by unconscious thinking, we use the idea of an active self – that aspect of self that acts in the world (Fig. 8.7).

## *Will: Decisions of Action*

Lastly in the older tradition of the psychological literatures on personality, there is a theme that self must be disciplined and deliberate and rational in action. Of the modern psychologists of the twentieth century, William James particularly emphasized the idea of will as a determination to act upon and even against events in the world.

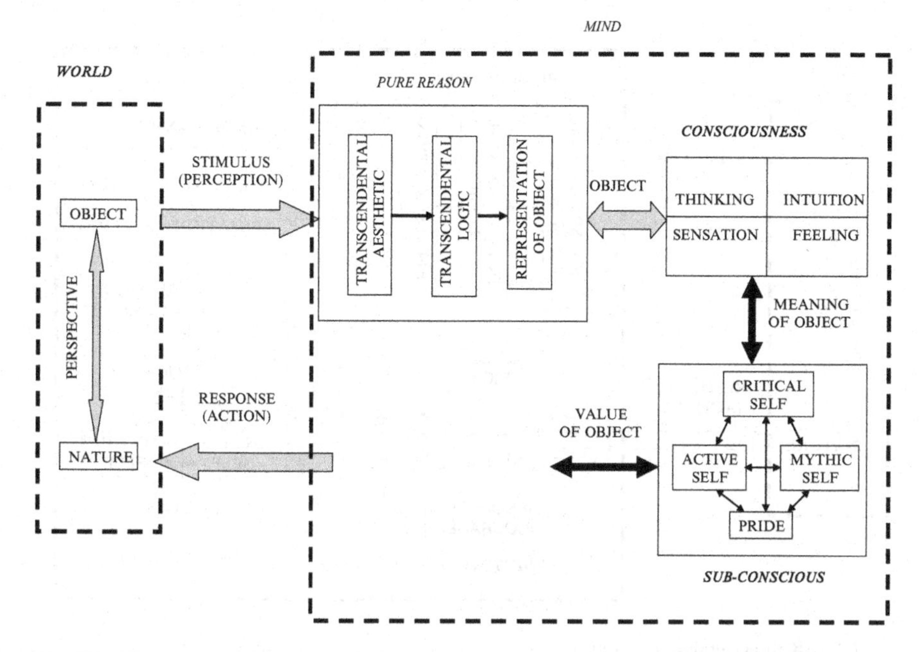


**Fig. 8.7** SR mind model: sub-conscious

However, one does not find in the psychological literature of the twentieth century much elaboration of this idea of will as a determination to act. In the second half of the twentieth century, what was once the traditional idea of will became elaborated under a new name of decision making – and in a new social science discipline of management science. We will use the terminology of "decision making" to describe the component ideas of a modern conception of will.

Decision making describes a commitment to action (will) as a decision to act. The logic in a decision to act involves selecting a means and an end for a task. The task is conceived as a concrete action for a purpose of a decision maker. The social context of a decision to act is the ethics of the action. And if the action encounters resistance in the world from competitors, then the decision maker must also display courage in the performance the action – an implementation of action even against resistance in the world.

Although the modern term for a determination to act is in management science called decision making, we will continue to use the older psychological term for this stage of cognition as will. In this way, we connect management science ideas to ideas of traditional psychology – thus expanding on the information model of cognition as a decision. In the S-R model of mind, will is the final stage of information processing in the mind that makes a commitment to action – a decision to act. The task and purpose for a decision can be formulated in conscious cognition. But subconscious ideas about the value and meaning of objects in the world to the person will also play a part in the cognitive formulation of the decision. Adding these functional characteristics into the S-R model then completes a functional model of the mind (Fig. 8.8).

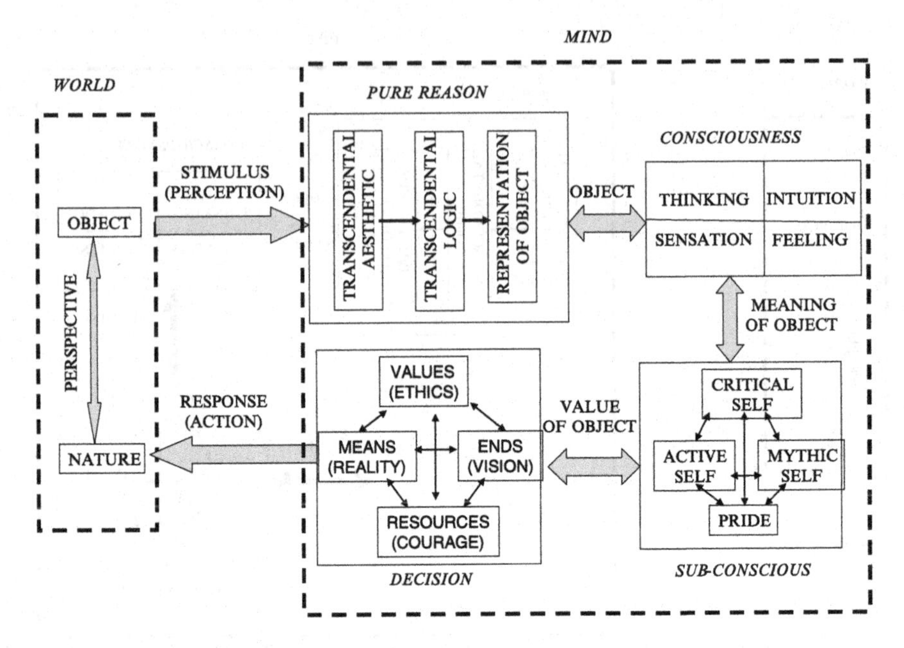


**Fig. 8.8** SR mind model: decision

This particular functional model we have depicted may not be the best functional model of the mind. But it does show how such a functional model can be constructed, using the scientific literature in psychology. Until the social science discipline of psychology begins to construct functional models of mind, scientific progress in the social science of individual behavior will lag, because psychology will remain merely a group of "schools" without an essential coherence in the underlying forms of psychological nature.

Model construction in science integrates theory across the different disciplines of science (and modeling is a topic that we will later examine).

## Summary

1. The five key ideas that together make up the paradigm of physical mechanism are: spatial description, temporal kinematics, force and energy dynamics, scales of explanation, and causal prediction.
2. Because of the invariance of physical laws as independent of any particular observer, the dynamics of physical laws can explain nature in cause–effect relations – causality.

3. Biological function is a description of the relevance of physical processes to an animate being.
4. The seeking out of materials and energy for living is the biological ground of animate purpose.
5. In evolutionary theory, function is the value of a biological mechanism to survival of the organism in its environment.
6. The instinctive behavior of individuals of a species is selected through the process of the species adaptation for survival in an environment.
7. But social laws in the perceptual spaces of the social science are never wholly independent of observers, and therefore causality does not empirically appear as explanation in the social sciences. (And we will discuss this further, later.)
8. To fully describe and explain life, science needs additional concepts to those of Mechanism, such as *purpose, intention, will, reason* – and these are central concepts in the paradigm of Function.
9. *Instincts* provide the primitive *hardwired* biological instructions for attaining the basic purposes of animate beings; whereas *reasoning* for attaining these purposes is learned by individuals and provides the *software* in biological instructions.
10. Willful action by a biological actor is performed in the present for future anticipated value to the actor.
11. The brain is a description of human intelligence in the paradigm of Mechanism, and mind is a description in the paradigm of Function.

## Notes

[1]Biographies of Boyle include those by Principe (1998) and Hunter (2009).
[2]Biographies of Lavoisier include those by Donovan (1993) and Poirier (1996).
[3]There are many biographies of Einstein, including those by Clark (1971) and Pais (1982).
[4]Biographies of Darwin include those by Desmond and Moore (1991).
[5]Biographies of Dewey include those by Boisvert (1997) and Martin (2003).

# Chapter 9
# Objectivity in Social Sciences

## Introduction

We continue with scientific method but now focus upon the social sciences. Historically, the physical and biological sciences are the most well developed methodologically and therefore provide the *standard of scientific excellence and objectivity* about nature. It is to their standard that the methodology of any scientific discipline must be measured.[1] The first issue for the social sciences is how can they be objective about studying social nature?

We recall that, in the example of Einstein's Special Relativity Theory, we saw how the methodological criterion of "objectivity" is central to scientific method in the physical sciences. Objectivity means that the physical knowledge about a physical object can be completely separate from knowledge about the observer. In Einstein's theory, the speed-of-light is an objective constant in the universe, measured exactly the same by any observer. To accomplish this standard of objectivity, all physical theories must be expressed in relativistic-invariant form. This means all physical theory is in a form in which the speed-of-light has the same constant value for any observer.

A scientific judgment about nature which does not imply value on the part of the observer is often said to be an empirical judgment, a fact of nature. In contrast, a scientific judgment about nature which implies value on the part of the observer is often said to be a normative judgment – a prescription about nature – the value of nature to the observer. In the physical sciences, *objectivity* has a kind of *absolute meaning* in terms of formal incorporation in theory, theory independent of the perspective of any observer. All observers of physical nature can be unbiased about the object, when theory is in an invariant form. In this form, the physical sciences have been called "value-free," unbiased by any values of an observer or applier of physical science knowledge.

> Objectivity is knowledge about the object, independent of any observer.

The methodological challenge for the social sciences is that they cannot do empirical studies in an entirely value-free method. The methodology of the social sciences is not a value-free but a "value-laden" methodology. The bias (values) of an observer can never be entirely absent in social science observations. It is difficult

F. Betz, *Managing Science*, Innovation, Technology, and Knowledge Management 9,
DOI 10.1007/978-1-4419-7488-4_9, 

in the methodology of social science to separate empirical judgments from normative judgments of social scientists. Social science observations of nature always have a normative judgment implied in any empirical judgment (value underlying fact). Thus, how is objectivity possible in social science? And where does this leave social science "empiricism" – social science as a science?

Whereas the value-free methodology of the physical sciences gives only facts (empirical judgments), the value-laden methodology of the social science must yield both empirical and normative judgments (facts and prescriptions).

## *Illustration: Max Weber on Social Science Methodology*

There are many topics in the social sciences and much profound knowledge therein to be found. However, the topic upon which we focus is methodology. Therein one finds fewer significant writings on this topic; and, in particular, one notes the important and influential views of Max Weber on methodology. Weber is considered one of the founders of the discipline of sociology. His views provide an important milestone in the development of social science methodology and still express contemporary concerns.

Weber described what is observable in social science: "From our viewpoint, 'purpose' (as observed in a social action) is the conception of an effect which becomes a cause of action (as observed).... Its specific significance (observation of action) consists in the fact that we not only *observe* human conduct but can and desire to *understand* it" (Weber 1897).

Observation of social phenomenon can see not only the action by an actor in a social phenomenon but see the motivation of the actor.

This is a form of the scientific paradigm of Function. In Weber's view of social science empiricism, the social scientist can observe, in social phenomena, not only the action but also the intention of the actor. Implicitly, Weber was identifying the paradigm of Function as a fundamental paradigm in social science.

In other writings of Weber we find the ideas of systems and of logic in social science description of social action and intention. For example, in *Fundamental Concepts of Sociology*, Weber defined the following basic terms in sociology:

1. "Sociology seeks to formulate type concepts and generalize uniformities of empirical *processes*. (System paradigm).
2. "Action is social when, by virtue of the subjective meaning attached to it by the acting individual(s), it takes account of the behavior of others and is thereby guided" (Function paradigm).
3. "Types of social action include: (a) rational orientation to a system of discrete individual ends (decision-making); (b) rational orientation to an absolute value (ethics); (c) affectional orientation (emotions); (d) traditional orientation (culture)." (Logic paradigm) *(http://www.sociosite.net/topics/weber, 2008).*

In retrospect, one can see that Weber was using not only the ideas in the scientific paradigm of Function but ideas from the paradigms of System and Logic.

Max Weber (1864–1920) was born in Erfurt, Germany. In 1882, he went to the University of Heidelberg to study law and then in 1884 to the University of Berlin. In 1886, he passed an examination (Referendar) to practice law. In 1889, he earned a doctorate in law from the University of Berlin, with a thesis on the history of medieval business organizations. He then completed a second thesis (Habilitationsschrift) on Roman agrarian history (with reference to law). This enabled him to teach at the University of Berlin as a lecturer (privatdozent). He married Marianne Schnitger in 1893. In 1894, he was appointed to a professorship at Freiburg University. In 1896, he moved to the University of Heidelberg.

In 1897, Max Weber had a quarrel with his father. And a few months later his father died, with the quarrel unresolved. Weber developed nervousness and insomnia, both severe enough to interfere with his teaching duties. In 1899, he took a leave of absence from his teaching and spent the summer and fall of 1900 in a sanatorium. He returned to Heidelberg in 1902 and resigned his professorship in 1903.

Next, he became an associate editor of a journal (Archives for Social Science and Social Welfare). After not publishing any papers between 1898 and 1902, Weber began again writing and publishing in 1904. Then he published his most famous essay, *The Protestant Ethic and The Spirit of Capitalism*. During the First World War, he directed army hospitals in Heidelberg. Afterwards, he helped draft the Weimar Constitution for the post-war German republic. He resumed teaching at the University of Vienna and then in 1919 at the University of Munich. He died in 1920. His major work on public administration (bureaucracy) was published posthumously in 1922 as *Economy and Society*.

Max Weber (http://en.wikipedia.org, Max Weber 2007)

## Empirical Observation in Social Science

As seen in Weber's view on methodology, social science empirical knowledge is descriptive of what exists in social nature – descriptive as factual statements of both action and intention. Both are observable in social phenomena: (1) what actions occur in a society and (2) what actions mean to the actors in a society – *action and intention*. Social science empiricism is methodologically possible by observing "social objects" (actors, action, intention) in the intellectual framework of the paradigm of Function.

Social science empiricism is based upon a description of social nature as "function" in participants' actions in a society.

Social science empiricism is also based upon a description of social nature as "system" and "logic."

Empiricism in social science and using the paradigms of Function and System and Logic, all together fulfill Weber's edict: (1) to observe empirical knowledge and (2) to separate empirical knowledge from normative judgments of the observer. The value judgments (Function) the social scientist can empirically describe are *those held by the observed participants in their organized interactions* (System). Moreover if these value-judgments are empirically observed to be made by rational rules by the participants, the observer can identify forms of "reasoning" (Logic) in their decision behavior.

Social science "laws" observed in social behavior characterize both empirical events and normative meaning of events.

When discussing the physical and biological sciences, one can find many historical illustrations of progress in the construction of empirically grounded theory. But this has not been the pattern in the social sciences. Instead of progressing with accumulating empirical evidence and integrating theory grounded upon that evidence, mostly the social sciences have divided in different schools within disciplines. Each school has contended for the mantle of scientific truth but without the necessary empirical grounds upon which to validate theory. And often instead of integrating theory, schools have contended with one another. (Earlier in illustrating a kind of functional model of the mind, we saw in the discipline of psychology that many different schools of psychology exist, each focused around seminal writings of a key psychologist. So psychology has not yet developed an integrated, empirically based model of the mind, as for example biology has accomplished in the DNA model).

The division of the social sciences into disciplines and further into contending schools has been probably not because of any bad intentions of the many communities of social sciences. All social science disciplines have aspired to truth, and thereby formed many scientific societies publishing many journals and also institutionally housing their communities in departments in universities. Instead the challenge to the social sciences is that it must methodologically create social theory grounded not only upon empiricism but also upon normative judgments. This is what Weber was referring to when he said that sociologists should "not only observe human conduct but can and desire to understand it." The "observation of human conduct" requires empirical judgments but the "desire to understand it" requires normative judgments. And this is why the social sciences are called value-laden sciences.

Social objectivity requires grounding theory in both empirical and normative judgments.

## *Illustration: Modern Economic Theory and the Global Financial Collapse of 2007–2008*

Since physical science theory requires grounding only in empirical observation, whereas social theory requires grounding in both empirical and normative observations, one can imagine why scientific progress is occurring more slowly in the social rather than in the physical/biological sciences. If we look at an interesting case wherein modern social science theory was challenged by a real social event, we can see that the failure of theory to anticipate or explain reality centers exactly upon this pinnacle of empirical and normative observation. This event is the case when modern economic theory failed in the empirical/normative test of the global financial market in 2007–2008. This historical social event challenged the social science "objectivity" about a major economic theory. Was the economic law of a "perfect market" an empirical or normative judgment by economic theorists? In 2007–2008, most economic theorists had not anticipated a monstrous financial/economic failure. Nor did economists even agree among themselves upon the description or explanation of the failure of the concept of an economic "perfect market" in real practice.

The two empirical facts were (1) the world's financial system collapsed in 2007 and (2) national governments "bailed them out" (subsidized the privately-owned banks in many countries of the world). But, in economic theory, if the financial market was a "perfect market," no bail out should have been necessary. The enormous scale of government banking subsidies then in several Western nations was the largest ever in economic history. The financial collapse had centered on a financial scheme, "mortgage-based" collateralized debt objects (CDO), a so-called financial derivative. Financial derivatives are contracts for the right to the payments of interest from loans (debt object).

The scheme had temporarily operated under special economic conditions then in the world in the decade of 2000, a combination of unusually low interest rates, rapidly increasing oil wealth, huge leveraged investments, global manufacturing outsourcing, and no proper government regulation of banking practices. These special conditions could be exploited for large-scale financial fraud because the instruments could be globally sold thanks to the new Internet connections of investment practice. That new derivative financial market had grown in 5 years from millions of dollars of contracts to trillions of dollars. It had been an enormous financial leveraging upon a relatively small asset base of a US mortgage market, attracting billions and billions of capital from around the world. Even the richest oil sheik in the world lost at least a quarter of his wealth in the debacle; and the Arab world noted at least $2.4 trillion dollars was lost – lost forever.

And to the participants and observers of the time, it all appeared complex and difficult to understand. This was because perhaps it was meant to be obscure. An example of an economic commentator at the time who used the term "fraudulent" in making normative judgments on the episode was the Nobel Prize economist, Paul Krugman, who summed up the scheme: "America emerged from the Great Depression (1930s) with a tightly regulated banking system, which made finance a staid, even boring business. Banks attracted depositors … used the money thus attracted to make loans, and that was that…. After 1980 a very different financial system emerged. In the deregulation-minded … era, old fashioned banking was increasingly replaced by wheeling and dealing on a grand scale…. And finance became anything but boring. It attracted many of our sharpest minds and made a select few immensely rich."

"Underlying the glamorous new world of finance was the process of securitization. Loans no longer stayed with the lender. Instead, they were sold on to others, who sliced … individual debts to synthesize new assets. Subprime mortgages, credit card debts, car loans – all went into the financial system's juicer. Out the other end, supposedly, came sweet-tasting AAA investments. And financial wizards were lavishly rewarded for overseeing the process."

"But the wizards were frauds … and their magic turned out to be no more than a collection of cheap stage tricks. Above all, the key promise of securitization – that it would make the financial system more robust by spreading risk more widely – turned out to be a lie. Banks used securitization to increase their risk, not reduce it, and in the process they made the economy more, not less, vulnerable to financial disruption. Sooner or later, things were bound to go wrong, and eventually they did. Bear Stearns failed; Lehman failed; but most of all, securitization failed" (Krugman 2008).

In this illustration, I am quoting at length from Pau Krugman. At the time of the crisis and as a recognized economist, he was a significant public voice over the crisis in his other role of a columnist for the New York Times. He also wrote about the divisions within the community of economics concerning the relevance of modern economic theory to the crisis. Krugman had received a Nobel Prize in economics from his colleagues, thus having the scientific prestige to command an audience among his peers. In Weber's methodological perspective, Krugman was both an observer (as newspaper comminatory) and participant (as professional economist) in the societal crisis, thus writing (for the historical record) both contemporary empirical and normative judgments.

For example, Krugman wrote the following (about his empirical and normative view) on the status of his discipline of economics: "It's hard to believe now, but not long ago economists were congratulating themselves over the success of their field. Those successes – or so they believed – were both theoretical and practical, leading to a golden era for the profession. On the theoretical side, they thought that they had resolved their internal disputes…. And in the real world, economists believed they had things under control: the "central problem of depression-prevention has been solved," declared Robert Lucas of the University of Chicago in his 2003 presidential address to the American Economic Association. In 2004, Ben Bernanke, a former Princeton professor who is now the chairman of the Federal Reserve Board, celebrated the Great Moderation in economic performance over the previous two decades, which he attributed in part to improved economic policy making. Last year, everything came apart" (Krugman 2009).

A methodological importance of theory is to understand a natural phenomenon enough to be able to anticipate empirical occurrences. Were economists capable of anticipating the global credit crisis, based upon their economic theory? Krugman wrote that: "Few economists saw our current crisis coming, but this predictive failure was the least of the field's problems. More important was the profession's blindness to the very possibility of catastrophic failures in a market economy" (Krugman 2009).

An exception was the economist Nouriel Roubin, who had forecasted the housing bubble and financial crisis. While at the time, his warning was ignored, but afterwards, he became well known and established a consultancy company, Roubini Global Economics. But, so to speak, his exception proved the rule, as Krugman explained: "During the golden years, financial economists came to believe that markets were inherently stable – indeed, that stocks and other assets were always priced just right. There was nothing in the prevailing models suggesting the possibility of the kind of collapse that happened last year" (Krugman 2009a).

We have emphasized how in science, when theory is contradicted by experiment, then theory is reexamined. This is what is meant by the scientific method creating "empirically grounded theory": theory constructed upon experimental evidence and validated by empiricism. If there is a conflict between empiricism and theory, it is theory that is reconstructed, not empiricism. But despite the empirical impact of the collapse of the world's financial market in 2007–2008, reexamination of economic theory apparently did not happen universally in the economic discipline. Then instead of a community of scientists (economists) setting out to reconstruct economic theory, instead the community of economists simply divided into two schools – two opposed camps of idealists and realists.

Krugman described this division: "Meanwhile, macroeconomists were divided in their views. But the main division was between those who insisted that free-market economies never go astray (idealists) and those who believed that economies may stray now and then (realists) but that any major deviations from the path of prosperity could and would be corrected by the all-powerful Fed. Neither side was prepared to cope with an economy that went off the rails despite the Fed's best efforts" (Krugman 2009).

Under the powerful empiricism of the collapse of the world's entire financial system, did all economists then become realists? Not according to Krugman: "And in the wake of the crisis, the fault lines in the economics profession have yawned wider than ever. Lucas says the Obama administration's stimulus plans are "schlock economics," and his Chicago colleague John Cochrane says they're based on discredited "fairy tales." In response, Brad DeLong of the University of California, Berkeley, writes of the "intellectual collapse" of the Chicago School, and I myself have written that comments from Chicago economists are the product of a Dark Age of macroeconomics in which hard-won knowledge has been forgotten" (Krugman 2009).

Paul Krugman (1953–present) was born in New York. He earned a B.A. in economics from Yale University and a Ph.D. from the Massachusetts Institute of Technology (MIT) in 1977. In 1982, he worked as staff in the government Council of Economic Advisers. In 2000, he became a professor of economics at Princeton University. He wrote on international economics, including international trade, economic geography, and international finance. In 2008, he received a Nobel Prize in economics for his studies on geographic analysis in economy theory. He has acted as an industrial consultant in economics and was a columnist for the New York Times in 2008–2010.

(http://en.wikipedia.org, Paul Krugman)

## Flyvbjerg on Idealism and Realism in Political Science

This illustration of economic theory is continued later; now we pause to observe how any social science discipline can divide into two schools, one of realism (empiricism) and the other of idealism (normative prescription). It has turned out that schools in a social science discipline have often occurred around this distinction. As a second example of realism/idealism divisions, we next look at Bent Flyvbjerg's views about methodology in political science (Flyvbjerg 2001). Flyvbjerg focused upon the writings of two influential political scientists of the late twentieth century: Jurgen Habermas and Michael Foucault. Flyvbjerg saw their differences as one of methodology, each describing differently the politics of a given societal era in *idealism* or *realism*.

Habermas described political activities by identifying the political ideals around which people gather, associate, and identify. Habermas called this a "discourse-ethics" of the politics. By the term "discourse," Habermas indicated that social ideals are discussed openly in the politics as a justification of political action. By the term "ethics," Habermas was indicating that the ideal of the discourse provided an ethical agreement around which the group associated.

But there is a reality about power in all political situations – the reality of how power is actually used, as opposed to how the power is justified. This is what Michel Foucault emphasized should be described (as an essential feature of in social science methodology). Foucault argued that in any political situation, even focused around a "discourse-ethics," there was also another side to power, a "realism" in politics – the "power analytics" of the situation.

For example, although equality of opportunity is an ideal of American democracy, the reality of politics (in America and elsewhere) is that the wealthy do have more opportunities in society and obtain better representation in politics than do the poor. Even in law, the American ideal of equal treatment depends in reality upon having sufficient wealth to hire the best lawyer or lobbyist.

Flyvbjerg wrote that between the two, Habermas and Foucault, modern political science methodology can capture both idealism and realism about power contexts: Flyvbjerg wrote: "The works of Habermas and Foucault highlight an essential tension in modernity. This is the tension between consensus and conflict.... Habermas is the philosopher of Moralitat (morality) based on consensus. Foucault ... is the philosopher of Wirkliche Historie (real history) told in terms of conflict and power" (Flyvbjerg 2001).

The morality of modern democracy is a discursive consensus for a democratic process, which defines the rules of governance in a constitution and provides for the exercise of government power by elected officials. This is Habermas's point about political morality as based upon consensus. But how such consensus actually operates is through conflict – struggle by parties for election, funding of elections by special interests, formulation of laws and enforcement to benefit special interests rather than the general civil society. This is Foucault's point about the actual operation of any real democracy in a society is through conflict and the gaining and exercise of power.

Flyvbjerg's argument is that both Habermas's and Foucault's perspectives on consensus and conflict in society are essential to the operations of a real democracy.

> The consensus about power in a group is constructed around an ideal expressed in a discourse-ethics of the group (idealism); while the reality of how power is really exercised in a group is expressed in the power-analytics of the group (realism).
>
> Discourse-ethics is the justification of power; while power-analytics is the exercise of power.
>
> Jurgen Habermas was born in Dusseldorf, Germany in 1929. He studied at the University of Gottingen and the University of Bonn, obtaining his PhD in 1954. He did his habilitation in political science at the University of Marburg. In 1962, he obtained a professorial appointment at the University of Heidelberg. In 1981, he published The Theory of Communicative Action.

(http://en.wikipedia.org, Habermas 2009)

> Michel Foucault (1926–1984) was born in Poitiers, France. He attended the Ecole Normale Superieure. He earned a license in psychology and a degree in philosophy in 1952. From 1953–1954, he taught psychology at the Université Lille Nord de France, where he published his first book *Maladie mentale et personnalité*, From 1954 to 1960, he taught in different universities in Sweden, Poland, and Germany. In 1960, he completed his doctorate and obtained a position in philosophy the University of Clermont-Ferrand. After the French student rebellion in 1968, the French government started a new university, Paris VIII. Foucault became head of its philosophy department. In 1970, he was elected to the Collège de France, as Professor of the History of Systems of Thought. His major works include: *Madness and Civilization, The Birth of the Clinic, Death and The Labyrinth, The Order of Things, The Archaeology of Knowledge, Discipline and Punishment, The History of Sexuality.*

Michel Foucault (http://www.foucault)

> Bent Flyvbjerg was born in Denmark in 1952. He is a geographer and urban planner. He has held professorial chairs at Aalborg University in Denmark and at Delft University of Technology in the Netherlands. In 2009, he moved to Oxford University in England. His books include: *Decision-Making on Mega-Projects, Managing Social Science Matter, Rationality and Power.*

*(http://www.sbs.ox.ac.uk/faculty/Flyvbjerg+Bent)*

## *Illustration (Continued): Economic Theory and Global Financial Collapse 2007–2008*

The division of schools in the social sciences has often occurred on this "fault line" of idealism-realism.

Even after the 2007–2008 economic crisis, the school of neoclassical economists remained resolutely idealistic. Krugman explained this as an esthetics in economic theory: "As I see it, the economics profession went astray because economists, as a group, mistook beauty, clad in impressive-looking mathematics, for truth. Until the Great Depression (1930), most economists clung to a vision of capitalism as a perfect or nearly perfect system. That vision wasn't sustainable in the face of mass unemployment. But as memories of the Depression faded, economists fell back in love with the old, idealized vision of an economy in which rational individuals interact in perfect markets, this time gussied up with fancy equations" (Krugman 2009).

The methodological problem in idealism in social theory is that of not being able to distinguish utopia from reality. Idealistic schools in social sciences apparently can ignore reality, as idealistic economists seemed to ignore real economic problems. About this idealism in economics, Krugman wrote: "Unfortunately, this romanticized and sanitized vision of the economy led most economists to ignore all the things that can go wrong. They turned a blind eye to the limitations of human rationality that often lead to bubbles and busts; to the problems of institutions that run amok; to the imperfections of markets – especially financial markets – that can cause the economy's operating system to undergo sudden, unpredictable crashes; and to the dangers created when regulators don't believe in regulation" (Krugman 2009).

But just how did this excessive dose of idealism come to dominate the US economic discipline after World War II? Krugman summarized: "The birth of economics as a discipline is usually credited to Adam Smith, who published 'The Wealth of Nations' in 1776. Over the next 160 years an extensive body of economic theory was developed, whose central message was: Trust the market. Yes, economists admitted that there were cases in which markets might fail, of which the most important was the case of 'externalities' – costs that people impose on others without paying the price, like traffic congestion or pollution. The basic presumption of 'neoclassical' economics … was that we should have faith in the market system" (Krugman 2009).

The last great dose of economic realism (empiricism) to hit economists in the US was the Great Depression of the 1930s. Krugman wrote that then the world depression of 1930 called the attention of economics to real context: "This (utopian) faith was, however, shattered by the Great Depression. Actually, even in the face of total collapse some economists insisted that whatever happens in a market economy must be right: 'Depressions are not simply evils,' declared Joseph Schumpeter in 1934. 'They are,' he added, 'forms of something which has to be done.' But many, and eventually most, economists turned to the insights of John Maynard Keynes for both an explanation of what had happened and a solution to future depressions. Keynes did not, despite what you may have heard, want the government to run the economy. He described his analysis in his 1936 masterwork, 'The General Theory of Employment, Interest and Money,' as 'moderately conservative in its implications.' He wanted to fix capitalism, not replace it" (Krugman 2009).

Krugman argued that while Keynes, as an economist, subscribed to the ideal economic theory that commodity markets should be self-regulating, yet Keynes recognized contexts to an economic theory: "But Keynes did challenge the notion that free-market economies can function without a minder, expressing particular contempt for financial markets, which he viewed as being dominated by short-term speculation with little regard for fundamentals. And he called for active government intervention – printing more money and, if necessary, spending heavily on public works – to fight unemployment during slumps" (Krugman 2009).

Thus the economists had merely divided into two schools: one of *realism* (e.g., Keynes) and the other of *idealism* (e.g. Friedman). Krugman emphasized this scholastic division: "Yet the story of economics over the past half century is, to a large degree, the story of a retreat from Keynesianism and a return to neoclassicism. The neoclassical revival was initially led by Milton Friedman of the University of Chicago, who asserted as early as 1953 that neoclassical economics works well enough as a description of the way the economy actually functions to be both extremely fruitful and deserving of much confidence" (Krugman 2009).

Krugman described the idealistic economic school of Friedman as turning away from any contextual role of government in the economy as a regulator. Krugman described the Friedman school as solely focused upon the financial factor of money supply: "Friedman's counterattack against Keynes began with the doctrine known as monetarism. Monetarists didn't disagree in principle with the idea that a market economy needs deliberate stabilization. 'We are all Keynesians now,' Friedman once said, although he later claimed he was quoted out of context. Monetarists asserted, however, that a very limited, circumscribed form of government intervention – namely, instructing central banks to keep the nation's money supply, the sum of cash in circulation and bank deposits, growing on a steady path – is all that's required to prevent depressions" (Krugman 2009).

This small modification of the idealism in the "perfect market" by *government regulation only of money supply* was seen by monetarist economists as a sufficient contextual interaction (between the economic system and political system of a society). But even with this small contextual modification to economic theory, the overall attraction of a pure "idealism" of the perfect market was still irresistible to economic theorists. Krugman wrote: "Eventually, however, the anti-Keynesian counterrevolution went far beyond Friedman's position, which came to seem relatively moderate compared with what his successors were saying. Among financial economists, Keynes's disparaging vision of financial markets as a "casino" was replaced by "efficient market" theory, which asserted that financial markets always get asset prices right given the available information. Meanwhile, many macroeconomists completely rejected Keynes's framework for understanding economic slumps. Some returned to the view of Schumpeter and other apologists for the Great

Depression, viewing recessions as a good thing, part of the economy's adjustment to change. And even those not willing to go that far argued that any attempt to fight an economic slump would do more harm than good. Not all macroeconomists were willing to go down this road: many became self-described New Keynesians, who continued to believe in an active role for the government. Yet even they mostly accepted the notion that investors and consumers are rational and that markets generally get it right" (Krugman 2009).

So Krugman had described two schools of economic theorists: one school of idealism (Friedman monetarists) and one school of realism (Keynesian government interventionists). Their division centered different views about the methodological status of a central economic law – a perfect market operating on an economic "Law of Supply and Demand" In a perfect market, prices are determined in an equilibrium where supply just meets demand of a product. Is this "Law" purely an empirical fact (empirical judgment) and all markets are empirically a "perfect market"? Or is this Law a prescription (normative judgment) about how markets ought to operate for the good of a society and should be regulated to be nearly a perfect market?

Within the economic community, debates about economic theory and economic reality continued. For example in April 2010, Chris Giles reported upon an economic conference held in Cambridge: "Many of the world's top academic economists agreed on Friday (April 9) that the financial and economic crisis had exposed fatal flaws in their subject and ideas were urgently needed to keep economics relevant. While this represented an unusual consensus, the eminent economic brains lived up to their stereotype by disagreeing on what policies, if any, should be adopted to prevent a repetition.... The participants were speaking at the inaugural conference of the Institute for New Economic Thinking, a think-tank sponsored by George Soros, the billionaire financier. They included five Nobel prize winners (in economics). Held at King's College, Cambridge ... the conference participants could neither agree on the cause of the crisis nor the necessary remedies" (Giles 2010a).

And these economists at the conference divided according to *realism and idealism* in economic theory: "One disagreement hinged on whether asset price bubbles lay at the heart of the crisis. Those who thought so argued for tighter regulation (realism)... (Others, such as) Michael Goldberg, of the University of New Hampshire, said it was wrong to suggest the price swings were necessarily a bubble and that they were more likely to be fundamental to the beneficial forces of capitalism (idealism)" (Giles 2010b).

In summary, both before and after the financial crisis, the two schools of economics differed about theory either as idealistic or realistic. Neoclassical economists held that markets ought to be perfect, and whether markets were really so was of little importance. Neo-Keynesian economists held that no real market is ideal and government regulation is needed to move real markets toward an ideal. Yet their arguments were not explicitly poised as such methodological issues. The community of economists simply argued who was right and who was wrong. When such methodological differences arise, communities do split into schools. And they often just argue, neither convincing the other.

Social science theory should be grounded both normatively and empirically, so to construct scientific theory grounded both in empiricism (realism) and in value (idealism).

## Idealism and Realism in the Social Sciences

As we noted in the illustration of economics and of political science, this kind of division of the schools into *idealism and realism* is a general pattern in all the social science disciplines. This is due to the methodological issue of empirical and normative judgments in social science observation.

And as we noted in social science observation, a participant's "action" can be observed empirically, realistically. But the observation of a participant's "intention" cannot be described so simply as either realistic or idealistic. This is because the "intention" of an action must be viewed from two perspectives in social science: (1) the perspective of the social "actor" and (2) the perspective of the social "observer." From these two perspectives, any observation of social "intention" must always be both empirical and normative. The observer's view of the "intention" of the observed participant is both empirical (what people do) and normative (what people think they should do). This mixture of empiricism and prescription of an observation of a social participant's intentions (social laws) is characteristic of observation in all social sciences.

It is the mixture of "realism" and "idealism" in the empirical observation of "social intention" which makes the methodological challenge of objectivity in social science theory so interesting and challenging!

Yet, while this mixture of idealism-realism in social science facts-intension is methodologically challenging, it does not negate "objectivity" in the social sciences. Instead, it provides the basis for the importance of the social sciences to modern civilization, an objective study of human values.

Value-laden social science can be the most important method for learning about the objectivity of value (universal values) in modern life.

Modern social science is essential to social progress in society – if it can successfully prescribe improved rationality for organized social activities – ethical rationality.

## Social Theory and Practice

Interestingly, when Max Weber discussed the methodological issue of the empirical and the normative in social science theory, he had noted the importance of practicing social theory as well as making social theory (theory application and theory construction). It was in 1903, when Weber became associate editor of a journal (*Archives for Social Science and Social Welfare*) and wrote: "We all know that our science (social science) … first arose in connection with practical considerations. Its most immediate and often sole purpose was the attainment of value-judgments concerning measures of State economic policy" (Weber 1903).

Weber noted an intimate connection between political policy and the origin of social science and this connection of science to practice continues. In the contemporary academic scene there is this connection of social science to practice – application of social science ideas to policies and consulting.

For example, an economist might advise a government on economic policy. A political scientist might consult for a political candidate on how to run a successful political campaign. A sociologist might contract to a company to suggest how best to run a new marketing campaign. A troubled client might go to a psychiatrist for treatment of a mental condition.

About empirical and normative knowledge, Weber wrote: "In the pages of this journal, especially in the discussion of legislation, there will inevitably be found social policy.... But we do not by any means intend to present such discussions as 'science'.... In other worlds, it should be made explicit just where arguments are addressed to the understanding and where to the sentiments. The constant confusion of the scientific discussion of facts and their evaluation is still one of the most widespread and also one of the most damaging traits of work in our field" (Weber 1903).

> Weber's editorial position illustrates how the methodological challenge to separate factual judgments from value judgments is basic to the idea of "objectivity" in sociology and economics, even from their early days.

Weber emphasized that social science needed to maintain: "the logical distinction between 'empirical knowledge' (knowledge of what is) and 'normative knowledge' (knowledge of what should be)... An empirical science cannot tell anyone what he should do, but rather what he can do and, under certain circumstances, what he wishes to do" (Weber 1903).

> The methodological challenge to the social sciences is to make and maintain a clear distinction between empirical observations and normative judgments – between what-is and what-ought-to-be.

One can find more discussions of methodology by Weber in his other essays. For example, in Weber's "Definition of Sociology," he wrote: "There is no absolutely 'objective' scientific analysis of culture ... 'of social phenomena' independent of special and 'one-sided' viewpoints according to which – expressly or tacitly, consciously or unconsciously – they are selected, analyzed and organized for expository purposes.... All knowledge of a cultural reality ... is always knowledge from particular points of view" (Weber 1897).

> Even in the selection of a social-science-object-to-study, there is some prior evaluative concern about the importance of the topic as chosen by an observer – a subjectivity in topic selection.

Weber emphasized that assuming there is no bias in observation is naïve: "... the naïve self-deception of the specialist, who is unaware that it is to the evaluative ideas with which he unconsciously approaches his subject matter, that he has selected form an absolute infinity, a tiny portion with the study of which he concerns himself" (Weber 1897).

Weber argued for a "self-awareness" about subjective bias in the observer. This does not yet provide any absolute "objectivity" in social science laws. But it does indicate a first step toward objectivity – an observer should be sensitive to an inherent evaluative bias in the observer's empirical observations.

We recall again that there is an absolute objectivity in the physical science paradigm of Mechanism, as all physical laws must be expressed in a form invariant (unchanging) of observer. The Special Relativity of Einstein requires all physical theory to be expressed in a form which keeps the speed-of-light the same constant for all observers. Such invariant theory – observation independent of any particular observer – objectivity of observation – is possible only for the physical sciences but not for the social sciences. This is how formally *value-free methodology* in the physical sciences must differ from the *value-laden methodology* of the social sciences. There can be no invariant theory in the social sciences.

In addition, Weber argued that the intentions-of-social-science-observers should be, at least, for empirical truth: "… the choice of the object of investigation and the extent or depth to which this investigation attempts to penetrate into the infinite causal web (of human action), are determined by the evaluative ideas which dominate the investigator and his (her) age…. (But) in the method of investigation, the guiding 'point-of-view' (of the observer) is of great importance for the construction of the conceptual scheme which will be used in the investigation. In the mode of their use, however, the investigator is obviously bound by the norms of our (social scientist community) thought. For scientific truth is precisely valid for all who seek the truth" (Weber 1897).

And a practical application of this principle can been seen by examining the funding sources for scientific researches. For example, in the second half of the twentieth century, there were notorious cases of "research" on the health effects of cigarettes – cited as proof that cigarettes did not "cause" cancer. These research studies were performed in universities and university hospitals as academic research projects funded by cigarette manufacturers. Some of these researchers allowed their research to be cited by cigarette manufacturers as proof that cigarettes were entirely innocent in lung cancer causation. This bias was enabled by the fact that cigarettes were only one of two causal factors in lung cancer (with viruses providing a second causal factor). But the researchers did not establish this proper kind of causal link to cigarettes in their research. The guiding scientific value of "seeking truth" was not maintained in that particular community of cigarette-and-health-issue researchers.

## Social Science Empiricism and Practice

We have seen that one of the guiding principles in the striving toward objectivity is a "value" held by a social science community to "seek the truth." This does not establish any absolute objectivity in social science – but only a temporary and contextually-dependent objectivity. But how can temporary and contextually-dependent theory lead to any universal social theory? The answer turns out to be "in stages." The first stage is in Weber's application of social science knowledge to practice – consulting.

We recall that in the taxonomy of all the activities of science, totality of science, there are categories of method and application. Now we need to look at the interactions between method and practice in science, as emphasized in Fig. 9.1.

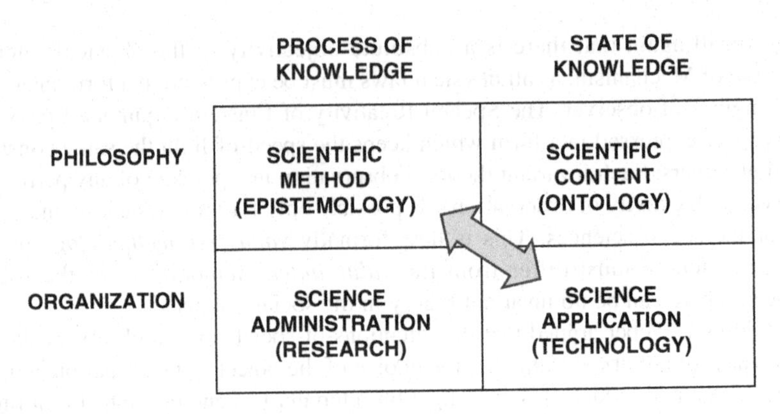


**Fig. 9.1** Philosophy and organization of science

We saw that Weber saw this connection between empirical judgments of scientific method and normative judgments of science application as a critical connection. In the physical and life sciences there is also a close connection between scientific method and application – between science and technology. The difference, however, is that physical science research can be performed in a "value-neutral" methodology, whereas social science research must be performed only in the "value-laden" non-laboratory context.

## Relationship of Physical Science to Technology

One can describe the relationship of physical science to technology in the following information model, connecting scientists to engineers to business managers (Fig. 9.2).

Inanimate physical things have neither mind nor humanity and so can be experimented upon in a laboratory, as the context of physical reality is independent of the validity of physical theory. Earlier, we noted that the principle of relativistic invariance in physics establishes the value-free nature of physics – physical laws expressed in form, independent of the perspective of an observer.

Thus in a laboratory setting, science and technology can be decoupled in time and place and purpose. This decoupling allows an absolute independence of objectivity in science from application in technology – separation of physical science from practice. We can see this by again looking at the information flow from science to technology, envisioned as science down to technology. In this perspective, one could draw a dotted line between science above and technology beneath to show a methodological division (separation) that is possible between physical

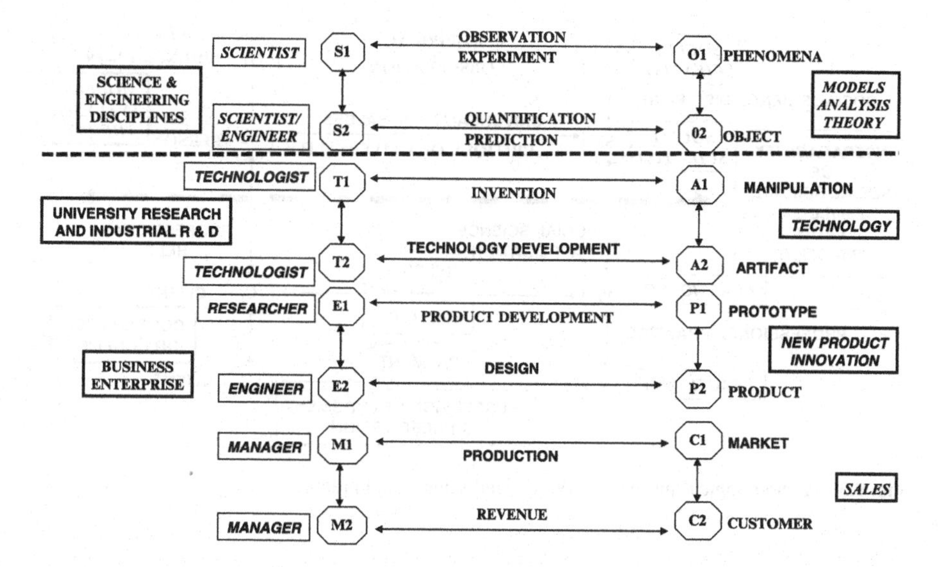


**Fig. 9.2** Roles in physical science and innovation

science and physical technology. Physical science about nature can be performed in a scientific laboratory, entirely separate from an industrial application in a technology. Physical nature can be investigated in a laboratory separate from the application of physical nature. This is an important methodological point. A physical scientist can be entirely objective about physical nature because experimentation on physical nature cannot alter physical nature.

> Since the application of the physical science of nature does not change physical nature, physical science can be methodologically separate from physical application.

## Relationship of Social Science to Practice

This complete methodological separation of physical science and its applications of technology are, in contrast, not possible in the social sciences. A social scientist will always hold a normative judgment (value judgment) about any empirically observed social behavior. (This is what Weber pointed out when he emphasized that even in the selection of a social science topic for study, an observer exercises a value judgment.)

> In the physical sciences, a physical scientist can be wholly objective about physical nature; but in the social sciences, a social scientist can be only partly objective about social nature.

> In the social sciences, the application of social science knowledge may always potentially alter social nature.

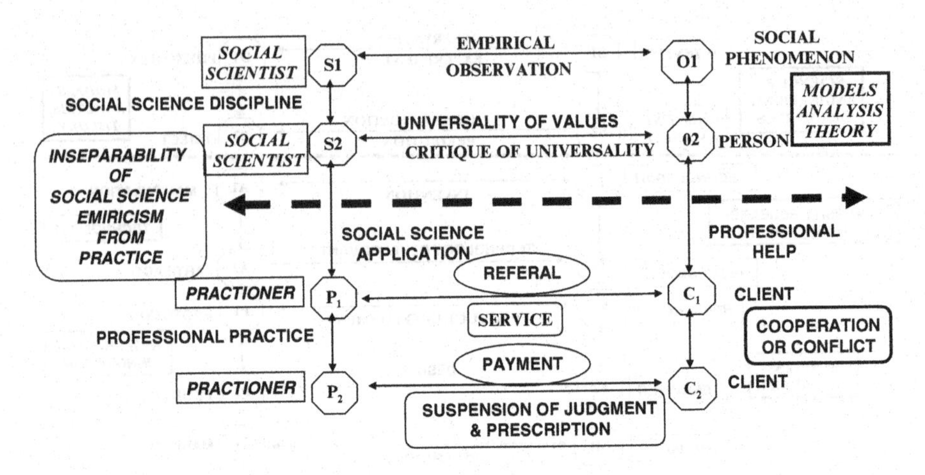


**Fig. 9.3** Methodological inseparability of social science and practice

Empirical social science and application of social science knowledge, practice, can never be methodologically entirely separated. We can sketch this idea of *methodological inseparability* in the social science in Fig. 9.3.

We can clearly see the relationship of theory to practice in a medical prescription from a medical doctor. If a sick patient wishes to get well, the doctor prescribes a medicine. The medical doctor is using biological knowledge and technology on health to cure a patient's disease. The medical doctor is making no judgment about whether or not the patient is worthy of health. Instead, the medical doctor temporarily adopts the intention of the patient (for health) to aid the patient in how to regain health.

In contrast to the physical sciences, there is for the social sciences no "laboratory" wholly independent of application in which social nature can be isolated. For example, a doctor cannot experiment with a drug on a patient. There are formal government procedures which medical drug companies must follow to prove a drug effective and licensed for medical application.

Since there is no "social science laboratory," social nature cannot be viewed independently of its value-implications for human society. As another example, there is no "national economy" to be experimented upon, independent of a real nation. Any change in economic policy in a nation is an economic experiment and simultaneously alters its economy.

Thus in social science, separation of empirical research from potential application is always only a partial separation. And this inseparability must be a factor in social science methodology. Accordingly in the social sciences, the value-implications (application) of any social science study are present as a context of the study.

> There is never a complete separation of social science empiricism from policy/practice application.

## *Illustration: US Academics and Consulting*

Successful professors now in leading business schools in the US and Europe often earn more money from business consulting than from professorial salaries. Many have established separate business-consulting firms. And some have temporarily in served government posts and then returned to their professorial positions in universities.

> Michael Porter, a professor at Harvard University, provides an example of the connection between research and consulting practice in business schools at American universities. Michael Porter wrote several books on management practice (i.e., business as a value-adding enterprise and on clustering of industries). Based upon these publications, he established a successful business consulting firm.
>
> Henry Kissinger provides an example of a successful Harvard professor in political science, who also served in the US government. From 1964 to 1972, Henry Kissinger became secretary of state under President Richard Nixon. Then after a decade of government service, Kissinger set up a successful international consulting firm in Washington DC – advising companies and foreign governments on international policy issues and situations.
>
> Robert Reich is another example of a Harvard economics professor who served in the US Government. From 1993 to 2001, Reich was the Secretary of Labor under President and then returned to academia.

## Social Science Consulting and Normative Judgments

Social scientists can describe (1) the action and purpose (Function) of social participants and (2) the rational rules (Logic) (3) in their organized activities (System). All this is empirical: (1) what the participant is doing, (2) why the participant is doing it, and (3) how the participants decide to cooperate to do it. One can observe societal participants in their behavior, reason, and organization.

While perhaps aware of their own observer bias, social scientists need not *immediately* make any of their normative judgments about a participant's behavior, reason, and organization. Weber's edict to partially and temporarily separate empiricism from normative judgments makes social science empiricism methodologically possible. But this is still only a partial separation of empiricism from normative judgment.

In consulting practice, the observed participant-values (intentions) can be temporally and partially separated from the value judgments by the social science practitioner on these intentions. And in such a partial separation there can be a temporary distinction between participants' values and an observer's values.

It is this kind of methodological partial-separation of empiricism from normative judgment that makes the kinds of business and policy and practitioner operations of social science professors useful to their clients. In the empirical description of what a client is doing (action) and in thinking what the action is accomplishing (purpose) may occur an unintended consequence for the client (unintended by the client's own intentions) and also unrecognized by the client.

In sociology, the idea that "the ends of our chosen means to action may not turn out as expected" is expressed by the term "unintended consequences." This term was popularized in sociology by Robert K. Merton as a kind of sociological phenomenological law, as a warning that a political intervention in the complexity of a society will likely produce unanticipated changes in society system, unintended consequences.

Robert King Merton (1910–2003) was born in Philadelphia, USA. He obtained a bachelor's degree at Temple University in 1931 and a Ph.D. in sociology from Harvard University in 1938. He became a professor at Tulane University and then in 1941 in Columbia University. He published The Sociology of Science: Theoretical and Empirical Investigations.

Robert Merton (http://en.wikipedia.org, Robert Merton 2010)

This interchange between social science empiricism and practice is actually observable in the careers of social scientists. For example, the scientific discipline of economics finds many of its graduates serving as economic forecasters (practitioners) in companies or as administrative officials in government. In the scientific discipline of psychology, the adjacent practitioner community is called "psychologists." In political science, some professors become political policy consultants for political candidates, and many graduates serve in government diplomacy posts.

Empirical observations by a practitioner – of the consequences of a client's action – may show that the client's actions may not be attaining the client's intention.

In such a case, a practitioner might suggest a different course-of-action to the client – a prescription for action.

If such a prescription is science based, it is more likely to prove technically effective than when not science based.

## Universality in Social Science as Empiricism/Practice

Max Weber had argued that "objectivity" in social science empiricism is possible, even given the complete inseparability of social science empiricism from practice: "… the science of sociology seeks to formulate generalized uniformities of empirical process. This distinguishes it from history, which is oriented to causal analysis and explanation of individual actions. Between the 'historical' interest in a family chronicle and that in the development of the greatest conceivable cultural phenomena which were and are common to a nation or to mankind over long epochs, there exists an infinite gradation of 'significance' arranged in an order which differs for each of us…. But it obviously does not follow from this that research in the cultural sciences can only have results which are 'subjective' in the sense that they are valid for one person and not for others" (Weber 1897).

Weber's methodological approach on how to deal with subjective bias in social science is this. If one can generalize the evaluative idea in the selection of a research topic over a "family chronicle" of all humanity ("nation or mankind"), then the empirical observations in the study can be relevant not just to a single perspective but over a common perspective to all humanity.

Empiricism in the physical sciences has had a similar challenge in always observing nature from some perspective. Any observation of nature in time and space occurs from some position and at some time of a physical scientist (perspective of the observer). Yet in the physical sciences, a theoretical requirement is that all physical laws must be formulated to be independent of any perspective – invariance of laws over observer perspectives.

In contrast to an "invariance of physical laws," Weber proposed a kind of "universality of social laws." Evaluative ideas (values) are implicit in the methodological choices of any observation in the social sciences. But can these evaluative ideas be generalized over all the "family chronicles" of humanity? Can there be a "universality" of evaluative ideas useful for all humanity? This kind of "universality" actually occurs in social science practice.

We have just discussed how in practice a consultant can empirically observe a client's action and intention and see when the outcome of action did not meet intention. Empirically a client's action may not be attaining what the client thinks it is attaining. So a social science consultant can describe the action and function of the action differently from how the client sees it. Then a consultant might be able to show to the client a disconnect between intention-and-results – when the client's action is not really attaining the client's purpose. Next a consultant could shift to a normative mode and tell the client how to modify action to better attain intention – change the action of the client by more efficient means and better ends.

> This shift by a consultant/practitioner from empiricism to normative advice can connect social science to practice – a temporary and partial objectivity in social science.

The consultant may temporarily accept the client's values (intentions) and rationally suggest more efficient means (changed action) to attain the client's valued ends (purposes). By temporarily assuming a client's values (purpose), a social science practitioner may aid the client to find an alternate and more efficient means (action).

> A consultant's temporary assumption of the client's values to prescribe more efficient and effective action to the client is universalizing a social science prescription from one client to another – toward a universal "family of clients" – so to speak.

This consultant's aid to a client is in the form of a prescription for action – so act to better achieve the intent of your action. As a prescription, the consultant is providing a tool, a decision aide, to improve the client's decision-making ability.

> In consultation, the practitioner is temporarily suspending the partial separation between empiricism and normative judgment.

> All social science practices consist of prescriptive aides to improve the decision-making capability of a client to better attain a client's purpose.

In the medical profession, good medical practice is based upon scientific understanding of biological nature (Mechanism and Function). But in that part of medical practice that is based upon social science, psychiatry, practice is not as well grounded scientifically. In the engineering profession, good engineering practice is based upon scientific understanding of physical nature (Mechanism). But in the profession of law, what is the science base for good legal practice? In the profession of politics, what is the science base for good governmental practice? In the profession of business, what is the science base for good business practice? These are the important issues which continue to connect social science and practice, from Max Weber's days to the present day.

> The more practice is based upon scientific knowledge, the more likely prescription will be technically effective toward attaining a client's purpose (And in chapter 11, we will address how social theory can be distinguished from political propaganda.).

However, when there is no significant attempt to universalize the ethical underpinnings of "social science knowledge" by the scientific communities, then such so-called knowledge might be used merely as propaganda. This fundamental challenge was so great in the twentieth century as to have diverted many social science efforts away from the objectivity of truth and only toward propaganda.

> The major methodological challenge of objectivity in the social sciences is to distinguish social science knowledge from political propaganda.

## Summary

1. Objectivity is knowledge about the object, independent of any observer.
2. In the physical sciences, a physical scientist can be wholly objective about physical nature; but in the social sciences, a social scientist can be only partly objective about social nature.
3. Observation of social phenomenon can see not only the action by an actor in a social phenomenon but see the motivation of the actor.
4. Social science empiricism is based upon a description of social nature as "function" in participants' actions in a society.
5. Social science empiricism is also based upon a description of social nature as "system" and "logic."
6. Whereas the value-free methodology of the physical sciences gives only facts (empirical judgments), the value-laden methodology of the social science must yield both empirical and normative judgments (facts and prescriptions).
7. Social science "laws" observed in social behavior characterize both empirical events and normative meaning of events.
8. Modern social science is essential to social progress in society – if it can successfully prescribe improved rationality for organized social activities, ethical rationality.

9. There is never a complete separation of social science empiricism from policy/practice application, so that the application of social science knowledge can always potentially alter social nature.
10. Empirical observations by a practitioner (of the consequences of a client's action) may show that the client's actions may not be attaining the client's intention. In such a case, a practitioner might suggest a different course-of-action to the client – a prescription for action.
11. If such a prescription is science based, it is more likely to prove technically effective than when not science based.

## Note

[1]This view of comparing the sciences in the positivist sociological tradition beginning with Auguste Compte (Pickering 1993). However, a comparison between sciences is not necessarily itself "positivism," particularly when it focuses upon the differences in methodology. Still the high standard of objectivity about nature is one to which all disciplines should aspire. Compte's influence on the development of sociology is noted in a later chapter (Chapter 11).

# Chapter 10
# Paradigms of Systems and Logic

## Introduction

Before we continue to examine the methodology of the social sciences, we need to understand details about the scientific paradigm of systems and logic. These along with the paradigm of function are essential to the social sciences. We recall that as an intellectual framework, paradigmatic ideas are used in an "a-priori mode," which means a previous format before the specific expression of any experiment and theory in a paradigm. First we look at the paradigm of Systems and then at the paradigm of Logic.

### *Illustration: Energy Systems*

We can illustrate the idea of a systems paradigm by examining the systems through which energy is provided to society. The provision of energy in an economy involves two kinds of systems, a natural system and an energy extraction system. Modern industrial societies are all based on the economic utility of energy extracted from nature. Figure 10.1 sketches how the energy we use on Earth ultimately derives from nature in the hydrogen fusion reactions of the sun.

Radiant energy from the sun powers the weather cycles on earth that transfer water in a hydrological cycle from the oceans to land and back to the oceans. Radiant energy from the sun plus rainfall from the hydrological cycle powers the growth of biomass. Biomass, ancient and modern, provides energy sources to society in the form of coal, petroleum, gas, wood, and biodiesel. In addition, the hydrological cycle provides energy in the form of hydroelectric power as rivers return water to the ocean. Also wind and wave motion can provide energy sources. (The exception to this sun-power energy source is nuclear energy. Nuclear energy uses the radioactive element of uranium. This was created in previous stellar furnaces; as all elements heavier than hydrogen were created in ancient, ancient stars.)

The description of a natural system (a system of nature) uses the physical and biological sciences for system description.

F. Betz, *Managing Science*, Innovation, Technology, and Knowledge Management 9,
DOI 10.1007/978-1-4419-7488-4_10, © Springer Science+Business Media, LLC 2011 

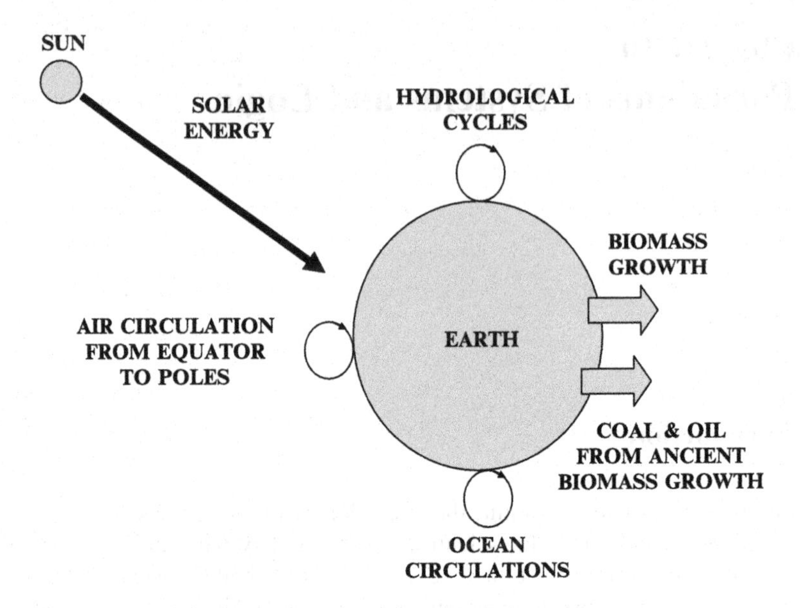


**Fig. 10.1** Natural system and energy systems

Figure 10.2 shows how a modern economy can use these cycles of nature to provide energy to society. This involves a sequence (chain) of industrial sectors to acquire energy, process energy, and distribute energy to consumers in an economy.

Energy-producing industrial sectors acquire energy through mining of coal and uranium, exploration and production of gas and oil, logging of timber, or wind or wave generation of electricity. The first industrial sector in the energy chain is the energy-extraction industrial sector of coal mining, petroleum exploration and production, timber, or wind/wave farming.

Other industrial sectors process energy for an economy in the form of (a) electrical utilities that produce electricity from burning, coal, uranium, or petroleum, (b) oil refineries that process petroleum into gasoline, diesel fuel and petroleum lubricants. Then another set of industrial sectors distributes energy through (a) electrical power transmission networks, (b) gasoline and diesel petroleum distribution stations, (c) fuel oil distribution services, and (d) natural gas distribution networks. Through this complicated scheme of natural cycles and industrial sectors, economies acquire energy from nature. Thus energy production involves two kinds of systems: (1) natural systems to create the sources of energy and (2) technological systems to extract, process, and distribute energy to an economy.

The description of a societal system (a system of society) uses the social sciences and the engineering sciences for system description.

Human societies use natural systems as resources for technological systems to create products and services for economic systems.

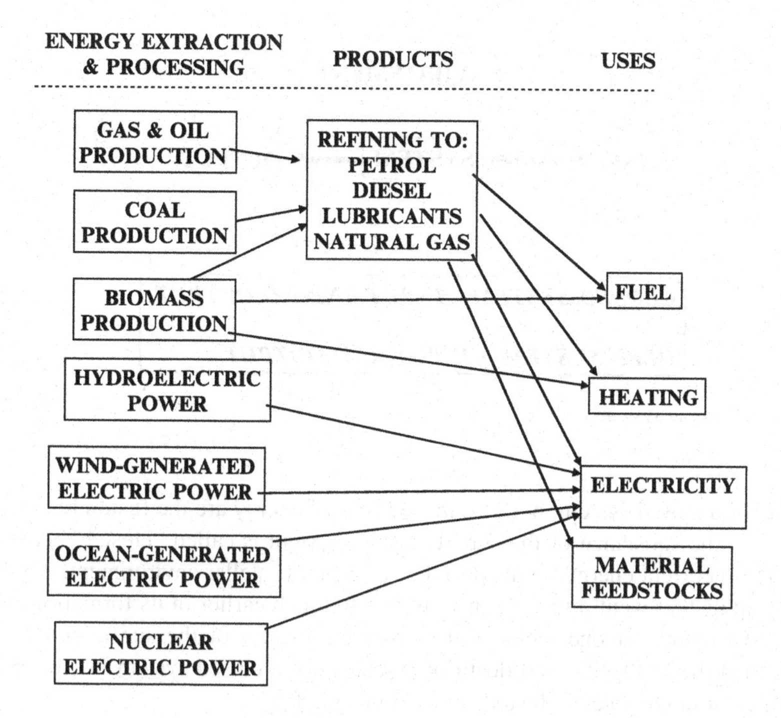


**Fig. 10.2** Energy systems

## Systems Paradigm

In all the disciplines of science and engineering, the concept of a "system" basically means to look at a "thing," with a view in seeing the thing as a "dynamic totality." A "dynamic totality" displays (1) an internal "change" and (2) interacts with an "environment."

> A 'system' is a dynamical object which can be defined as (1) an entity, (2) composed of a complex structure (3) performing a process, (4) which is controlled, and (5) operating within an environment.

The complexity of a structure requires the system to have parts connected to each other to express the totality of the structure. The process of the system requires that the connection of the parts to each other produces some activity characteristic of and controllable by the system as a whole. The environment of a system depicts the external conditions within which the system's process operates.

> The system idea contains six key sub-ideas: entity, complexity, structure, process, control, and environment.

As shown in Fig. 10.3, there are two general forms of systems: closed and open systems.

A closed system does not have significant outputs from its environment but may have significant inputs. The previous illustration of solar energy onto earth is an

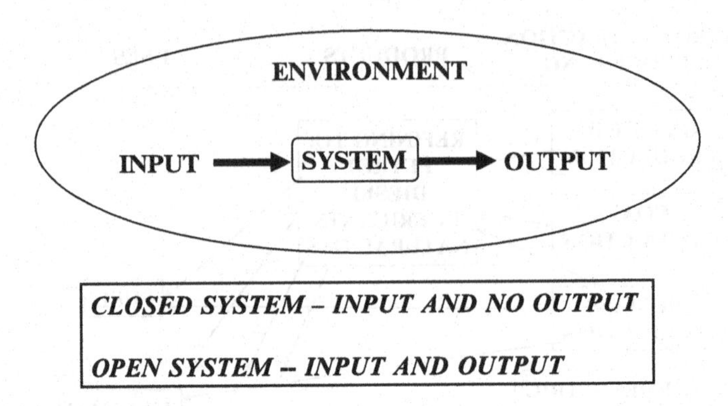


**Fig. 10.3** Dynamic systems

example of a closed system, because the sources of energy are the fusion reactions burning up the substance of the sun. The solar system is called "closed" because there are no current energy inputs into the sun from its stellar environment. All the energy inputs that went into the sun were much, much earlier in its formation in a stellar dust cloud. (If one includes the complete history of the sun as a system object, then the birth, life, and death of a star is not a closed system but an open system, open in the galactic history of the universe.)

A system is defined as closed or open by its conceptual boundary.

Change in a closed system is described by transformations between internal states of the system. Change in an open system requires description of both inputs and to and from its environment and outputs into its environment. In the previous illustration, the natural solar energy system of sun shining on the earth is described as a "closed system" (since the energy source is internal to the sun from hydrogen fusion); and, in contrast, the energy technology system is an open system, transforming natural-resource inputs (such as oil production) into energy outputs (such as petroleum products).

## Systems Theory

The importance of the concept of a "system" can best be understood by examining its philosophical roots in two early ontological concepts about the nature of the world: the world as "Becoming" or the world as "Being" (as "Permanence" or as "Change"). In 1873, Friedrich Nietzsche identified these two concepts as central to the history of philosophical cosmology, pointing back to two philosophers, Heraclitus and Parmenides, who lived before Socrates.[1] Socrates is important because his student Plato advocated his doctrine that the true nature of the world lies in the forms of nature underlying appearance. In modern terms, we now call "underlying forms" as "theory," and natural "appearances" as "observation/experiment."

Plato argued that the philosophic character of the world was expressible in forms of *permanence*, unchanging forms, just as Parmenides had argued.[2] A modern example of this is the physical law of the conservation of energy and mass. In any interaction between two physical things, their total energy and mass are conserved, unchanged in the interaction. In modern science, Parmenidean laws expressing what forms of nature do not change are important, such as the conservation of energy and mass, conservation of electrical charge, etc. Thus the philosophical concept of "Being" is essential to modern scientific ontology. However, laws of conservation are not sufficient to describe physical nature, as physical states of nature do change, two particles collide, waves intersect, stars radiate, etc.

In contrast to Parmenides' idea of the world as permanance, the Greek philosopher Heraclitus had argued that the world was essentially one of change. Later the philosopher Nietzsche agreed with Heraclitus's view. Nietzsche pointed out that Parmenidan/Socratic/Platonic idea of "Being" was insufficient to describe all the underlying forms of nature, because *change* in nature is also basic to nature, nature as "Becoming." An example of this is the concept of "growth" in biology. A tree changes during its lifetime, from seed to sapling to mature tree to dead tree to rotting wood. All organisms change during their life. Understanding the forms of change, "Becoming" is also essential to modern scientific ontology. Nietzsche argued that the underlying form of all sentient organisms was a "will-to-power." This was Nietzsche philosophy of Becoming (and which we touched upon in the previous chapter of the scientific paradigm of Function).

The importance of the scientific paradigm of systems is that as an intellectual framework it emphasizes the depiction of any natural thing in both permanence and change. The permanence of a natural thing is descriptively captured in the expression of "totality," while the change in any natural thing is descriptively captured in the expression of its "transformation."

The Systems paradigm enables scientists to describe both the permanence in the "totality" of existence of any observable object as well as its "change" in the transformations of its system.

Heraclitus lived in Ephesus (now on the coast of modern Turkey) about 535 to 475 BCE. His principle doctrine was that things were always in flux, constantly changing. His ideas come down to us as fragments and stories about him quoted in later writings.[3] Parmenides lived about 510 to 450 BCE in a Greek city Elea (on the coast of modern Italy). His work was in the form of a poem, called "Nature," fragments of which have survived. Parmenides' approach to cosmology as one of forms of permanence influenced Plato.[4]

Instead of these ancient terms of "Being" and "Becoming," the modern term for such ideas in science is now called "systems theory" and was popularized by Karl von Bertalanffy (1901–1972). He was an Austrian professor of biology at the University of Vienna from 1934 to 1948. He moved to England and to Canada and finally the United States. In 1968, he published a book called *General System Theory*, which was a seminal book for the modern system schools. (von Bertalanffy 1968)

Karl von Bertalanffy (http://en.wikipedia.org, Karl von Bertalanffy, 2007)

Other persons interested in the idea of systems were not only in biology but also in computer science, economics, sociology, engineering, and management science. Taken all together, these constitute a "school of thought" in the history of ideas, a school focused upon the idea of a system as central within many disciplines. Systems have been traditionally important in biology and engineering; but also after the advent of the computer, computer science has particularly emphasized a systems view, necessary to the conceiving the architecture of programs.

But there is no real systems "theory," in the sense of theories in the physical or chemical or biological sciences. Instead there is this intellectual school around the idea of "system" as *a perspective,* as a *scientific paradigm.* For example, this difference between the concept of "system" as a "theory" or as a "perspective" was noted by Ervin Laslo in a preface to von Bertalanffy's book: "The original concept of general system theory was Allgemeine Systemtheorie….'Theorie' has a much broader meaning in German than the closest English word 'theory' …. When von Bertalanffy spoke of Allgemeine Systemtheorie, it was consistent with his view that he was proposing a new perspective, a new way of doing science…. Von Bertalanffy created a new paradigm for the development of theories." (preface to von Bertalanffy 1968)

> We use the term 'system' in this sense as an intellectual 'perspective', a scientific meta logic, a scientific paradigm.

## *Illustration: Enterprise Systems*

In the field of management science, for example, a particularly important kind of a system is an organizational system for production, an "enterprise system." All businesses can be represented as an open system of value-adding operations within a marketing environment. This idea of a business as a "system" was popularized in the 1980s by Michael Porter, who called this an "enterprise model," a value-adding open system. In thinking about competitive strategy, Michael Porter argued that it was important to view a firm as a kind of "open system" which transforms inputs into outputs, adding "economic value" in that transformation (Porter 1985).

As shown in Fig. 10.4, a firm's inbound logistics purchases resources as materials and components and supplies and energy – from the market environment of the firm. Then the firm's production operation transforms the resources into the products of the firm. Products are stored in inventory until after sales, they are distributed to customers. This transformational sequence connects the firm with its customers in adding economic value to the resources the firm procures. Also, Porter pointed out that above this transformational chain, several other functions of the firm provide *overhead* support to the firm, including the activities of: (1) procurement, (2) technology development, (3) human resource management, and (4) firm infrastructure. One sees that Porter's description of a business is as an open system, with its environment as markets for purchasing of supplies (as inputs) and selling its products (as outputs).

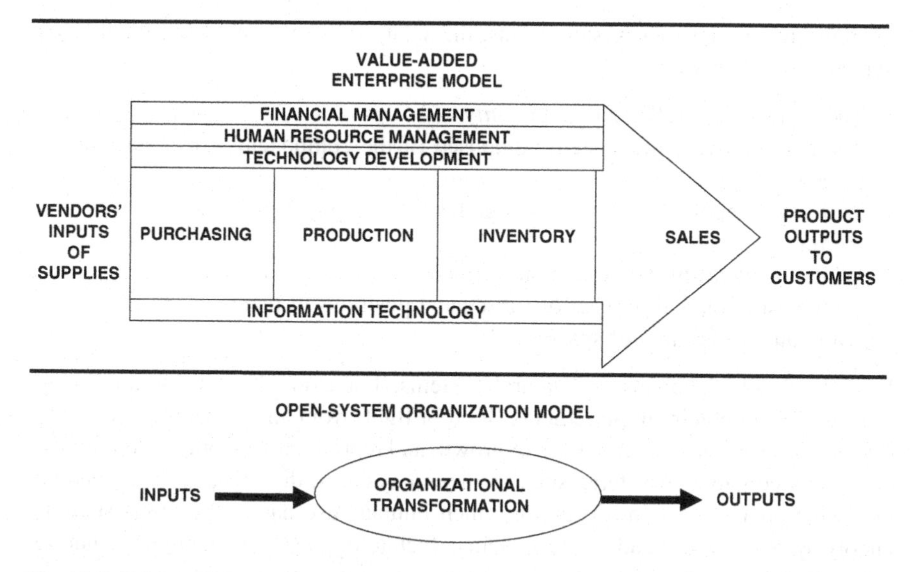


**Fig. 10.4** Value-added and open-systems descriptions of a business enterprise

Michael Porter was born in Ann Arbor, Michigan, U.S.A. He attended Princeton University, received a mechanical engineering bachelor's degrees in 1969. He enrolled in Harvard University Business School, receiving and MBA in 1971 and a Ph.D. in Business economics in 1973. He then joined the faculty of Harvard University, where he teaches and consults.

Michael Porter (http://drfd.hbs.edu/fit/public/facultyInfo)

## Kinds of Systems

There are many kinds of systems, such as natural systems, technology systems, organizational systems, or functional systems.

Natural systems occur in nature, such as environmental systems, meteorological systems, hydrological systems, material or energy cycles, ecological systems, and biological systems, and so on. We recall that Copernicus's model of the sun-centered planetary system is a kind of system, a solar system. Newton developed new mathematics (calculus) and put together a new physical mechanical model for representing the solar system. Thus, the solar system was modeled finally as a dynamical system with:

- The system as a stellar thing, an *entity*
- The *complexity* of the system composed of the components–of a star and its planets
- With its components *connected* together by the physical laws of gravity and inertia
- While the *states* of the system are formed
- By the astronomical *process* of subsequent positions of plants in orbits
- All in an *environment* of space

In biology, all organisms are natural systems. For example, a bacterium is an organic system obtaining ingredients (food or light) from an environment and processing these as nutrients for states of growth and reproduction. Complex biological organisms contain subsystems such as: neural systems, digestive systems, and so on. Functional systems are ways in which humans use natural systems, such as energy systems, water and waste systems. Technology systems manipulate nature for human function. Technology systems are used within functional systems. For example, technologies of petroleum extraction and of refinement into fuel are essential technologies of the current energy systems used in the nations of the world. Where water power is available as rivers, technologies of dams and electrical turbines can be used in hydrological systems to contribute to energy systems. Even some wind power contribution to energy systems is possible where local meteorological system conditions provide relatively constant winds to rotate wind power propeller structures.

Organizational systems, such as government agencies or businesses, implement technology systems to operate as functional systems. For example, in the 1930s, the US government's Tennessee Power Authority Agency built dams in the eastern part of the United States to supply electrical power to rural areas. Corporations establish businesses that use technology systems in produced and selling products and services to the market economies as systems of trade and retail.

Thus we see that the different types of systems interact in modern societal structures. As an example in the twenty-first century, the interaction between natural systems and organizational systems resulted in global warming.

The idea of system is central to the social sciences. For example in management science, the term "socio-technical" systems was used to indicate systems in which technologies and human interacted to provide system processes. In Fig. 10.5, we can see a list of different kinds of socio-technical systems.

In the systems paradigm, we see is that inanimate (mechanism) objects and animate (function) objects must also be scientifically expressible as systems, because they can be observed as static (stasis) objects and as changing (dynamic) objects. A Systems paradigm enables a dynamical description of the natural world.

> A complete scientific description of physical objects requires a priori scientific frameworks of Mechanism and System.

**ENVIRONMENTAL SYSTEMS**
**TECHNOLOGY SYSTEMS**
**PRODUCT SYSTEMS**
**PRODUCTION SYSTEMS**
**OPERATIONS SYSTEMS**
**INFORMATION SYSTEMS**
**COMMUNICATION SYSTEMS**
**TRANSPORTATION SYSTEMS**
**POWER SYSTEMS**
**ENVIRONMENTAL SYSTEMS**
**ORGANIZATIONAL SYSTEMS**
**BUSINESS SYSTEMS**
**GOVERNMENTAL SYSTEMS**
**ECONOMIC SYSTEMS**
**EDUCATIONAL SYSTEMS**
**HEALTH SYSTEMS**
**MILITARY SYSTEMS**
**JUDICIAL SYSTEMS**

**Fig. 10.5** Types of socio-technical systems

A complete scientific description of animate objects requires a priori scientific frameworks of Mechanism and Function and System.

Next we will see that a complete scientific description of sentient animate objects requires a priori scientific frameworks of Mechanism and System and Function and Logic.

## Logic as the Language for a Language: Linguistic Meta-regression

Now we will examine the Logic paradigm, and what we will see is that animate objects (function) are dynamic systems communicating with one another in social (functional) interactions. Some animate objects are sentient (thinking) beings, with their interactions facilitated by language (linguistic forms of communication).

Traditionally and in modern times, "logic" has been taught as topics in philosophy and in statistics. Philosophically, logic is a form of inference with: (1) deductive inference as drawing conclusions from premises and (2) inductive inference as drawing premises from conclusions. For example, in philosophy, deductive logic is taught as propositional calculus, inferring the truth content of a combined sentence from element sentences, such sentences combined by the prepositions of NOT, AND, OR, IF-THEN. Mathematically, logic is also a form of inferences: (1) inferences from premises to conclusions as mathematical deductions, solutions, or (2) inferences from conclusions to premises as mathematical probabilities, statistics.

We will use the term of logic in this way, only naming such linguistic use in small letters, "logic." But we will also use the term in a larger way, as large as an a-priori scientific framework, which we will indicate as a capitalized word, "Logic." The reason for this is that the idea of logical inference plays a broader large role in modern science than simply philosophical or mathematical "inference." As an "a-priori" framework, the paradigm of logic provides both a "transcendental aesthetic and logic" for describing and explaining *sentient behavior* in intelligent social organisms. It is this larger meaning of "Logic" as a paradigm that we now explore.

Historically, the term "logic" arose in philosophy as a term for describing the a-priori structure of a language in common spoken and written languages (e.g., German, French, Turkish, Arabic, English, etc.), such an a-priori linguistic structure is called a "grammar" of the language. Linguistic grammar structures a language by first structuring sentences for expression. Grammar differentiates words of the language into types: nouns, verbs, adverbs, adjectives, prepositions, and connectives. Terms are classified as nouns or adjectives. Grammar differentiates nouns into proper nouns (names) and common nouns (things). Relations are classified as verbs, adverbs, propositions, and connectives. Grammar differentiates verbs into tense – past, present, and future. Sentence order specifies the order of terms and relations into a meaningful predication of a noun.

In addition to sentence structure, languages have forms for relating sentences to each other, and this is called "inference." Inference provides forms for constructing larger units of meaning above single sentences, such as sentential connectives (e.g., "and," "or," "if-then") or paragraphs of communication, or themes in communication, or deductive reasoning in communication, or inductive reasoning in communication. Grammar distinguishes sentences as declarative, inquisitive, or imperative for argument and inference.

As we earlier noted grammar provides a meta-language (a language-for-a-language) to structure expression and inference and argument in a language. In a traditional philosophical jargon, one can call a language-about-a-language as a linguistic "meta-regression" – regressing language one conceptual step up to a language-of-language. We saw that notion of a "linguistic meta-regression" fits nicely with the idea of a scientific paradigm as an a-priori conceptual framework for scientific expression.

Thus the scientific paradigm of mechanism can be thought about as a linguistic meta-regression for scientific expression of physical nature; and the scientific paradigm of function can be thought about as a linguistic meta-regression for scientific expression of living/animate nature; and the scientific paradigm of system can be thought about as a linguistic meta-regression for scientific expression of dynamical nature. Next, we will examine how the scientific paradigm of Logic can be thought about as a linguistic meta-regression for scientifically expressing sentient beings (social animals, humans, and computers).

## *Illustration: Fortran – The First Programming Language*

As an example of a linguistic invention, we review the history of the first high-level programming language for computer, Fortran. After the invention of electronic computers, a technical need was for a language in which to program computers, a language bridging human speech and computer speech. Programming the first stored-program, electronic computers in the 1950s was very difficult. This required programming directly in the only language computer machines really understand, writing in binary mathematics of ones and zeros.

For example, Lohr commented about those early days before the first high-level programming language, Fortran, was invented: "Professional programmers at the time worked in binary, the natural vernacular of the machine then and now – strings of 1's and 0's. Before Fortran, putting a human problem – typically an engineering or a scientific calculation – on a computer was an arduous and arcane task. It could take weeks and required special skills. Like high priests in a primitive society, only a small group of people knew how to speak to the machine" (Lohr 2001, pp. 20).

With our human minds, we had to talk to those computers with their machine minds – but in a language computers understood. The idea of a higher-level-language in which to write programs was a good idea, a language closer to the human mind than to the machine. The computer could compile the higher-level language program into a binary language program. In the early 1950s, John Backus, an IBM employee, suggested a research project to improve computer programming capability: "In late 1953, Mr. Backus sent a brief letter to his boss, asking that he be allowed to search for a better way of programming. He got the nod and thus began the research project that would eventually produce Fortran" (Lohr 2001, p. 20).

Backus put together a team to create a language in which to program calculations on a computer, using a syntax more easily learnable and humanly transparent than was binary programming: "They worked together in one open room, their desks side by side. They often worked at night because it was the only way they could get valuable time on the IBM 704 computer, in another room, to test and debug their code… The group devised a programming language that resembled a combination of pidgin English and algebra. The result was a computing nomenclature that was similar to the algebraic formulas that scientists and engineers used in their work." (Lohr 2001, p. 20)

This was the essential point of Fortran – that it looked to the human mind of the scientist or engineer as a kind of familiar algebraic computation. Learning to program in Fortran could use a familiar linguistic process which the scientists/engineers had learned as students – how to formulate and solve algebraic equations.

For example, writing a program in Fortran first required a scientific programmer to define the variables which would appear as operands in the program – scalar variable ($A$), vector variables ($A_i$), or matrix variables ($A_{ij}$). Then an algebraic operation could be written that looks like an equation (e.g., $A_i = B_i + C_i$). Any algebraic operation could be automatically repeated by embedding it in repetition loop (called a "DO" loop in Fortran – e.g., For i = 1 to 10, Do $A_i = B_i + C_i$)

One can see that the successful trick of developing computer programming language lay in making it more familiar to humans. Fortran was the first high-level language developed specifically to talk to computers. Many others followed.

John Backus (1924–2007) was born in Philadelphia, Pennsylvania, USA. In 1949 he graduated with a master's degree in mathematics from Columbia University. In 1950, he joined IBM. In 1954, he assembled the team to create FORTRAN. He also contributed to standards in the ALGOL language. In 1963, he became an IBM Fellow (given the freedom to do whatever research he wanted to do for the rest of the length of his employment).

John Backus (http://en.wikipedia.org, John Backus, 2007)

The history of computer programming is replete with examples of how languages can be structured by a logical grammar for a specific purpose. Language structured by logic underlies modern information technology. For example, after Fortran was released in 1957, ALGOL was invented in 1958 as was FLOW-MATIC (both languages for programming control). In 1959, a syntactic language for linguistic inferences, LISP, was invented to assist writing artificial intelligence programs. Pascal was invented in 1972 to improve the structuring of programs. The first object-oriented program language (to further improve structuring of programs) was Smalltalk in 1980. In 1985, C++ was a modification to C to provide a widely used object oriented programming language. Visual BASIC was invented to add ease of graphics in programming. Java was invented in 1995 for performing operations independent of computer platform (and transmittable across the Internet).

All programming languages are specialized in linguistic syntax to facilitate communication between the human mind and the computer mind.

## Types of Languages Developed in Science

As we saw in the invention of Fortran, science has been encouraging the development of specialized languages particular for computer and information technology. But even before the computer, science was developing specialized languages, particularly for quantitative expression, as in the development of mathematics. Used in scientific expression, mathematics provides a language of pure form, a syntax, for quantitative expression. We recall that Galileo's physical laws could be expressed in quantitative form. For example, the first law of Galileo stated the velocity ($v$) of a physical object was constant, unless acted upon by a force; or quantitatively: $v=c$ (where c is a constant speed). The second law of Galileo expressed how an external force ($F$) acting on a physical object changed its velocity ($v$) by changing the acceleration ($a$) of the object proportional to its mass ($m$); or quantitatively: $F=ma$.

Mathematics as a syntactical language (language of quantitative form) has been an essential tool for modern science. We saw in Newton's quantitative model of the Copernican solar system, Newton used Descartes new mathematical topic of analytical geometry and invented differential calculus to quantitatively express this gravitational model:

$$\mathrm{F_g = gMm / r^2 = ma}$$

Besides mathematics and programming languages, what are all the kinds of languages which science has created? A philosopher named Charles Morris asked this question in the 1930s. His answer classified the types of specialized languages in science in terms of syntax, semantics, and pragmatics. In his book, *Foundation of the Theory of Signs*, Morris argued that words were "signs" which stood for "thoughts," stating that: "every thought is a sign." (Morris 1938) In language, syntax is the form of an expression, semantics is the content, and pragmatics is the purpose for the expression.

Charles Morris (1903–1979) was born in Denver, Colorado, U.S.A. He gained a bachelor's degree in engineering from the Northwestern University. He obtained a Ph.D. in philosophy at the University of Chicago. He belonged to the Pragmatism school of philosophy, which included John Dewey.

Earlier, the author used these distinctions to identify different kinds of specialized formal languages developed in science and added a fourth as control languages. (Betz 2003)

A *syntactic specialized language* is a formal language focused upon aiding the form of expression. In science, mathematics provides the syntactic language for quantitative expression of physical theories. For example, in Isaac Newton's creation of Mechanics, he had to invent the new mathematical language of differential calculus to express quantitatively the idea of "instantaneous velocity" ($v=\mathrm{d}x/\mathrm{d}t$).

A *semantic specialized language* is a formal scientific language focused upon aiding the elucidation of a field of experience. The complete set of terms (such as "mass" or "energy" or "place" or "velocity" or "acceleration") and theories (such as $F=ma$) in Newtonian Mechanics can all be viewed together as a language specialized to express the content of physical nature. And these semantically specialized terms can be further written in mathematical syntactical forms. For example, the physical concept of "mass" is expressed mathematically as a scalar number "$m$." The physical concept of "velocity" is expressed mathematically as a vector $v$ – needing to be expressed quantitatively (syntactically) with three scalar numbers, $v_1$ and $v_2$ and $v_3$ as: $\mathrm{V=(V_1, V_2, V_3)}$

The scientific discipline of physics is written in two specialized languages – semantic in physical content and syntactic in quantitative mathematical expression.

A *pragmatically specialized language* is a formal language focused upon aiding an area of purposeful activity. Any computer software program developed for an *application* is constructed as a pragmatically-specialized language. Familiar examples are word processing programs (e.g., MS Word) or spread-sheet programs (e.g., Excel) or graphics programs (e.g., PowerPoint). An application program creates specialized linguistic terms (functional terms) for the tasks in an application and is programmed as operational commands. For example, in a word processing program, linguistic terms such as "Cut" and "Paste" enable one to move script around in a text as an editing task.

A *control specialized language* is a formal language focused upon expressing the writing of algorithms to control a process or system. For example, a computer operating software program (such as Unix) is constructed as a special control language for controlling computational operations in a computer. Another example is the programming language of "Algol," developed in the 1950s to facilitate the programming of algorithms in controlling processes and procedures.

Syntactic specialized languages enable a sophistication of communication about the forms of communication – such as mathematics provides forms for expressing the quantitative aspects of phenomenon.

Semantic specialized languages enable sophistication of communication about things in an experiential field of existence – such as things of physical phenomenon, chemical phenomenon, or biological phenomenon.

Pragmatic specialized languages enable sophistication of communication about valuable and purposeful tasks – such as the task of writing as aided by an application software word processing program.

Control specialized languages enable sophistication of communication about the control of valuable and purposeful systems of activities – such as modeling and simulation programs used in engineering.

## *Illustration: Specialized Languages in Information Systems*

The usefulness of this classification of scientific languages one can be seen in the many different language-specialized programs that are needed to operate a modern information system. As an example of an information system, we will use the innovation scheme, which we described earlier. A national innovation system requires communications among and between scientists, engineers, and managers. In Fig. 10.6, we show the kinds of specialized languages needed in the programs operating as an information system for an innovation network (with the type of computer language named for each connecting line in the chart.

### Syntactic Specialized Languages

The observation connection between scientist and natural things ($S_1$ to $T_1$) is facilitated by syntactic languages, such as mathematics which enables the expression of

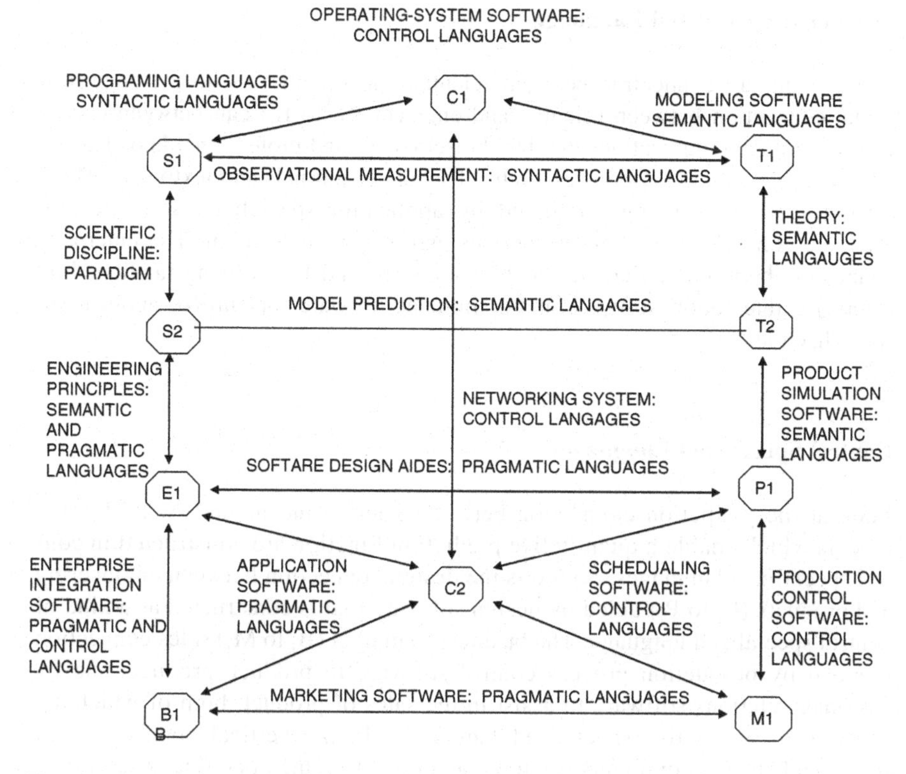


**Fig. 10.6** Operating system software

the quantitative aspects of things. The programming connection between scientist and computer ($S_2$ to $C_1$) is facilitated by syntactic programming languages, such as Fortran, Basic, C++, etc. The industrial standards connection between engineering and computer ($E_1$ to $C_2$) is facilitated by syntactic standards in computerized communications. The search connection between business and computer ($B_1$ to $C_3$) is facilitated by syntactic browser languages.

## Semantic Specialized Language

The vertical connection between things of nature ($T_1$ and $T_2$) is expressed in scientific theories of nature, and such theory is expressed as a semantically specialized language. The manipulative connection between things of nature and products ($T_2$ and $P_1$) can be described in semantically specialized languages of technology. The value connection between market and products ($M_1$ to $P_1$) can be described in semantically specialized languages of utility.

### Pragmatic Specialized Language

The disciplinary connection between scientist and scientist ($S_1$ to $S_2$) or professional connection between scientist and engineer ($S_2$ to $E_1$) can between can be facilitated in communications that build disciplinary and professional knowledge – such as professional conferences, archival scholarly journals and textbooks. These activities are now being facilitated by application-specialized software that facilitate conferences or on-line journals and distance education. The enterprise connection between engineering and business ($E_1$ and $B_1$) is being facilitated by enterprise integration software and/or intranet software that utilize application-specialized languages.

### Control Specialized Languages

Look at the prediction connection between science and nature ($S_2$ to $T_2$) and models which enable a quantitative prediction function are constructed in control specialized languages. So too is the design connection between engineering and products ($E_1$ to $P_1$) aided by design software that is constructed as a kind of control specialized language. The business to market ($B_1$ to $M_1$) sales connection is aided by production process control software to produce products and by customer relations software to assist in the sales of product, both of which are constructed as control specialized languages. In the vertical connections of Fig. 10.6 between computers ($C_1$ to $C_1$ and $C_1$ to $C_2$), the operating system connection and the network system connection are facilitated by software written as control languages.

In the lines from scientists to computers (the connection of $S_2$ to $C_1$), one can see the importance of software for programming, *programming languages,* for humans (such as a scientist) to communicate with a computer. In the connection within a computer ($C_1$ to $C_1$), an operating system software, operating system language, is necessary for the internal working of the mind of a computer. Also in the connection of one computer to another ($C_1$ to $C_2$), a networking system software, networking system language, is necessary.

For an operating system or a network system to be shared and widely used, standards must be established on the syntax of the system, as indicated in the engineering to computer connection ($E_1$ to $C_1$). And the connection of humans such as business to computer networks ($B_1$ to $C_2$) requires another kind of software, a search capability, as for example a browser language. There are several kinds of software languages necessary for computer communications: programming, operating, networking, standards, and browser languages in specialized software.

Computer science and information technologies use different scientifically specialized languages in developing all the kinds of software needed to operate information networks.

## Illustration: Logic in the Computer

We can further see how the traditional idea of *logic* has been expanded to a larger *paradigm of Logic* by looking at the invention of the computer (and its attendant scientific discipline of computer science). The first successful general purpose electronic computer was invented by John Mauchly and J. Presper Eckert. Mauchly was a physicist and Eckert an electronic engineer. While doing research during World War II at the University of Pennsylvania in the United States, Mauchly conceived the idea of an electronic computer, which he and Eckert proceeded to build as the first general purpose electronic computer, called the ENIAC.

But the story of the invention begins before Mauchly and Eckert. It begins in the topic of logic, with the mathematicians Goedel, Turing, and Von Neumann. As T. A. Heppenheimer commented on the computer invention: "Fascination with the power of technology has long been a part of human culture. The computer changed the mythic power of humanity to acquire and process information. Central to its invention were the humans – Von Neumann, Goedel, Turing, Mauchly and Eckert – and tragically, each did suffer a Promethean end." (Heppenheimer 1990).

The scientific knowledge upon which the stored-program computer was invented was provided by the mathematician John Von Neumann. Von Neumann met John Mauchly and Presper Eckert, just when they were working on the first computer. John Von Neumann was born in Hungary in 1903, the son of a Budapest banker. He was precocious. At the age of 6 he could divide eight-digit numbers in his head and talk with his father in ancient Greek. At the age of 8, he began learning calculus. He had a photographic memory. He could take a page of the Budapest phone directory, read it and recite it back from memory. When it was time for university training, he went to study in Germany under the great mathematician, David Hilbert.

John Von Neumann (http://en.wikipedia.org, Von Neumann, 2008)

David Hilbert believed that all the diverse topics in mathematics could be established upon self-consistent and self-contained intellectual foundations. Hilbert, in a famous address in 1900, expressed his position: "Every mathematical problem can be solved. We are all convinced of that. After all, one of the things that attracts us most when we apply ourselves to a mathematical problem is precisely, that within us we always hear the call: here is the problem, search for the solution; one can find it by pure thought, for in mathematics there is no ignorabimus (we will not know)" (Heppenheimer 1990, p 8).

As a graduate student, Von Neumann worked on the problem of mathematical foundations. But in 1931, Kurt Goedel's famous papers were published, arguing that no foundation of mathematics could be constructed wholly self-contained.

If one tried to provide a self-contained foundation, one could always devise mathematical statements that were formally undecidable within that foundation (incapable of being proved or disproved purely within the foundational framework). This disturbed Von Neumann, as it did all other mathematicians of the time.

But also Goedel had introduced an interesting notation of any series of mathematical statements or equations as encoded as numbers. This notation would later turn out to be a central idea for Von Neumann's idea of a stored program computer. All mathematical statements, logical expressions as well as data, could be expressed as numerically encoded instructions. However, the first person to take up Goedel's "instruction" idea was not Von Neumann but Alan Turing.

In 1937, Turing was a 25-year-old graduate student at Cambridge University when he published his seminal paper: "On Computable Numbers." Turing used Goedel's idea for expressing a series of mathematical statements in sequential numbering, Turning proposed an idealized machine that could do mathematical computations. A series of mathematical steps could be expressed in the form of coded instructions upon a long paper tape. The machine would execute these instructions in sequence as the paper tape was read. His idea was later to be called a "Turing machine." Turing had described the key idea in what would later become a general-purpose programmable computer. Although many people had thought of and devised calculating devices, these had to be externally instructed or could solve only a specific type of problem. A machine which could be generally instructed to solve any kind of mathematical problem had not yet been built.

After finishing his graduate studies in Germany, Von Neumann immigrated to America to escape Nazi persecution in Germany. He joined the faculty of Princeton University in 1930. In America in 1936, Turing's and Von Neumann's paths temporarily crossed. Turing came to Princeton to do his graduate work. He was thinking about the problem of his idealized machine; and he worked with Von Neumann, exposing Von Neumann to his ideas. Von Neumann offered Turing a position as an assistant after Turing received his doctorate. But Turing chose to return to Cambridge. There he published his famous paper. This is how Turing's ideas were in the back of Von Neumann's mind.

The story of the computer then jumps to Mauchly and Eckert (whom Von Neumann was later to meet and then contribute to their invention). Mauchly's idea for a computer happened in 1941, when he learned of some computational electronics invented by another physicist, John Atanasoff at Iowa State University. Mauchly visited Atanasoff, who showed him an experimental electronic adder. The heart of Altanasoff's adder was an electronic flip-flop circuit. The flip-flop circuit connected two vacuum tubes together – so that when an electronic signal arrived at the grid of one tube, it turned its state to conducting (on) while at the same time the other connected tube was turned to nonconducting (off).

Mauchly saw he could construct a binary number system with Altanasoff's flip-flop circuits. A signal applied to the grid of a tube could turn on the tube, and thereafter it would stay on and conducting (state of 1). Another signal might then be applied to the tube, and it would turn off (state of 0). In either state, the flip-flop circuit would be stable, until a new signal arrived to flip it. This circuit used a pair of tubes properly hooked together. Mauchly outlined in the memorandum how one could use a set of flip-flop circuits to express a number system, binary numbers and so construct a reconfigurable calculating machine. This would be different from mechanical calculating machines, which could only perform mathematical operations (addition, subtraction, multiplication, division). An electronic device could not only perform these mathematical operations but also logical operations (e.g., "and," "or," "if-then," which enable a wider range of operations – programmable operations).

In the fall of 1942, Mauchly wrote a proposal for building a computer and submitted it to the US Army Ordinance Department. Colonel Paul N. Gillon, Colonel Leslie E. Simon and Major H. H. Goldstine were in charge of funding research for future military applications. They approved the idea and gave an R&D contract to the University of Pennsylvania to build the machine, which Mauchly would call the ENIAC. Mauchly would be the research engineer and Eckert the chief engineer. J. G. Brainerd was project supervisor; and several others made up the team, including: Arthur Burks, Joseph Chedaker, Chuan Chu, James Cummings Leland Cunningham, John Davis, Harry Gail, Robert Michael, Frank Mural, and Robert Shaw. It was a large undertaking, and required a large research team (Brainerd and Sharpless 1984).

John Mauchly (1907–1980) was born in Ohio, USA, but grew up in Maryland. In 1925 he entered Johns Hopkins University in electrical engineering. In 1932 he obtained a PhD in physics. He then taught at Ursinus College from 1933 to 1941. When the war began, Mauchly moved to the Moore School of Engineering at the University of Pennsylvania. The Moore School was a center for calculation techniques in ballistics for the Army. In 1942, Mauchly wrote a memo proposing a general purpose computer using vacuum tube electronics. An army lieutenant Heman Goldstine read the memo and suggested Mauchly write a formal proposal to the Army. In 1943, the Army gave the Moore School a contract to build a computer.

J. Presper Eckert (1919–1995) was born in Philadelphia USA. He enrolled in the University of Pennsylavnia and graduated in electrical engineering in 1940. In 1941, Eckert became a laboratory assistant in the Moore Engineering School. Eckert became the project engineer for building the ENIAC.

Eckert and Mauchly in 1948 (http://en.wikipedia.org, John Mauchly, 2007)

When constructed, the ENIAC took up the whole of a large air-conditioned room, whose four walls were covered by cabinets containing electron-tube circuits. The circuit cabinets weighted 30 ton and drew 174 kW of power. The major problem was tube failure. Mauchly and Eckert had calculated that of the 17,468 tubes they were using in ENIAC were likely to have on the average one failure every 8 min.

This would have made ENIAC useless, not able to compute anything that took 8 min or longer. Cleverly, they decided to run the tubes at less than one half their rated voltage and one fourth of their rated current – which reduced the failure rate to one tube about every 2 days.

(http://en.wikipedia.org/wiki/ENIAC, 2008)

Eckert and Mauchly had also invented a memory unit consisting of an electro-acoustic delay line in a longitudinal tube filled with mercury. (This was the weak point of the technology, the active memory, which later Jay Forrester would solve in next computer invention of the Sage computer.) Also the ENIAC was very difficult to program. To set up a new problem for calculation, one had to physically connect circuits with patch cords between jacks, and all the cabling ran up to a total of 80 ft in length. The patching task itself could take at least 2 days.

Next the computer story jumps back to Von Neuman. In August 1944, Von Neumann met the Army program manager, Major H. H. Goldstine on a train platform in Aberdeen, Maryland. Von Neumann was working on the US Manhattan project and had gone to the Army's Aberdeen Proving Ground. Major Goldstine, who was in charge of the Army's ENIAC project had been to the Proving Ground and was also waiting for the train. Major Goldstine had before the war taught mathematics at the University of Michigan, and he recognized the famous mathematician. Goldstine introduced himself and found that Von Neumann was a warm, pleasant man. Von Neumann amiably chatted with Goldstine, asking him about his work. Later Goldstine said of the meeting: "When it became clear to Von Neumann that I was concerned with the development of an electronic computer capable of 333 multiplications per second, the whole atmosphere changed from one of relaxed good humor to one more like the oral examination for a doctor's degree in mathematics" (Heppenheimer 1990, p 13).

The ENIAC was already working, and Eckert and Mauchly were thinking about a better successor, which they intended to call EDVAC (Electronic Discrete Variable Automatic Computer). At Goldstine's urging, the Army was considering awarding the University of Pennsylvania $105,600 to build the EDVAC: "Into this stimulating environment stepped Von Neumann. He joined the ENIAC group as a consultant, with special interest in ideas for EDVAC. He helped secure the EDVAC contract and spent long hours in discussions with Mauchly and Eckert" (Heppenheimer 1990, p 13).

The contract was let in October 1944, and Von Neumann completed his famous paper in June 1945: "First Draft of a Report on the EDVAC." This was one of the most influential papers in what was to become "*computer science*" – *science underlying the technology of the computer*. Goldstine circulated the draft with only Von Neumann's name on the title page: "In a later patent dispute, Von Neumann declined to share credit for his ideas with Mauchly, Eckert, or anyone else. So the 'First Draft' spawned the legend that Von Neumann invented the stored-program computer. He did not, though he made contributions of great importance" (Heppenheimer 1990, p 13).

> The concept of a "stored-program" provided an "a-priori transcendental logic" for the "pure reason" of the computer.

Mauchly and Eckert planned to build a successor to ENIAC called EDVAC (which would have been the first computer to have a stored program capability) but Penn would not assign them the intellectual rights to the invention. They left the university to build computers on their own. They obtained a $75,000 contract from the Census Bureau to develop a computer and used this contract to start their new company, the Eckert-Mauchly Computer Company. However, Eckert and Mauchly never prospered in their business venture. Remington Rand bought out their business, by assuming the business's debt. Nor did they gain a patent on the computer, because Von Neumann had published a paper on the stored-program computer idea before Eckert and Mauchly filed for a patent.

What likely happened was this. When Von Neumann met Mauchly and Eckert, the former had not yet thought of the "stored program" concept, and Von Neumann proposed this to them. Thus Von Neumann could have thought of this concept as his own contribution and so written his paper without their names as co-authors. However from Mauchly's and Eckert's view, they might have thought that Von Neumann only had the idea while they actually implemented the idea in hardware. They assumed that they shared actual ownership in the implemented concept. When Mauchly and Eckert applied to patent the concept of a stored-program computer, Von Neumann's previous publication of the idea prevented US government approval under US patent law. Still the stored-program electronic computer began a major new technology and stimulated economic growth as a new computer industry; and history owes a great debt to all – to Mauchly and Eckert and Von Neumann.

## Logic in the Computer

The concept of the stored program computer used the basic technical idea that a binary number system could be directly mapped to the two physical states of a flip-flop electronic circuit. This circuit was one in which one tube was constantly conducting electricity (on) until a signal arrived to turn it off. Then it remained off (nonconduction) until another signal might arrive to turn it on again. The logical concept of the binary unit "1" could be interpreted as the on (or conducting state; and the binary unit "0" could be interpreted as the off (or not conducting state) of the electric circuit. In this way the functional concept of numbers (written on the binary base) could be directly mapped into the physical states (morphology) of a set of electronic flip-flop circuits. The number of these circuits together would express how large a number could be represented. This is what is meant by "word length" in the digital computer.

Binary numbers not only encode data but also the instructions that perform the computational operations on the data. One of the points of progress in computer technology has been how long a word length could be built into a computer. For example, in the micro-computers (computers on a chip), the word length began at 8 bits in 1975, increased to 16 bits in 1980, and to 32 bits in 1988. Mainframes used 64 bits and 128 bit word lengths. Each increase in word bit length would lead to faster processing speeds and larger memory access.

The design of the computer also used a hierarchy of subsystems. The lowest level of the computer morphology mapped to logic was the mapping of a set of bi-stable flip-flop circuits to a binary number system. At an intermediate logic level between the binary number logic and the program logic of the computer was an intermediate level of logic in arithmetic operations.

At the highest logic level, Von Neumann's stored program concept was a clocked cycle of fetching and performing computational instructions on data. This is now known as the Von Neumann computer architecture, sequential instruction operated as a calculation cycle, timed to an internal clock.

1. Initiate running the program
2. Fetch the first instruction from main memory to the program register
3. Read the instruction and set the appropriate control signals for the various internal units of the computer to execute the instruction
4. Fetch the data to be operated upon by the instruction from main memory to the data register
5. Execute the first instruction upon the data and store the results in a storage register
6. Fetch the second instruction from the main memory the program register
7. Read the instruction and set the appropriate control signals for the various internal units of the computer to execute the instruction
8. Execute the second instruction upon the recently processed data whose result is in the storage register and store the new result in the storage register
9. Proceed to fetch, read, set, and execute the sequence of program instructions, storing the most recent result in the storage register until the complete program has been executed
10. Transfer the final calculated result from the storage register to the main memory and/or to the output of the computer

The modern computer has several hierarchical levels of schematic logics mapped to physical morphologies (forms and processes) of transistor circuits:

1. *Binary numbers* mapped to bi-stable *electronic circuits*
2. *Boolean logic operations* mapped to basic *electronic circuits*
3. *Mathematical operations* mapped (through Boolean constructions) to *electronic circuits*
4. *Program instructions* mapped sequentially into temporary *electronic circuits*

## Logic in the Modern University

One sees in the illustration of the invention of the computer how the traditional topics of logic became broadened, stretched, and enlarged to a vaster notion of logic. Logic in the computer was not only inference (calculation) but also an algorithmic procedure, how to perform any calculation. Inference, algorithm – how many ideas now reside in the big idea of Logic as a scientific paradigm? We can see this bigger idea of Logic by looking at all the places in a modern university wherein logic now is taught; and we next review as the different kinds of logic in a university.

### *Logic in Ordinary Languages*

We start first with university language departments, such English or French, Chinese, Arabic, German, Spanish, etc. Therein logic is still taught as grammar. As noted earlier, linguistic grammars structure and classify sentences. Also further organization in inference occurs through the structuring of an essay in paragraphs, subheadings, or chapters. Also expositions have a thematic structure, such as topic, plot, episodes, climax, denouement, etc.

In ordinary languages, a linguistic logic provides the formats for grammar and argument.

### *Logic in Philosophy*

In philosophy departments, logic is taught not as a grammar but as a form of inference as a "propositional calculus." In propositional calculus, statements or "propositions" are the units of discourse. They are related by linguistic connectives: NOT, AND, OR, IF-THEN. The four operations on any two propositions (A, B) are: (NOT A), (A AND B), (A OR B), (IF A THEN B),

Further, any statement can have a value of either true or false. In the propositional calculus aspect is that one can test for the "truth or falseness" of any conjunctive proposition (A,B) with the following rules:

1. The negated statement (NOT A) is true when A is false.
2. The statement (A AND B) is true, if and only if A is true and B is true.
3. The statement (A OR B) is true if either A is true or B is true or both A and B are true.
4. In the statement (IF A THEN B), B is true when A is true.

In philosophy departments, logic is focused upon the truth value of propositional inference.

## Logic in Mathematics

In modern mathematics department, logic is taught as a mathematical topic called "set theory." Set theory provides a modern foundation (or meta-language) for modern mathematical topics. A set of things is a collection of objects which share a common property. Early in the twentieth century, the mathematicians, Whitehead and Russell, provided a modern foundation of mathematics using set theory as a meta-language for structuring mathematical languages. Afterward, all modern mathematical topics are taught with the basic terms of each topic so defined in the mathematical meta-topic of set theory.

This is important because defining the fundamental terms (primitive concepts) in terms of the concepts of set theory means that probability can be consistently used with any other mathematical topic in which its fundamental terms are also defined in terms of sets. An a-priori requirement of mathematics is internal self-consistency. A formal meta-language assures that all languages grammatically structured by the same meta-language will be logically consistent from language to language in their grammatical structure.

> For any language, logic provides the proper forms of the language's grammar and self-constant inferences.

## Logic in Computer Science

We saw in computer science, that logic is algorithmic in form. An algorithm is a procedure for making a calculation. For example, the procedure we were taught in school to determine the square root of a number is an example of an algorithmic procedure. Procedures for calculation (algorithms) are one form of mathematical inference. Also in the computer science departments of the late twentieth century, logic was taught in the form of Boolean algebra. Boolean algebra provides a true-false look-up table for conjoined sentences by the set conjunctions of: and, or, exclusive or, negation. Logic also plays a role as standards for communication, such as internet protocols. Thus logic in computer science provides three roles of: inference (algorithms) and grammar (Boolean algebra) and standards (Internet protocols).

> For any language, logic provides the proper forms of the language's grammar, inference (and algorithms), and standards of communication.

## Logic in Science

In departments of the physical and life sciences, as we have been discussing all along in this book, logic is taught as a method, the scientific method for observing and explaining nature. And the central point of the scientific method is to construct empirically grounded theory. The logic of method consists of research techniques and their interaction to result in empirically grounded theory. Scientific method provides the forms of the processes for experimentation and analysis in scientific description and explanation.

In addition to proper forms of grammar, inferences, standards, logic also can provide the proper forms of perception and analysis in the language.

## Logic in Management Science

In management, logic is taught as a form of reasoning about action, planning, and decisions. Planning is thinking out future action, making concrete what actions one needs to perform in the present to bring about a future state that one desires. Action is the basis of all productive organizations, whether commercial, governmental, military, etc. We recall that the logic of thinking about action occurs in the conceptual dichotomy of means and ends. The means of an action is the way the action is carried out by an actor and the ends of an action is the purpose of the action to the actor. Humans reason and communicate about possible futures as means and ends:

- In order to satisfy both instinctive and learned needs as ends
- To plan action as means to bring about such futures

The reasoning and communication about future action is called planning; and the means-ends description for future action from such reasoning is called a plan. Action-as-a-logic involves the important idea that all action, although planned, will occur in a present-time of the action. The quality of the logic of planning is measured by its effectiveness in facilitating action to bring about a planned future and reduce the risk in action toward that future. Action implemented according to a plan is the operations of the plan.

In addition to proper forms of grammar, inferences, standards, perception, logic also provides proper forms of planning and operations.

## Logic in Engineering

In schools of engineering, logic is taught as design principles in the different engineering disciplines. All engineered systems can be designed in components and connections to construct a product system, a hard good (such as a computer, automobile, building, etc.).

Designed objects are then produced in production systems.

In addition to proper forms of grammar, inferences, standards, perception and analysis, planning and operation, logic also can provide design principles.

## Logic as a Scientific Paradigm

Now we can summarize all the component ideas in a scientific paradigm of Logic. The many processes of reasoning now require that a paradigm of Logic should cover and provide the a priori forms for structuring:

1. Grammar, inference, and standards
2. Perception and analysis,
3. Planning and decision-making
4. Design
5. Control

Grammar, inference, standards, perception, analysis are all logic modes to describe and explain the world as to the quality, quantity, and relation of things in the world.

Design and control are logical modes to describe the world in relation to real or imaginary and to relate the imaged to the real through design and control.

In planning, a future action (an imagined idea) is detailed as a plan (virtual object) and implemented as an operation (a real object).

Similarly in design, an imaginary object (an idea) is laid out as a design (a virtual object) and then implemented.

This modern scope of logic has arisen from the development of schools of science and engineering and business in modern universities. The scientific paradigm of Logic provides an intellectual a priori framework for the thinking and communication of the sentient beings and communities of scientists and engineers and managers. It provides an intellectual framework for the logical forms in the thinking and communications of modern science and technology.

## Summary

1. The system paradigm contains six key sub-ideas: entity, structure, process, complexity, control, and environment.
2. Types of system are: natural systems, technology systems, organizational systems, or societal systems.
3. In reasoning, modern logic provides proper forms as: (1) grammar, inference, and standards, (2) perception and analysis, (3) planning and operations, (4) design and transformations, (5) control and production.
4. A linguistic grammar provides the meta-theory for structuring a language in sentential and inferential forms.
5. Scientific languages can be specialized as syntactic, semantic, pragmatic, or control languages.
6. The scientific paradigm of Logic provides logical forms for grammar, inference, standards, perception & analysis, planning & decision-making, design, and control.

# Notes

[1]Friedrich Nietzsche encouraged modern attention to philosophers before Socrates in his book (Nietzsche 1873). He particularly noted Heraclitus as expounding a philosophy of "Becoming" and Parmenides a philosophy of "Being."

[2]Contemporary discussions about Heraclitus include: Dilcher (1995), Kirk (1954).

[3]Contemporary discussions about Heraclitus include: Dilcher (1995), Kirk (1954).

[4]Contemporaty discussions about Parmenides include: Scott (1986), Coxon (1986), Curd (1998).

# Chapter 11
# Theory in the Social Sciences

## Introduction

Science is of a "piece," the whole piece being threaded together by methodology. The social sciences are different from the physical/biological sciences. But the skein of scientific method ties them all together. Although social science differs from the physical/biological sciences, it is not completely separable. Science is one. This is the perspective we have adopted in this book.

We have examined the physical/biological sciences in detail to extract scientific method, because historically these disciplines have been the most successful in science in constructing empirically grounded theory. They provide a standard for social science, a standard to measure up to but not exactly emulate. The methodology we have been reviewing encompasses all physical, biological, mathematical, and social sciences, while yet allowing major differences between them.

As we have emphasized, one difference between scientific disciplines is that the scientific paradigm of Mechanism is central to that of physics and biology but not to mathematics and social science. In contrast, the scientific paradigm of Logic is central to mathematics and socials science but not to physics and biology. Moreover, the scientific paradigm of Function is only shared by biology and social science, and not used in physics or mathematics. However, the scientific paradigm of Systems is used by all the disciplines.

A second difference is in phenomenological laws. We have emphasized how only the physical and biological sciences have causal explanations, phenomenological laws of cause and effect. Mathematics and social science do not use causality in explanation, instead using prescriptive or thematic explanations.

A third difference is in objectivity. The physical science disciplines can achieve theory of absolute objectivity, knowledge of a physical object independent of context and observer (the principle of invariance in the formulation of physical theory). In contrast, social science disciplines can achieve theory only partly and temporarily nearly objective. Social science theory is context dependent and influenced by the values of the observer. Physical science attains "value-free" and empirically grounded theory. Social science attains "value-loaded" and both empirically- and normatively-grounded theory.

F. Betz, *Managing Science*, Innovation, Technology, and Knowledge Management 9,

DOI 10.1007/978-1-4419-7488-4_11, 

Let us now examine societal theory, social sciences theories about social sentient beings. In particular, we will here focus upon social theory for the discipline of sociology. We do this because the methodology for theory (which is both empirically and normatively based) has been clearly articulated. This occurred in a sociological school called "structural functionalism," which traces back to Max Weber. A later American scholar of the school in the second half of the twentieth century was Talcott Parsons. Structural functionalism focused upon observing in a society/group its shared patterns (structures) and their relevance (function) to participants.[1]

As a discipline, sociology has focused upon such social interactions. The term is derived from the Latin term "socius" for "companion," indicating that sociology is the knowledge of companionship, social interactions. The first to popularize this term was Auguste Comte (1798–1857), suggesting that social ills could be solved by scientifically understanding human nature, as a positive advance in knowledge beyond theology and metaphysics. (It was from Comte's use of the term "positive advance" that the later philosophical school of the Vienna Circle took as its name "logical positivists.")

Institutionalizing sociology in academia began when Émile Durkheim (1858–1917) established the first European department of sociology at the University of Bordeaux in 1895. And in 1896, Durkheim started a sociology journal L'Année Sociologique. In America, sociology courses were begun in universities in 1875, and in 1884 the first sociology department was established at the University of Chicago. In England in 1904, a sociology department was established in the London School of Economics and Political Science. In Germany in 1909, Ferdinand Tönnies and Max Weber founded the German Sociological Association, and Weber in 1919 established the first German sociology department at the Ludwig Maximilian University of Munich.

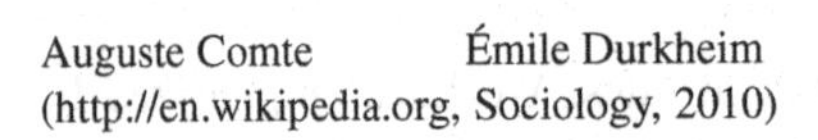

Auguste Comte Émile Durkheim Ferdinand Tönnies Max Weber
(http://en.wikipedia.org, Sociology, 2010)

## *Illustration: History of Political Reason in the US Constitution*

We review the intellectual history of the emergence of the social/political reasoning embedded in the US Constitution. We will look at it as an illustration of the development of a social theory. As a social theory (wherein its principles are inscribed in that legal document of the US Federal Government), this reasoning "evolved" socially in the history of English and American societies: evolving not as biological principles (e.g., Darwinian) but as rational (logical) principles (e.g., political science).

Moreover, this theory of political governance as a "constitution of governing principles" is certainly empirical, since it arose in the natural phenomena of human history. And this theory is certainly a kind of "phenomenological law," for it provides the principles for determining legal status of laws and governmental authority in USA. But those who composed the social theory were not social scientists but political and legal practitioners, politicians and lawyers. As we earlier reviewed, social theory is tested for objectivity and validity in practice; so social theory can be constructed not only by social scientists but also by practitioners.

In 1776, the US Constitution was formulated and passed by a Continental Congress of 13 American states. Where did this sociological idea arise of an ideal type of governance as a "constitutional government"? This idea of basing a national government upon a written constitution was a novel political idea in the 1700s. But its historical roots (thematic pattern) for this societal idea went back to the 1200s in Europe. In the history of the political ideas that informed the US framers of the constitution in 1776, there were two sources of their ideas: (1) the Magna Carta in the British governmental tradition and (2) Rousseau's idea of a "social contract."

The Magna Carta was written in the historical epoch of English society in 1215. King John of England had incurred the rebellious wrath of English Barons. John had lost control by England over his ancestral territories in Normandy, France. This disgraced him militarily to his English barons. And adding injury to insult, John repeatedly increased taxes on the English barons, to replace lost taxes from his former Norman barons. In addition, he angered the bishops of England by interfering with the Pope's appointments in England, and as a result John was forced to submit to the Roman Pope's edict – further showing him as a weak and ineffectual ruler to his English barons.

By 1215, the English barons had enough, and they entered London in force on the 10th of June. The barons then forced concessions from John, restraining the King's power. These were written as the "Articles of the Barons." John agreed and attached his Great Seal to the document in a meadow at Runnymede on June 15. In return, the barons renewed their oath of fealty to John. A month later, the royal chancery recorded the agreement in a formal document called the Magna Carta (Great Charter).

Throughout later epochs of English history, this Magna Carta was periodically reissued – as a continuing affirmation of the King's political power by the Law in English government. This was an "ideal type" principle of order, which had emerged in English history as a rational political idea, the political ideal of a written law which can limit the absolute power of government. The Magna Carta became a British political tradition between a King and a Parliament, bound under Law. This was the central idea for the principle of order in the political rationality – toward an ideal type of constitutional government.

Next in the historical epoch of the evolution of societal rationality, the English philosopher John Locke (1632–1704) popularized the further idea of "natural rights" of individuals in a society. This extended the idea of the rights of barons (in the Magna Carta) to rights of any citizens in a society. Jacques Rousseau (1712–1778) then extended the idea of citizens' rights into that of a "social contract" between a people and their government, in *Du Contract Social* (Of the Social Contract).

The idea of a social contract became the intellectual foundation of a new government in 1776, written as the Constitution of the United States of America. James Madison (1751–1836) was a principal author of the document. Madison had been influenced by English tradition of the Magna Carta and by the writings of John Locke and Jacques Rousseau.

The first attempt at formulating a new government for the former colonies was the Articles of Confederation. Madison argued for a stronger form of central government than a mere confederation of states. In 1787 at the Constitutional Convention; in Philadelphia, Madison formulated a three-sector form of government (Legislative, Executive, Judicial) with powers prescribed as a Constitution.

We see that these Federalist ideas came from several historical epochs – societies in England, France, and America. We also see that the logic (principles of order) for political reasoning was developed throughout these epochs – an empirical basis of the developing social theory. First, there was a principle of bounding a king's power by a written agreement, and this was historically expressed in a Magna Carta. Next, this principle became an English political tradition – a logic with the force of historical precedent extending over English history through time – from epoch to epoch. And this universalized the value of "law limiting the exercise of political power" over all future time for English society. Next in the new American nation, this principle was universalized as a political idea for that or any society in any time as a "social contract." And this principle was tested in practice by founding a new government entirely based upon a written social contract – a constitutional government.

Methodologically, one can view this social "history of an idea" as a kind of empirical case study. It is a natural case about how participants in a society normatively extracted a natural law of social nature, a societal phenomenological law:

1. Participants in a societal epoch designed a social structure to provide a governmental function (structure-functionalism).
2. The functions of government were to be controlled by rational principles (paradigms of Function and Logic).
3. These rational principles were willed as universal (universalization of ethics).

## Societal Phenomenological Laws

Let us again review the research technique of phenomenological laws in science. Scientific "Laws" are the expression of the regularities of patterns in natural processes discovered in observing between natural objects and their relationships. And we recall that the mode (kind) of a phenomenological law depends upon the scientific paradigm in which the object is described. In the physical sciences, the paradigm of mechanism provides a modality of relationship between two objects interacting by physical forces as a cause–effect relation. This means that a cause (A) must precede in time an effect (B), so that:

1. If B has occurred, then A must have previously occurred, A is necessary to B; and
2. Whenever A occurs, then B follows, A is sufficient to B

A causal law is characterized by both necessity and sufficiency (N&S) in the relation of the cause to the effect. And, as shown again in Fig. 11.1, we can distinguish four kinds of modalities of relationships in science – by taking the two logical relational requirements of necessity (N) and sufficiency (S) – in all possible logical combinations.

We recall that the paradigm of Mechanism uses the causal relationship (N&S) between physical objects. Physical forces are both necessary and sufficient for physical objects to change their motions in a space-time framework. The paradigm of Function uses the prescriptive relationship of (N&$\underline{S}$) between functional objects. For example in biology, connecting a function to a mechanism in biology requires that the mechanism object be necessary for the functional value to a biological thing, but not sufficient. The paradigm of Systems uses causal laws (N&S) for natural systems and prescriptive laws (N&$\underline{S}$) for socio-technical systems. The paradigm of Logic uses both (1) prescriptive relationships (N&$\underline{S}$) and (2) accidental relationships ($\underline{N}$&S) to both establish logical necessity and logical sufficiency in the construction of logical topics.

But what can one say about thematic relationships? They are neither necessary nor sufficient to explain observed phenomena; and yet we saw in the societal history of the principles of constitutional government that the evolution of the principles could only be explained thematically. In thematic explanation, an idea (theme) becomes observable as a social factor in explaining how participants thought. Thematic explanation is essential to the normative judgments in social theory. In thematic explanation, the contexts of the ideas provide the necessity and sufficiency in the explanation.

In summary, theoretical laws establish regularities in the relationships between phenomenal objects within the paradigms of the field of nature. Laws use different modalities of relations according to different paradigms in which a law is formulated.

| *RELATIONSHIP* | *NECESSITY* | | *SUFFICIENCY* | |
|---|---|---|---|---|
| CAUSAL | NECESSARY (N) | & | SUFFICIENT(S) | N&S |
| PRESCRIPTIVE | NECESSARY (N) | & | NOT-SUFFICIENT ($\underline{S}$) | N&$\underline{S}$ |
| ACCIDENTAL | NOT-NECESSARY ($\underline{N}$) | & | SUFFICIENT (S) | $\underline{N}$&S |
| THEMATIC | NOT-NECESSARY ($\underline{N}$) | & | NOT-SUFFICIENT($\underline{S}$) | $\underline{N}$&$\underline{S}$ |

**Fig. 11.1** Modality of relationships in science

The field of nature being observed determines the selection of paradigms used to perceive the phenomenal field. Causal relationships in natural phenomena occur only in the physical sciences (physics and chemistry) and never in any other of the sciences (biological, cognitive, or societal – and therein only in the physical or chemical processes that underlie functions in biology, cognition of societal action).

Physical science laws and theory use causal explanation (N&S) and are thus "contextually independent" (since sufficiency of explanation is contained within a causal law).

All social science laws and theory use prescriptive (N&S) and/or thematic(N&S) explanations, being thus "contextually dependent" (requiring additional explanation for sufficiency).

In the previous example of the intellectual evolution of the constitutional theory of governance, explanation of its social evolution was thematic (N&S) and so required a historical context to add to the thematic explanation how (necessity)and why (sufficiency) the idea (theory) had meaning in the particular societies and times. Even though using only thematic explanation, still the constitutional theory emerged historically as an empirically grounded theory. Also, the theory was created by political and legal practitioners and later elaborated upon by political scientists.

Earlier, we had emphasized that the empirical test of a social theory is through its use in practice. But whose practice? The issue is whether the theory can be valid for any society – or only the English society or only the US society? Can a logic in a political idea be applied to govern society and universalized as an ideal political value for different societies at different times and in different places? To address this issue, we turn next to Max Weber's methodological idea of "ideal-type social theory."

Explaining either thematically or prescriptively, yet social theories must be grounded both in empirical and normative judgments.

## Ideal-Type Social Theory

Let us now look at what kind of social theory can be grounded both in empiricism and in normative judgments. This methodological issue was addressed back in the early twentieth century in the new discipline of sociology, Max Weber suggested that social theory could be expressed as a kind of rational ideal – an "ideal-type social theory." This type of theory arises from addressing the following methodological issue. In a historical epoch in society, how can the ideas of that period be empirically described – both in societal events and in rationalizations (ideas) interpreting events – both action and meaning?

Max Weber suggested social theory should describe both a society's history (empiricism) and reasoning (prescription) in that society, writing: "We have in abstract economic theory an illustration of those synthetic constructs which have been designated as 'ideas' of historical phenomena. It offers us an 'ideal picture' of events on the commodity-market under conditions of a society organized on the principles of an exchange economy, free competition and rigorously rational conduct" (Weber 1897).

For Weber, an "ideal type" in social theory is an abstraction of the "principles-of-order" empirically observed in a society (e.g., the economic principles for a commodity-market).

Weber proposed a social science methodology of (1) the idea of historical epochs as an empirical basis in the social sciences and (2) the idea of "ideal types" as a form of social theory. And we saw in the illustration of the US Constitution an illustration that Weber's methodology does actually fit some historical cases of the evolution of ideas – in this case, the thematic pattern in ideas from the English Magna Carta to the US Constitution.

Weber elaborated on how exactly economic theory was a form of an "ideal type": "An ideal-type of a commodity market's relationship to the empirical data consists solely in the fact that where the market-conditions are *discovered to exist in reality*, we can make the characteristic features of this relationship clear and understandable *by reference to an ideal-type*.... In its conceptual purity, this mental construct (of an ideal type) *cannot* be found empirically anywhere in reality. It is a *utopia*"(Weber 1897).

An ideal type (in a Weberian form of social theory) is a descriptive abstraction of the *principles of order* that can be empirically observed in a historical social situation. Ideal types are abstractions and do not exist completely in reality. But such ideal-type abstractions *do* express principles of order, which can *actually operate* (at least partially) in a society. An ideal type social theory is not merely an empirical description of *what* people are doing in a society and *why* they think they are doing that – but also *how* they thought they should think.

Ideal-type social theory is a generalization of the "principles-of-order" that a society thinks it should be operating in a given historical situation.

An "ideal type" in social theory is an abstraction of the "universal intentions of a society."

Universal intentions in an Ideal-type are expressed as "principles-of-order to guide social behavior."

Weber's example of an economic commodity market as an ideal type explains not only the *what* (commodity market) and the *why* (utility of economic exchanges in a market-organized society) but also the *how* (supply equals demand) – as perceived by its participants. Furthermore, Weber saw that if this ideal type of a commodity market not only operated in historical epochs of particular societies, it might also operate in a present or future society – for economic benefit of present or future societal inhabitants.

The principles-or-order empirically observed in societal-historical-epochs might be universalized over the human family of societies as a prescriptive injunction.

What might be universalized – universally projected for all societies or all times – are principles of societal order.

Weber argued that a generalization of an "ideal type" inherently occurs in any historical study that attempts to explain historical events from the perspective of the historical participants: "Every conscientious examination of the conceptual elements of historical exposition (empiricism as societal history) shows that the historian, as soon as he attempts to go beyond the bare establishment of concrete relationships and to determine the *cultural significance of … (historical) events* must use concepts … in the form of *ideal types*." (Weber 1987).

Weber proposed to describe social theory as the "cultural significance of historical events to participants." This kind of social theory could be expressed as a model of *social rationality*, which was seen as "ideal" to those participants. Weber called an "ideal-type social theory" an utopian idea – an utopia. Weber's use of the term utopia came from a tradition in Western literature of writing about what a perfect society would be like and calling this perfect society an utopia. The term comes in English literature from the title of Thomas Moore's book, Utopia.

In the example on the US Constitution, this document expressed principles of order for a perfect society, in the perspective at that time of that Continental Congress of the former American colonies. These principles of order became the ideal rules for governing succeeding generations in USA for the next two centuries. The social reasoning expressed the ideas for a utopian government, which then was a new political rationality on the stage of the "evolution" of human social organization. It was not Darwinian but Weberian. What was unique was that this new government was based upon a constitution – a written expression of societal principles of order. In Weber's terms, this idea of government operating upon a written constitution was an "ideal-type" theory.

We will call such Weberian principles of societal order as an *ideal of reason* in a societal system – *as perceived by the participants of that society at that time*. This idea of an ideal-type social theory to model a society is centered upon the idea of principles of order. This term indicates the use of logic in societal reasoning – reasoning of participants in a society about their choices of behavior – decision making by participants. This is an idea of reasoning within a society. It is an idea that connects traditional philosophic ideas to social science ideas.

For example, Max Weber's seminal study on bureaucracy codified a new social science approach to the proper forms of modern organizations. Particularly, Weber argued that bureaucracies showed a form of organizational rationality in developing procedures that properly applied rules to organizational activities. This idea of "organizational rationality" was an ideal view (normative judgment) of how bureaucrats should behave. Later, some social scientists criticized Weber's ideal type of bureaucracy. They argued that this idea of bureaucratic rationality could also become irrational if too rigid or if misused in empire building. This idea of organizations as both rational and irrational became the standard in teaching organizational theory in business schools as the prescription that organizations should be rational. Later, what Weber described as "empirically observable" in sociology became a prescription to improve organizations, as an application of the social theory of bureaucracy.

## *Illustration: Ideal-Type Theory of Societal Systems*

As an illustration of how to construct an "ideal-type" social theory, having both normative (ideal) and empirical (realistic) grounding, we next use Weber's empiricism of social relations to construct a systems theory of a modern society. We will show how a systems theory of societal systems can be constructed on Weber's theory that in any social interaction, participants can hold four kinds of expectations about that interaction: (1) utility or identity and (2) reciprocity or authority (Weber, 1947).

By the *dichotomy of utility or identity*, Weber meant that in any societal interaction, each party to the interaction will anticipate either:

1. *Utility in a relationship*: As a useful value for a participant in the interaction (such as buying or selling goods)
2. *Identity in the relationship*: As an identification of one party with the other party as belonging to some same group and sharing the values of the group (such as belonging to the same family or same political party)

By the *dichotomy of reciprocity or authority*, Weber meant that in any societal interaction, each party will also anticipate as a basis for the interaction either:

1. *Reciprocity in the relationship*: As a mutual and equal advantage for each party in the relationship, such as fairness in treating one another
2. *Authority in a the relationship*: As one of the parties in the relationship for making decisions about the relationship (such one being a judge and the other a plaintiff or one being a mayor of a city and the other a citizen

To build a theory of society based upon these empirical/normative observations, we first need to construct a perceptual space for observing social/sentient objects in nature, an a-priori transcendental aesthetic for social nature. We recall that Weber's approach to objectivity in social science used the paradigms of Function and of System. Function looks at relationships of value between objects, and system looks at dynamical descriptions of objects. For Weberian observations of society then, a perceptual space must then have at least one Functional dimension and one System dimension. For this, one can use fundamental distinctions found in the sociological school of "Functionalism-Structuralism." In this school, functional descriptions of a society need to observe and distinguish between individuals and society; whereas structural descriptions of a society needs to observe and distinguish between groups and processes. One can construct a societal perceptual space for a function/structure sociological school as a two-dimensional perceptual space, with dimensions of society–individual and of process–group (Fig. 11.2). And in that perceptual space, one can see that Weber's two dichotomies of social interactions depict two sets of social processes that can occur within social groups of a society.

Within the discipline of sociology and in the USA, the structure–function school of sociology was popularized by Talcott Parsons. He used the term "structural

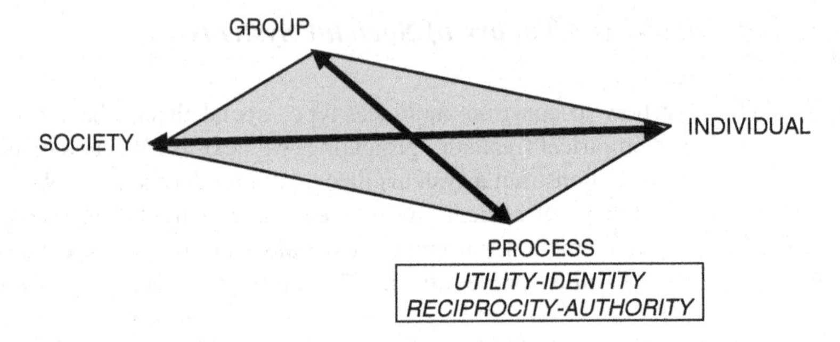


**Fig. 11.2** Structure/function societal perceptual space

functionalist" in his social theory of action. By the term "social structure" Parsons indicated the patterns in the social arrangements of life, and by the term "functionalism" Parsons indicated the relevance of the social patterns (structure) to the participants in the society. Parsons also formulated social theory in what he called "action theory."[2]

Talcott Parsons (1902–1979) was born in Colorado, USA. In 1924, he obtained a bachelor's degree from Amherst College. He attended the London School of Economics and then the University of Heidelberg, from which he obtained a Ph.D. in sociology and economics in 1926. He wrote his thesis on the concept of capitalism, examining the works of Max Weber, and began translating Weber's works into English.[2] In 1926–27, Parsons had returned to the USA and taught at Amherst College and then began teaching at Harvard University in 1927.

Talcott Parsons (http://en.wikipedia, Talcott Parsons 2010)

The perceptual space with dimensions of individual-society and group-process provides an observational a-priori framework for (1) describing the structural patterns in a society (groups-processes) and (2) the functional relevance of these structures to societal participants (individual-society).

Although Weber described his two dichotomies (utility-identity and reciprocity-authority), he did not construct a structural–functional societal taxonomy, based upon them. A taxonomy is a classification scheme, based upon taking combinations of dichotomies, two at a time.

In a previous work (Betz 2001), the author constructed a taxonomy of groups who specialize around pairs of these interactions (Fig. 11.3). A structural–functional taxonomy of groups can be formed from Weber's two dichotomies, by setting utility-identity across the top of a taxonomy and reciprocity-authority down the side of the taxonomy. Across the top are *structural factors* in the formation of groups, *utility or identity,* as expectations in forming a group. Down the side are *functional factors* in the processes of groups, *reciprocity or authority*, in the operations of

| | *UTILITY* | *IDENTITY* |
|---|---|---|
| *RECIPROCITY* | ECONOMIC INTERACTIONS | CULTURAL INTERACTIONS |
| *AUTHORITY* | TECHNOLOGICAL INTERACTIONS | POLITICAL INTERACTIONS |

**Fig. 11.3** Societal taxonomy of social interactions

groups. Next, one can identify the kind of social interactions in each structural/functional group in the taxonomic box.

The group pattern, in which participants anticipate structural benefits of *utility* and functional benefits of *reciprocity*, occur in *economic interactions*. Therein two participants each expect from their interaction both usefulness (utility) and also that utility should be reciprocal in mutual benefit (fairness). For example, two participants in a market as buyer and seller both expect to benefit from the sale (product for the buyer and price to the seller) and that the sale should be fair (a competitive price for a quality product). The economic theory of a 'perfect market' is an 'ideal-type social theory.

The group pattern, in which participants anticipate the structural benefits of *identity* and the functional benefits of *reciprocity*, occur in *cultural interactions*. Therein two participants each expect to share a mutual identity in their interaction and also expect actions that are reciprocal in mutual benefit (fairness). For example, two participants in a church as priest and congregant both expect each to believe in the same religious faith (as members of the same church or synagogue or mosque) and that shared religious practice will enhance each other's service of the religion.

The group pattern, in which participants anticipate the structural benefits of *identity* and the functional benefits of *authority*, occur in *political interactions*. Therein, two participants each expect to share a mutual identity in their interaction but also expect actions to be decided by the one participant superior in societal authority and followed by the inferior participant. The participant superior in authority is said to hold political power over the other participant. For example, a political office holder such as a judge in a court of law can sentence another participant in a trial (having been brought into court as an arrested offender) to a sacrifice of life or freedom of property. The judge has legal power over the defendant in a trial.

The group pattern, in which participants anticipate structural benefits of utility and functional benefits of authority, occur in *science–technologyl interactions*. Therein two participants each expect from their interaction a usefulness (utility) and also that utility is based upon an action (technical process) that can effectively create the utility – a methodological authority that guarantees the technical effectiveness of the useful action. For example, as a business person might hire an engineer to design a factory to produce the business person's product. (One example is that of a chemical engineer hired to design chemical processes for producing chemicals. In this interaction, the engineer's useful action in designing a factory is based upon his methodological authority of engineering knowledge.)

As an illustration of these different sectors in real societal events, we can recall that in the previous illustration of the Global Financial Collapse of 2007–2008:

1. The functional collapse of the financial structure due to the failed market in derivatives occurred in the sector of *Economic Interactions*.
2. The failure of governmental regulation to prevent fraud of the explosive derivative market occurred in the sector of *Political Interactions*, followed by government "bailout" of banks by the many different governments.
3. The mechanism of the collapse was facilitated by the information technologies in the *Science & Technological Interaction*, which made possible the packaging of all that information into thousands of derivatives and instantaneous sales around the world to hundreds of buyers of the derivatives.
4. In the *Cultural Interactions* of affected countries, citizens impacted by the economic recessions and by the massive public debts became very angry at the banks (and at the large bonuses bank managers continued to pay themselves after the bailouts).

A note on technique: for theory construction, we are using the technique of dichotomies and taxonomies. A dichotomy is a conceptual means, in a universe of discourse, to separate two contradictory concepts – one thing and its opposite (in that conceptual universe). A taxonomy is a set of categories defined by the ideas in two dichotomies. Taxonomies are constructed from dichotomies that characterize the property of an object into contrasting aspects of the property.

Categories in a taxonomy can then be used to classify natural objects, which are empirically distinguished by properties of the dichotomies. Taxonomic models formalize the content and boundaries of natural objects. The dichotomies define the content and boundaries. Formal taxonomies provide a technique to distinguish objects in a perceptual space with properties defined by the dichotomies.

One can see where a taxonomy of groups fits in a structural/functionalism perceptual space, as shown in Fig. 11.4.

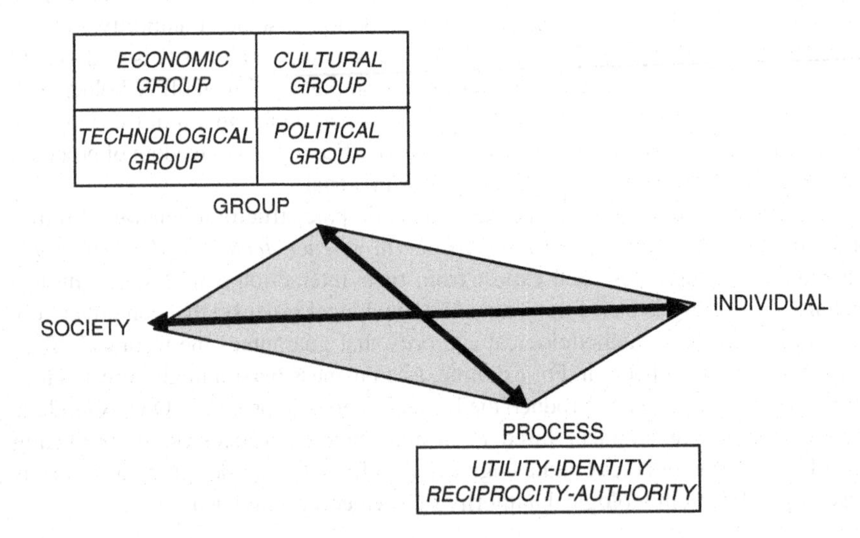


**Fig. 11.4** Structure/function societal perceptual space

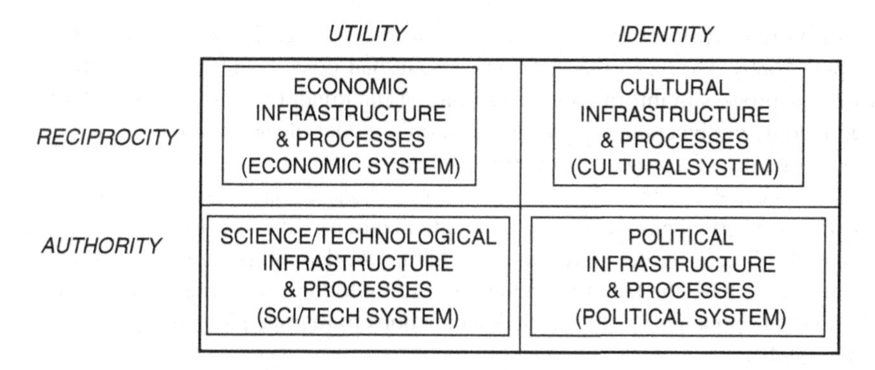


**Fig. 11.5** Societal taxonomy

Next in the idea of a group, sociologically one can add the idea of a social infrastructure and its processes (structure-function). When a structure and its function is described both as an infrastructure and processes, then one can describe this kind of totality as a system, a social system (Fig. 11.5).

To empirically ground this theoretical taxonomy does require observing real societies to see if such institutional arrangements do exist in modern societies. And certainly, in a glance at US society, one can see (1) an economic system such as industries and services, finance and markets, (2) a technological system as schools and universities, industrial R and D, government science and technology, (3) a political system such as political parties, elections, government institutions, and (4) cultural systems such as families, religions, entertainment media. All these institutional patterns, groups, and processes can be observed in a modern society.

But can one normatively ground this theoretical taxonomy? How can this social theory be observed also as an "ideal-type" theory? For this normative grounding, an observer must see not only what societal participants *do in their actions*, but also what they *intended by their actions*.

Thus for normative grounding of societal theory, we must methodologically examine how to describe and explain the "patterns-of-social-intentions" (functional relevance) in the collective behavior of participants in social structures.

And we will call patterns-of-social-intentions as "reasoning" in a society.

## Principles of Social Order

We are examining a Weberian taxonomy of societal systems in order to see how to construct Weberian idea-type social theory. We will next see that the social processes (interactions) in the subsystems of the society (economic, cultural, technological, political) can be described in an *ideal-type* social theory as the *principles of order*, as reasoning about societal function in a societal structure.

Let us remind ourselves about how the term "reason" has been used in the traditions of philosophy. We recall that, Immanuel Kant had made "pure reason" as the first essential process in information processing of the mind, to sense an external world and recognize objects in that world. In Kant's model of the pure reason, the principles of order for object perception and recognition fell into the categories of "transcendental aesthetics" (a-priori sensory data formats) and "transcendental logic" (a-priori cognitive object conceptualization). Specifically about "human rationality," Kant wrote: "Everything in nature works according to laws. Rational beings alone have the faculty of acting according to the conception of laws, that is, according to principles … the deduction of actions from principles requires reason…" (Kant 1949).

This Kantian idea of "reason" was from a long tradition in philosophy about "rationality." For example, a compact summary of this philosophical tradition was made by Paul Diesing: "Reason appears as creativity in some of the works of Plato, Hegel, and Whitehead. It appears as the application of rules in natural law theorists such as Aquinas, Locke, and Kant. It appears as calculation in Hobbes, Bentham, and the Utilitarians" (Diesing 1962).

It is in this philosophical tradition (about rationality as being the "deduction of actions from principles) that we will use the term reason.

We saw that paradigm of Logic provides an intellectual framework for observing and expressing the concept of "principles of order" as a logic in which reasoning should operate. We recall that the paradigm of Logic includes proper forms for reasoning such as: (1) grammar, inference, and standards, (2) *perception and analysis*, (3) planning and operations, (4) design and transformations, and (5) control and production.

But there are two levels or reason, one at the individual level and one at the societal level. For individuals, Kant's model of "pure reason" described the principles of order needed by an individual mind (1) to sense things in the world and (2) to formulate mental ideas of objects of those things in the world. For societal reasons, one needs to look at the social structures (organized processes) that individuals share in reaching decisions in that society. The idea of reason in a social order is expressed by the principles of order that guide organized decision making in a society.

In the Weberian idea of social theory as an "ideal-type", the thinking of participants in a society in a given historical epoch can explain that societal reasoning as particular logical principles-of-order.

## Reasoning in Social Structures

We look further at the idea of "reasoning" within social structures. To find such a thematic survey of types of societal reasoning, we can use an interesting book by Paul Diesing (1962). Diesing was trained as a philosopher and empirically reviewed the different literatures of the social sciences. He tried to identity the types of social reasoning described within each scholarly discipline of the social sciences and their application: sociology, economics, political science; law, and management science. From his analysis of the types of reasoning described in these literatures, he generalized upon the kinds of reasoning in modern societies (as described in the social science literatures). He argued that social reasoning consisted of: "Decisions … made according to principles, and organized structures embody principles-of-order; accordingly principles which can be thought of as rational." (Diesing 1962).

This idea of – organized structures embodying principles-of-order – is an idea which connects (1) the idea of rationality in an individual (e.g. Kant's pure reason in the human mind) to (2) the idea of rationality in a social order (e.g. Weber's ideal-type of a social order).

Diesing wrote: "We (can) distinguish two phases of reason, the rationality of organizations and the rationality of decisions. … When the relations of which all societies are composed are ordered according to some principle, the society is rational in some sense." (Diesing 1962).

The idea of reason can be used at two levels in social science observations: (1) at the level of individuals and (2) at the level of society.

Diesing's idea of "principles-in-organized-structures" is similar to Weber's concept of an "ideal-type" wherein either one explains the observation of rationality in a society at both an individual level of thinking and at a societal-level of thinking.

Paul Diesing received a doctorate in political philosophy from the University of Chicago. He then became a professor in the Department of Political Science at the University of Illinois in Urbana, where he finished his study on societal rationality. He went to the State University of New York at Buffalo, where he continued to teach until his retirement as a professor emeritus in the Department of Political Science. He wrote about social science research methodology and political philosophy. Diesing identified five kinds of logic in reasoning described empirically in the social science disciplines: (1) political rationality, (2) legal rationality, (3) cultural rationality, (4) technical rationality, and (5) economic rationality.

## *Political Rationality*

Political structures operate as governments passing and enforcing laws. So to describe political rationality, one needs to look both at the social science discipline of political science and at the practice of law. Within the law literature, Diesing identified two sets of principles of order for expressing rationality in a legal system as (1) rights and duties of individuals and (2) public expression of rights and duties (Diesing 1962). A legal system is a set of public rules (laws) constraining the behaviors of individuals in a social system – such as (1) defining property and contract forms in the economic system and (2) defining individual rights and obligations of cultural roles in the social system.

Also, the application of laws to individual cases of behavior occurs in courts of justice in a legal system. The rationality of court procedures in the application of laws to cases is one of deductive inference. The court must identify which law applies to a legal case and also establish the fact that the individual action fits this case.

A legal system in a society exists within a broader context of the political system of a society. The description of legal rationality must be included into a broader rationality of political processes and governments. In reviewing the political science literature, Diesing identified that the way in which decision structures are constructed and maintained is the focus of political rationality (Diesing 1962). Political rationality is about gaining and keeping power as the right to make decisions.

The operation of political decisions occurs within political processes that facilitate (1) political discussion, (2) shared beliefs and values, and (3) action commitments of a group. Respect and fear are the rational criteria upon which is based the power of authority in a political decision-making structure. Within a governmental structure of a political system, office holding, either by appointment or election, is a source of authority in political decision-making structures.

The idea of political/legal processes combines the ideas of both constructing and operating governments. And in any historical epoch in any society, these two processes are intimately intertwined.

> Political rationality is prescriptive – using the criteria of attaining power and authority to construct and maintain a political process as an organized-structure for the right of decision-making in the system.
>
> And within a political context, the legal structures of a government operate. Legal rationality is prescriptive – choosing a judicial process as an organized-structure for the enforcement of law – so that laws can be publicly applied as rules to individual cases of law.

We have emphasized the value-loaded nature of social theory, and if we look at political reasoning as the intention to gain and wield power, what "values" are held highest in this rational process? Empirical and normative observation of political histories can show that the value of "loyalty" is most valued by politicians. Loyalty is the support, commitment and obedience by a follower to obey the instructions of a leader. Accordingly, the sociological/political science study of political events can study the social nature of loyalty among sentient social beings.

## *Cultural Rationality*

Diesing searched the literatures of the disciplines of sociology and anthropology – to identify principles of order that these disciplines have observed as patterns in social relations. He summarized a social relation as an enduring pattern of shared experience, conceptually thought of in a set of beliefs, obligations, expectations, and ideals about the behaviors within the relationship (Diesing 1962). He also noted that in sociology, social relationships are empirically observed as pairs of social roles. He also noted that social roles become cultural roles – socially organized structures – when a society institutionalizes a set of role conceptions as all members sharing such concepts of roles and socially reinforcing the role concepts. A social structure is an organized structure consisting of the complete sets of institutionalized cultural roles.

This sociological concept of "cultural roles" connects theory in sociology to theory in psychology. In social psychology, an individual's personality is socially constructed by the roles learned in life. Diesing then noted that within sociological empiricism, the idea of "cultural roles" is to interact with rational rules that serve to *integrate* a social system and tie together its different roles that participants play in a society. Diesing identified a cultural principle of order (logic) in a social system as this *integration of the cultural roles*. As a social process, integration involves individual efforts to integrate conflicts between cultural roles. The cultural institutionalization selects successful role conceptualizations that reduce cultural conflict. Also, in addition to the rational principle of "integration" of social roles, there is another principle of order in cultural systems that seeks to solve conflicts between cultural roles in an equilibrium, a societal balance.

An illustration of this in USA was the civil rights movement of the 1960s. The cultural conflict over racism was legally removed in the passage of Civil Rights Act of 1968 by the US government. This law formally integrated all US citizens as to their legal rights as citizens. It legally outlawed racial segregation in the US social system. An equilibrium was reached between the principle of order, equal rights, and the treatment of all groups in America.

Diesing suggested that sociologists can use the idea of "social rationality" for the logic of a process of reasoning in an *cultural system of a society*.

> Cultural rationality within a cultural system uses the logical principles of "integrative action" and "cultural equilibrium" to effect cooperation and resolve conflicts in social roles.
>
> *Cultural rationality is* prescriptive – *to minimize conflicts between roles through integrative action and to balance actions as an equilibrium between conflicts.*

If we look at cultural reasoning as the intention to integration and equilibrium, what "values" are held highest in this rational process? Empirical and normative observation of political histories can show that the value of "equity" is most valued in cultures. Equity is the just and fair treatment of members in a culture and the provision of equal access to resources and opportunities. Accordingly, the sociological/political science study of cultural events can study the social nature of equity among sentient social beings.

## *Economic Rationality in Society*

Diesing described the rationality of societal thinking that he empirically observed operating in the social science discipline of economics. In economic rationality, decisions are made by societal participants in their interactions that involve the buying and selling of goods and/or services. These economic interactions occur as economic exchanges in "markets" as an organized structure of an economy. In the market are transfers of "economic value" between sellers and buyers. Economic value is usually called "utility." A financial subsystem enables the "pricing" of a given good or service in a given market.

> Economic rationality is the principle-or-order (logic for decision-making) in an ideal type of commodity-market system, with rational principles-of-order for processes of "exchange" and "allocation."

And in the discipline of economics, one important economic theory (as an ideal-type theory) is the law of supply and demand'. This law is proposed as the rational principle for establishing correct pricing in a perfect economic market – a price determined in the balance between supply and demand. This is to say that the economically most rational price of a good/service occurs in the market when the supply of the good/service just equals the demand. This economic law of "supply and demand" is a principle of logic about how economic rationality should occur in an economic exchange.

> The economics discipline prescribes that a perfect economic market should operate as organized structure with the ordering principle of "supply & demand."

If we look at economic reasoning as the intention to exchange and allocate products, what "values" are held highest in this rational process? Empirical and normative observation of economic histories can show that the value of "profit" is most valued in cultures. Profit is the gain from trade after the costs of a trade is subtracted. Accordingly, the sociological/political science study of economic events can study the social nature of profit among sentient social beings.

## *Technical Rationality in Society*

Also in social science disciplines, there is another form of empirically identified societal rationality, which Diesing called a "technical rationality" (Diesing 1962). Technical rationality is a principle of order for describing rational action as an efficient choice of means to an end. Technical rationality primarily operates in the engineering science disciplines in engineering schools and in business schools in the disciplines of operations and management science.

Technical rationality is a form of reasoning described in the paradigm of function. We may recall that in the paradigm of Function, the basic dichotomy of *means–end* describes the *action* of an *intelligent organism*. Means is a choice of action taken by an organism to attain an end of value to the organism (e.g., hunting as a means of predators to obtain food as a valued end). In this means-–end dichotomy, the measure of the "efficiency" of a means is the ratio of how much resources are consumed by the means in producing the quantity of outcome in an end. This is to say that for any means, efficiency = outcomes/resources.

Management science is a relatively new social science discipline which began after the Second World War. The principles of order (rationality) in management science is that of "technical rationality" – assisting a client in the choice of a means to an end, which optimizes efficiency. For example, one theory in management science – called the "expected-value theorem" – calculates the probability of a set of means for producing an end as the sum of efficiencies of the means toward the end.

> Technical rationality is "prescriptive" in using the principle-of-order of efficiency in decision-making – chose a course of action which is most efficient to attain your end.

If we look at technical reasoning as the intention to efficiently and effectively attain a goal (end), what "values" are held highest in this rational process? Empirical and normative observation of political histories can show that the value of "truth" is most valued in scientific and technological communities. Truth is the objective and empirical knowledge of the nature of realty. Accordingly, the sociological/political science study of scientific/technological events can study the social nature of truth among sentient social beings.

## Societal Systems Model as Ideal-Type Theory

Combining Weber's social dichotomies with Diesing's approach to societal reasoning as principles-or-order, one can then construct a theoretical taxonomy of ideal-type societal systems, as in Fig. 11.6.

One can see that the different principles of order for reasoning in the different societal systems also stimulates different values in their reasoning. The principle value to be attained in economic reasoning is profit. The principle value to be attained

| | *UTILITY* | *IDENTITY* |
|---|---|---|
| ***RECIPROCITY*** | ECONOMIC SYSTEM *(ECONOMIC RATIONALITY)* **PROFIT** | CULTURAL SYSTEM *(SOCIAL RATIONALITY)* **EQUITY** |
| ***AUTHORITY*** | TECHNOLOGICAL SYSTEM *(TECHNICAL RATIONALITY)* **TRUTH** | POLITICAL/LEGAL SYSTEM *(POLITICAL/LEGAL RATIONALITY)* **LOYALTY** |

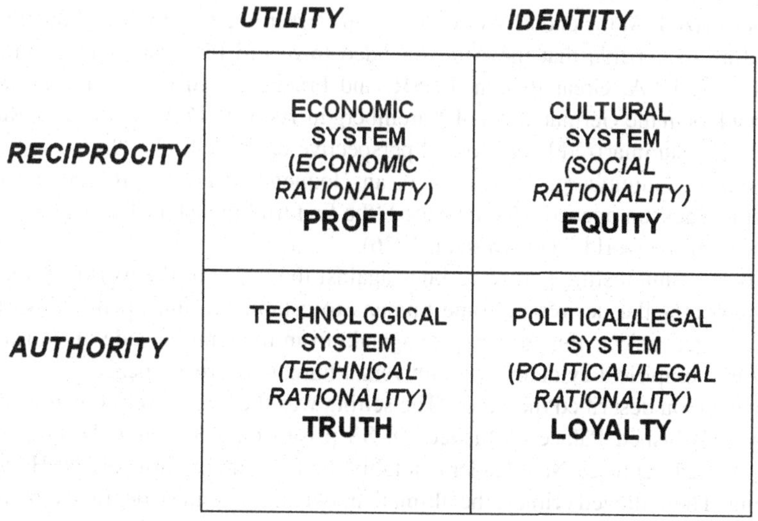

**Fig. 11.6** Societal systems model

in cultural reasoning is equity. The principle value to be attained in political reasoning is loyalty. The principle value to be attained in technological reasoning is truth.

> The value-loaded disciplines of the social sciences can add to human understanding the nature of and the prescriptive (empirical and normative) form of some of humanities deepest values.

But what about the nature of human values? Are they merely subjective, changing from person to person, society to society, time to time? Or is there something universal in human values, generalizing over individuals, societies, and all time? This is the challenge of "universalization" in the normative grounding of social science, which we had noted that Weber earlier urged as a methodological challenge (in his words of "families of humanity").

## *Illustration: Nuremberg Trials*

But how can principles of reason experienced in one society be generalized to principles of reason in other societies? We next look at a historical instance of societies trying to empirically and normatively establish universal values, ethics in society. A horrible but instructive example is that of the case of the Nuremberg trials after World War II in the middle of the twentieth century. Occurring in 1947–1948, the Nuremberg Trials subsequently influenced the growth of international law.

It was in 1933, when Adolf Hitler seized power and quickly subjected Germany to the political tyranny of a Nazi dictatorship. In any society under a dictatorship, the ethical choices of individuals become limited – be a brute or be brutalized, terrorize or be terrorized. At the end of World War II, the terror imposed by Nazi officials was viewed as so horrible that the Allies decided to record this historically in a legal format. The USA, Great Britain, USSR, and France conducted criminal trials of Nazi leaders in the German town of Nuremberg in 1945–1946. Later in 1996, Robert Shnayerson summarized his historical perspective on these trials: "In the war-shattered city of Nuremberg, 51 years ago, an eloquent American prosecutor named Robert H. Jackson opened what he called 'the first trial in history for crimes against the peace of the world'" (Shnayerson 1996).

This is an interesting phrase "crimes against the peace of the world." It empirically indicates the desires of some humans to establish a universal ethics of the world for peace. Was this merely propaganda from the winning side of World War II? Or is it empirical evidence for a universal human desire for peace?

Shnayerson described the time: "The setting was the once lovely Bavarian city's hastily refurbished Palace of Justice, an SS prison only 8 months before. In the dock were 21 captured Nazi leaders, notably the fat, cunning drug addict Hermann Goering. Their alleged crimes, the ultimate in twentieth-century depravity, included the mass murders of some six million Jew and millions of other human beings deemed 'undesirable' by Adolf Hitler. 'The wrongs which we seek to condemn and punish,' said Robert Jackson, 'have been so calculated, so malignant and so devastating, that civilization cannot tolerate their being ignored because it cannot survive their being repeated'" (Shnayerson 1996).

These are strong ethical terms: "calculated and malignant and devastating." Also there is the essential presumption that there does exist a group called "civilization"? And why would ignoring such crimes against peace threaten the survival of that civilization? These were some of the historical assumptions being articulated by the Allied participants (victors) of World War II.

It was a forgone conclusion among the victors that the vanquished must be punished. It was terrible, dangerous, and hard war. But how to punish them? Was the punishment merely to satisfy the presence of a universal desire for revenge by the victorious? Or was there a more universal, futuristic reason for punishment – a more ethical reason? Policy among the Allies was divided on the course of punishment.

In the US government, one official, Treasury Secretary Henry Morgenthau Jr, argued that captured Nazi leaders should be summarily executed and Germany reduced to an agricultural state. Another, Secretary of War, Henry Stimson thought that such a solution would violate the US belief in law (1) against the presumption of innocence of individuals until proven guilty and (2) collective punishment of everyone for specific crimes of a few. And the President of the USA, Franklin Delano Roosevelt, asked Murray Bernays (a lawyer serving in the army) to find a compromise solution. Bernays suggested an international court be established to try individuals accused of crimes. Shnayerson described his idea: "Bernays suggested putting Nazism and the entire Hitler era on trial as a giant criminal conspiracy. In a single stroke, this would create a kind of unified field theory of Nazi depravity … He also suggested picking a handful of top Nazi defendants as representatives of key Nazi organizations like the SS. If the leaders were convicted, members of their organizations would automatically be deemed guilty. Result: few trials, many convictions and a devastating expose of Nazi crimes (Shnayerson 1996).

President Roosevelt died, and the next US President, Harry Truman implemented the Bernays' plan. Truman appointed Robert Jackson, then a US Supreme Court Justice, to run it. Jackson organized an International Military Tribunal to hold war crimes trials in Nuremberg, with one military judge each from USA, Great Britain, USSR, and France.

Robert Jackson acted as prosecutor when the trial began: "This war did not just happen. The defendants' seizure of the German state, their subjugation of the German people, their terrorism and extermination of dissident elements, their planning and waging of war …, their deliberate and planned criminality toward conquered peoples – all these ends for which they acted in concert" (quoted by Shnayerson 1996).

Jackson then began producing documented evidence. The Allied military had found files of the Nazi propagandist, Alfred Rosenberg, hidden in a castle (47 crates of files). They found tons of diplomatic papers hidden in caves in the Hartz mountains. They recovered hundreds of works of art looted from occupied countries in Goring's estate. They found Luftwaffe records stored in a salt mine in Obersalzberg. They found notes made by officials of Nazi governmental meetings. And they had American movies documenting the liberation of concentration camps at Bergen-Belson, Dachau, and Buchenwald. These movies showed the starving survivors as nearly skeletons. They showed the stacks of naked corpses, with bulldozers having shoveled such stacks of victims into mass graves. They had records of the Nazi genocide program and minutes of its meetings to plan the program.

Shnayerson summarized: "The scale of Hitler's madness was almost beyond imagination. The documents showed that after conquering Poland in 1939, he ordered the expulsion of nearly nine million Poles and Jews from Polish areas … the SS unleashed hundreds of Einsatzgruppen – killer packs assigned to spread terror by looting, shooting and slaughtering without restraint … these SS action groups murdered and plundered behind the German Army as it advanced eastward" (Shnayerson 1996).

One of the historically important functions of the Nuremberg trials was to acquire and record documentary evidence of the Nazi policies and aggression and genocide.

In January 1946, Jackson began bringing witnesses. The first was Otto Ohlendorf, former commander of an Einstazgruppe in Russia. Jackson asked questions and Ohlendorf answered:

Q. How many persons were killed under your direction?
A. Ninety thousand people.
Q. Did that include men, women, and children?
A. Yes.
Q. Did you have any scruples about these murders?
A. Yes.
Q. And how is it they were carried out regardless of these scruples?
A. Because to me it is inconceivable that a subordinate leader should not carry out orders given by the leaders of the state.

(Quoted by Shnayerson, Robert, 1996. "Judgment at Nuremberg", *Smithsonian Magazine*, October, pp. 124–141.)

This is the ethical issue. Are subordinates ethically responsible for carrying out "evil acts" under "evil policies" by their official superiors. This is the ethical connection between the acts of an individual under the policies of a society.

In the first trial, 24 Nazis were tried and judged. Those involved in the founding of the Nazi Party were charged with conspiring to launch World War II and related atrocities. Others were accused of planning aggressive war. Eighteen were charged with wartime crimes and crimes against humanity, such as genocide.

During the trial, one judge, Donnedieu de Vabres argued that the defendants acted not so much in complicity but in bondage to a megalomaniac: "The conspiracy charge was then restricted to eight of the defendants who knowingly carried out Hitler's war plans from 1938 onward" (Shnayerson 1996).

Also, the Judges ruled that guilt could not be assigned for only belonging to a Nazi organization. Any trial for other participants must be run in evidence of personal responsibility for crimes: "But since the Nuremberg judges ruled them all innocent until proven guilty, relatively few were ever tried – the prosecutorial job was too formidable." (Shnayerson 1996).

The 24 Nazi leaders received the following verdicts:

Herman Goering – Commander of the German Air Force – death sentence.
Karl Donitz – Admiral of the German Navy – prison sentence.
William Keitel – Head of Hitler's Military Command – death sentence.
Alfred Jodl – Keitel's second in Command – death sentence.
Erich Raeder – Admiral of the Germany Navy before Donitz – death sentence.
Ernst Kaltenbrunner – Highest surviving SS leader – death sentence.
Martin Borman – Nazi Party Secretary and Hitler's chief of staff – death sentence.
Albert Speer – Minister of Armaments – prison sentence.
Julius Streicher – Nazi Head of Franconia and publisher of Nazi paper – death sentence.
Hans Frank – Nazi Governor of occupied Poland – death sentence.
Arthur Seyss-Inquart – Nazi Governor of occupied Netherlands – death sentence.
Wilhelm Frick – Nazi Minister of Interior, author of Nazi Race Laws – death sentence.
Hans Fritzsche – Deputy Leader of Nazi Propaganda Ministry – death sentence.
Alfred Rosenberg – Nazi Minister of Occupied Territories – death sentence.
Fritz Sauckel – Head of Nazi slave labor program – death sentence.
Julius Streicher – Publisher of Nazi newspaper – death sentence.
Robert Ley – Head of the German Labor Front – committed suicide before trial.
Rudolf Hess – Hitler's deputy – prison sentence.
Baldur von Schirach – Head of Hitler Youth – prison sentence.
Joachim von Ribbentrop – Nazi Ambassador – death sentence.
Konstatin von Neurath – Previous Minister of Foreign Affairs – prison sentence.
Franz von Papen – Chancellor of Germany before Hitler – acquitted.
Gustav Krupp – Major industrialist and Nazi supporter – not tried due to ill health.
Hajalmar-Schacht – President of Reichsbank and Economics Minister – acquitted.

The names are listed here to see the kind of personal responsibility the Court regarded as a crime against humanity, even when carried out as official duties in the Nazi party and government. One can see in this list that the first trial focused upon Nazi leaders (1) in the German Military, (2) in the Nazi Party, (3) in the Nazi government, and (4) Nazi industrial supporters. Of 12 sentenced to death, ten were hung. Goering poisoned himself the evening before his scheduled execution. Borman had not been captured and was sentenced in absentia – but he was already dead, with his remains being discovered a decade later.

Adolf Hitler, the Head of the Nazi Party and the German Government, was not tried because he committed suicide. Goebbels, Nazi Propaganda Minister, had committed suicide, along with his wife and five children. Heinrich Himmler, Head of the SS, had been killed during the Russian capture of Berlin. Eichmann, Head of the Nazi Jewish extermination program, escaped to Argentina (but was captured in 1960 by Israel officials, tried in Israel and executed). Joseph Mengele, Nazi Doctor who performed inhuman experiments on people also escaped to Argentina.

The legacy of the Nuremberg trials was important to establishing an international legal tradition, post-twentieth century. For example, in 2009, Henry King, Jr. (one of the Nuremberg war crimes prosecutor and then a Professor at Case Western Law School) wrote: "A milestone passed quietly by on Sunday, October 1, 2006 – the 60th anniversary of the judgments rendered by the International Military Tribunal at Nuremberg against the key Nazi figures that led the world in the chaos of World War II… It is right and proper that we reflect on this seminal event in legal history, an event that became the cornerstone to modern day international criminal jurisprudence" (http://jurist.law.pitt.edu/forumy/s006/tree-fell-in-forest-nuremberg.php, 2009).

In this illustration, the focus of the Nuremberg Trials was that membership in the Nazi regime structured ethical choices of Nazi officials – choices which included use of slave labor, conduct of genocide, and other war crimes – crimes against humanity. The issue of the Nuremberg trials was: Did or did not an official of a government (Nazi government) bear personal responsibility about the ethics of the actions, which that official performed in accordance with official government policy? Did the Nazi military, government, and party officials who implemented these policies ethically have only a loyalty to Hitler or also a larger ethical responsibility to humanity?

Through laws and policies, governments establish ethical situations in which officials must administer and implement. If the laws and policies are unethical to a standard more universal than a government, then the acts of that government are unethical acts. Therefore, society determines the ethical situation for individual action. But since government laws and policies have been earlier made by individuals in positions of power (i.e., Hitler as Fuhrer of the German Reich), then societal ethics are constructed by individuals. Thus the ethics of a society are formulated and implemented within the interactions between individuals and society. Descriptions of ethical situations in societal history occur in the interactions between individual behavior and societal conditions prescribing ethical choices.

And this value issue about politics and culture (loyalty and equity) is a profound and general challenge for the governance of all modern societies. How and why can societies of decent and cultured individuals be quickly changed to monstrous societies of terror and genocide by modern ideological dictators? In the twentieth century under ideological dictatorships of communism and fascism, government terror was imposed in Russia in 1917, in Italy in 1922, in Germany in 1933, in China in 1948, in Cambodia in 1975, in Rwanda in 1990, and in Serbia in 1995.

In Weber's sense of a *principle of order for the family of humanity,* the ethical focus of the Nuremberg Trials was the issue of what is a "common good" for all people in all places at all times. Certainly Hitler's Nazi policies of slavery and genocide were not for the common good of the Polish people or European Jews or people in the nations of Norway, Denmark, Belgium Holland, France, or Russia. One can think of these situations in the context of ethics, a sociology of ethics. In a given historical situation, how are the practices of ethics by individuals constrained by the social context? Conversely, how have some individuals changed the practice of ethics in a given society at a given time?

The ethical context of the interaction of the individual and society structures the normative judgments (ethical decisions) – performed or experienced by individuals in the society of that epoch.

## Universality in Societal Reasoning

One can see in the example the importance of grounding social theory both empirically and normatively. It is the *normative grounding* that provides for the development of a universal human ethics, as *empirically-grounded* (observed) ethical nature is generalized and universalized (over all the "families of humanity" now and into the future).

A social science observer tries to capture in a social-theory (ideal type) that the participants in a society in an historical epoch thought their principles of order (rationality) were proper guides to behavior. How can social theory as an ideal type apply to a different society and a different era? Could such principles of order be used for rationality (effective logic) for decisions in a different society and/or different time? Here is where practice and science can interact methodologically – over the universalizing of rationality in an ideal type of societal system. The application in practice/policy of principles of order (rationality) abstracted in an ideal type of a societal system can test the generalization of such principles. Can they really work as a logic of decision making (rationality) in practice in a real society? And if they do not, what is the context for their non-workability, failure to provide order?

This is in accord with Weber's suggestion that social science objectivity (only partially-objective at any time and in any study) can be aimed toward increasing objectivity in application to the "family of humanity." And this means that trying to use the principles of order (rationality found empirically in a historical study of an ideal type) for present guidance of contemporary action is a way to test the universality of the rationality (obtained from the ideal-type social science study).

An ideal-type social theory can methodologically connect social science empiricism to practice.

An ideal-type theory derived from history may provide an illustrative form of normative judgments which might be applied over the "family of humanity."

The validity of the reasoning that is empirically observed in a societal model can be tested in application. Using a societal model as an "ideal type," consulting practices derived from it can test the extent to which the principles-of-order (logic) in a given historical epoch from which the model was abstracted is useful – to future decision-making by participants in future times and different places of society. This future-application of societal rationality provides a means of social science to generalize its empiricism & normativism over the "Family of Humanity." The differences between empirical judgments and normative judgments may be resolved over the empirical observation of historical epochs of societies and future application of principles of order as reason in society.

> Societal laws (rationality) can prescribe how a participant in a system should decide upon an action, based on a principle-of-order, necessary for rational behavior and sufficient when ethically generalized over all contexts of human experience.

## Summary

1. Methodologically, a Weberian ideal-type social theory can satisfy the requirement to ground social theory both empirically and normatively.
2. Ideal-type social theory can be constructed in a perceptual space of two dimensions: society–individual and group–process.
3. Empirically, social interactions in a modern society can be categorized in four societal subsystems of economic, cultural, political, and science–technology interactions.
4. Normatively, the kinds of rational principles valued in each subsystem differ: "profit" in economic systems, "equity" in cultural systems, "loyalty" in political systems, and "truth" in science/technology systems.
5. Ideal-type-theory construction addresses both the empirical and normative judgments in social science observation.
6. History provides the empirical contexts for testing the normative prescriptions of social theories over time and over different societies.

## Notes

[1] Structure functionalism is not so much a tight "school" in the discipline of sociology but more like a major theme running through many sociological perspectives. For example, one description of this can be found in (Collins 1994).

[2] Parson's seminal works appeared as: Parsons (1937, 1939, 1951). Parson's translations of Weber's works appeared as: Weber (1905), *translated by Parsons in 1930 and* Weber (1921), translated by Parsons and Alexander Henderson in 1947.

# Chapter 12
# Models

## Introduction

Scientific models play an important role in theory construction. This has been frequent in the history of science, in all disciplines. We saw the importance of Newton's gravitational model of the Copernican solar system in constructing the theory of physical mechanics. We saw the importance of Watson and Crick's model of DNA in the construction of the theory of molecular biology. We next look in detail at how modeling as a research technique plays the critical role of connecting empiricism to theory, and experiments to theory construction and validation.

As a historical illustration, we review the quantum modeling of the atom, upon which the theory of quantum mechanics was constructed early in the twentieth century. Quantum mechanics is a difficult theory to understand mathematically, but not as difficult philosophically. And the philosophical issues are interesting. What is light – a wave or a particle or both? What are electrons – a wave or a particle or both? Why cannot the exact position and velocity of a quantum particle be measured to any accuracy? What is material substance at that tiny, tiny atomic level – matter or energy or both? We review the philosophical developments in quantum theory (but not its mathematical development) in order to see the direct connection that models provide between experiment and theory.

This is a difficult chapter to read because it goes down into a physical world that we can neither directly see nor experience but only observe through instruments – the instruments in physics that experimentally probe the very tiny atomic world. Yet we need to read this historical illustration because it highlights a major methodological point. Theory is grounded by experiment, not directly but through the technique of modeling.

> In the scientific method, a model of a natural object connects experiment to theory; and theory is validated by experiment through measurements on a model.

F. Betz, *Managing Science*, Innovation, Technology, and Knowledge Management 9,
DOI 10.1007/978-1-4419-7488-4_12, 

## *Illustration: Heisenberg and "Adequate Concepts"*

To get a feeling for those exciting times of the new quantum physics, one can read the essays of Werner Heisenberg (one of the key participants in creating the new quantum paradigm): "The history of physics is not only a sequence of experimental discoveries and observations, followed by their mathematical description; it is also a history of concepts. For an understanding of the phenomena; the first condition is the introduction of adequate concepts..." (Heisenberg 1983, p. 19).

This is one of the hallmarks of a paradigm shift: "introduction of adequate concepts." Theories are constructed within a set of adequate concepts, and these concepts constitute the "paradigm" (or intellectual framework) within which the theories are constructed.

When the concepts in a paradigm are no longer adequate for new theory to explain new phenomena, the paradigmatic concepts need to be changed, and this is a "paradigm shift."

There were puzzling experimental results about atoms that classical mechanics could not explain, as Heisenberg summarized: "There were mainly three phenomena, which had to be connected. The first was the strange fact of the stability of the atom. An atom can be perturbed by chemical processes, by collisions, by radiation or anything else, and still it always returns to its original state, its normal state. This was one fact which could not be explained satisfactorily in earlier physics (the classical Newtonian physics)" (Heisenberg 1983, p. 20).

In classical physics (that explained the gravitationally bound solar system), if an asteroid hits any of the planets, then things change permanently in the system. For example, once in the early history of our planet Earth, a large planetoid collided with earth, blowing off material to coalesce to make the earth's moon and plunging into the earth to create earth's hot and rotating core to make the earth's magnetic field. Even now, hundreds of millions of years later, the earth still has a revolving moon and an oscillating magnetic field. Changes in classical mechanics to systems are permanent.

But why at the atomic scale in physics should atomic systems be different? Why should they be stable under collisions, that is, be only perturbed by collisions and return to stable states afterward? This was the strange fact of the *stability of the atom*.

Heisenberg summarized a second strange fact of the new physics about atoms: "Then there were the spectral laws (atomic spectra), especially the famous law of Rydberg, that the frequency of the lines in a spectrum could be written as a difference between terms, and that these terms had to be considered as characteristic properties of atoms" (Heisenberg 1983, p. 20). What these spectral laws indicated was that when the stable atoms were perturbed by heating the atoms, the light radiated by the hot atoms came off not as a continuous spectrum of light frequencies but as a discrete spectrum of a series of sharp spectral lines – and these line spectrum were characteristic of the structure of the atom (characteristic properties).

This stood as a fact about atoms – in stark contrast to the behavior of how charged orbiting particles in classical physics behaved. If electrons were in orbits around atomic nuclei as described by classical physics, then orbiting electrons would radiate off light (electromagnetic waves) continuously as the electrons circled the nuclei. An electron orbit would shrink in size, as the orbiting electron continuously radiated away its kinetic energy as emitted light. Then the light spectrum from an atom in classical physics would not appear as a continuous spectrum and not show the discrete spikes of light frequency as actually seen in atomic spectra. Light is connected to electrons in Maxwell's electromagnetic field as moving electrons have electric and magnetic fields around them and light consists of a traveling wave of pulsing electromagnetic fields.

Thus, classical Newtonian and Maxwellian physics could neither explain (1) the stability of electron orbits around the atoms nor (2) the discrete spectra of light radiating from heated atoms. This was the problem facing the physicists of Heisenberg's generation.

Then there was that third strange fact about the structure of the atom, as Heisenberg wrote: "And finally there were the experiments of Rutherford, which had led him to his model of the atom" (Heisenberg 1983, p. 19). Electrons did orbit the atomic nuclei. And while this was impossible by classical physics – yet so it was – as Rutherford had demonstrated!

Thus, not just one set of experimental phenomena had to be explained with new, more adequate concepts (paradigm ideas) but a total of three sets of experiments about atoms: "So these three groups of facts had to be combined, and as you know, the idea of the discrete stationary state (of electrons in atomic orbits) was the starting point for their combination" (Heisenberg 1983, p. 21).

By these three experimental facts, those early atomic physicists were confronted with the need for new basic ideas (adequate concepts). The problem went deeper than mere theory. It went down into the intellectual framework, down into the very paradigm of classical physics – space, time, and matter! How were these paradigm challenges addressed by the scientific community? Heisenberg wrote about the time in 1925: "Planck's quantum theory (which we soon review), in those days, was really not a theory, but an embarrassment. Into the well-founded edifice of classical physics it brought ideas that led, on many points, to difficulties and contradictions, and hence there were not many universities where there was any desire to tackle these problems seriously" (Heisenberg 1983, p. 39).

> The deep problem was – could an atomic-scale particle travel as both a particle and as a wave? And the experimental facts were – indeed, an electron traveled both in wave and in particle motion.

For example, in an instrument called the Wilson cloud chamber, one can see the actual track of a traveling electron as a straight line of ionized bubbles in the chamber. Moreover, one can apply electric and magnetic fields onto the chamber and measure the actual velocity and mass of the electron making a track through the cloud chamber. Therein one observes directly that an electron does actually travel like a particle, with definite mass and velocity and charge.

Yet, in contrast, one can also directly observe electrons traveling as waves! For example, in another instrument, a slit experiment, one can send electrons through a narrow slit and see their subsequent impact upon a fluorescent screen (after traveling through the slit). The pattern one sees on the screen is a diffraction pattern – alternating light and dark patterns above and below the plane of the slit. But a diffraction pattern can only be formed if electrons travel as "waves!"

Thus, different kinds of experiments establish the real fact that electrons travel both as particles and waves. This was a contradiction in classical mechanics – because therein a particle is a form of a massive object, whereas a wave is a form of energy transfer. Heisenberg wrote of the groups addressing this contradiction: "Other than in Copenhagen (Bohr's institute), Bohr's theory (the Bohr atom) was primarily taught and developed by Sommerfeld in Munich, and it was only in 1920, with the appointment of Franck and Born, that the Gottingen faculty finally decided to join this scientific movement" (Heisenberg 1983, p. 19). Heisenberg joined Born's group in Gottingen as a new post doctoral assistant in 1924.

In 1925, these were the three physics groups in the world, trying to sort out a new paradigm for quantum mechanics: "If we compare the three centers, Copenhagen, Munich, and Gottingen, where the subsequent development primarily took place, we can relate them to three lines of work in theoretical physics …" (Heisenberg 1983, p. 38).

1. "The phenomenological school (Sommerfeld in Munich) was attempting to unite new observational findings in an intelligent fashion (and) to present their connection by means of mathematical formulae that appear to some degree plausible from the standpoint of current physics."
2. "The mathematical school (Born in Gottingen) endeavored to represent natural processes by means of carefully worked-out mathematical formalism, which also satisfies to some extent the mathematicians' demands for rigor."
3. "The third school (Bohr in Copenhagen), which may be called conceptual or philosophical, tried above all to clarify the concepts by means of which events in nature are ultimately to be described" (Heisenberg 1983, p. 38).

Arnold Sommerfeld Max Born Niels Bohr
(http://en.wikipedia.org, Arnold Sommerfeld 2007; Max Born 2007; Niels Bohr 2007)

Werner Heisenberg (1901–1976) was born in Wurzburg, Germany. He obtained a doctorate in physics at the University of Munich in 1922. In 1923, Heisenberg studied as a postdoc with Niels Bohr in Copenhagen. He went back to Germany as a research assistant to Max Born at the University of Gottingen and completed his Habilitation (license to teach in German universities) in 1924. In 1925, Heisenberg invented a new approach to quantum mechanical calculations as a form of matrix mechanics. As a young man, Heisenberg was at the heart of the development of quantum mechanics, the paradigm shift, extending mechanics down to the atomic level.[1]

(http://en.wikipedia.org, Werner Heisenberg 2007)

## *Illustration: Quantum Theory: Max Planck*

Paradigm shift and theory construction of quantum mechanics did not happen instantly or easily. It took many scientists over three decades to puzzle out the nature of the atom and to understand it with a new quantum theory – from Rydberg to Planck to Einstein to Rutherford to Bohr to Heisenberg to Born to Dirac. But in the end, the quantum theoretical model of atomic-scale phenomena was verified by many different kinds of atomic experiments. Here is the story, beginning with the research of Max Planck.

Planck's work in 1900 occurred before Rutherford's atomic experiment in 1904. Max Planck introduced the idea of a quantum of energy to explain the radiation of light from hot gases. The experimental data Planck analyzed (induced from) was about how the intensity of light varied as a function of temperature in emission of heat from a hot body – "blackbody radiation." "Blackbody radiation" occurs in an experimental setup in which a container of gas is heated, to give off light as "radiant heat" passing through a small hole. By the term "quantum," Planck meant that energy was radiated in packets, quanta, not continuously in size but in discrete sizes, a quantum of energy.

There had been several analytical attempts in the late 1800s to theoretically derive a mathematical formula that fit the empirical shape of the radiance (intensity of light emission) as a function of frequency of the radiance. One formula was derived by Rayleigh and Jeans and then another by Wilheim Wien. But both did not fit the experimental data, as shown in Fig. 12.1. Next, Planck derived a formula that

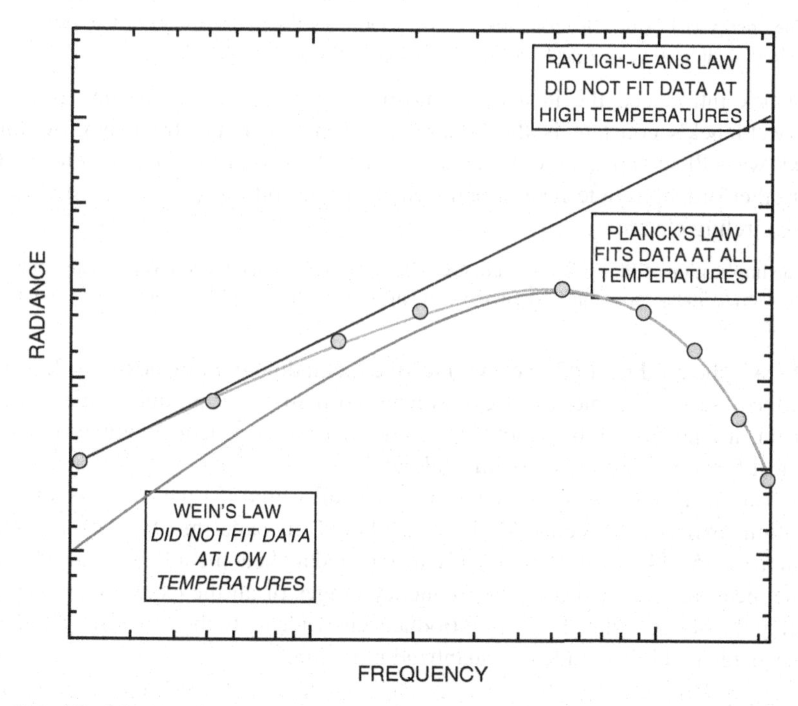


**Fig. 12.1** Black body radiation, experiment and analysis (http://www.en.wikipedia.org, Max Planck 2008)

matched the experimental data; he derived his mathematical model from statistical thermal physical theory but needed a strange assumption to make his model fit the data – quantized electronic radiators of the radiation.

Using the classical electromagnetic theory of Maxwell to describe the emission of light by the atoms of a gas, Planck had calculated the radiation from the atoms as kinds of oscillators in which the frequency of oscillation would be continuous from low frequencies to high frequencies. (This is what Rayleigh–Jeans had done, and this derivation produced a formula that did not match experimental data at high temperatures.)

Planck had the theoretical boldness to conjecture something new about the physics of the electronic oscillators (atoms) in the heated gas. But what? Then Planck imagined: what if the oscillators of the atoms did not oscillate at any and all frequencies but only at discrete frequencies? When Planck put this discreteness of oscillation by the atoms into the calculation of radiation, the resulting formula matched the empirical data! The formula depicted empirical reality, and therefore, something must be real in the atom, which corresponded to Planck's assumption. *Atoms really do radiate light as if they were discrete electromagnetic oscillators.*

Mathematically, Planck expressed his philosophical assumption about discreteness (quantization of the energy of radiation) as a phenomenological physical law between energy and frequency. (The frequency of an oscillator is how rapidly it is oscillating, as for example, how rapidly a pendulum on a pendulum clock swings back and forth.)

> The energy ($E$) of the electronic oscillators determines the frequency ($f$) of the radiated light in quantized energy packets: $E=hf$, where $h$ is Planck's constant.

Planck introduced the idea of a "quantum of energy": ($E=hf$) into physical theory. Planck's constant is the quantity of energy per unit-frequency of light. (Next, we will see how, 11 years later in 1912, Bohr used Planck's idea to model the Rutherford atom in terms of quantized electron orbits, as discrete energy levels of the circling electrons.)

> The implication of Planck's law on blackbody radiation was that the atoms oscillated only in discrete states, not at any frequency – quantization (discreteness) of the energy states of an atom!

As sketched in Fig. 12.2, we can trace the methodology in Planck's work, using both the research techniques of experiments to inductively produce experimental data on atomic radiation, which are analyzed next in a deductively formulated physical formula to fit the experimental data.

Planck analyzed (7) experimental data that measured (8) the intensity of light radiation, using instruments (4) in "blackbody" in experiments (3) on atomic phenomena (5). Planck mathematically formulated a phenomenological law (6) that fit the measured data (8) over the frequency ranges of all the experiments (3). To develop the law (6), Planck had to introduce a new idea into the scientific paradigm of mechanism (11) – the idea of quantized radiation.

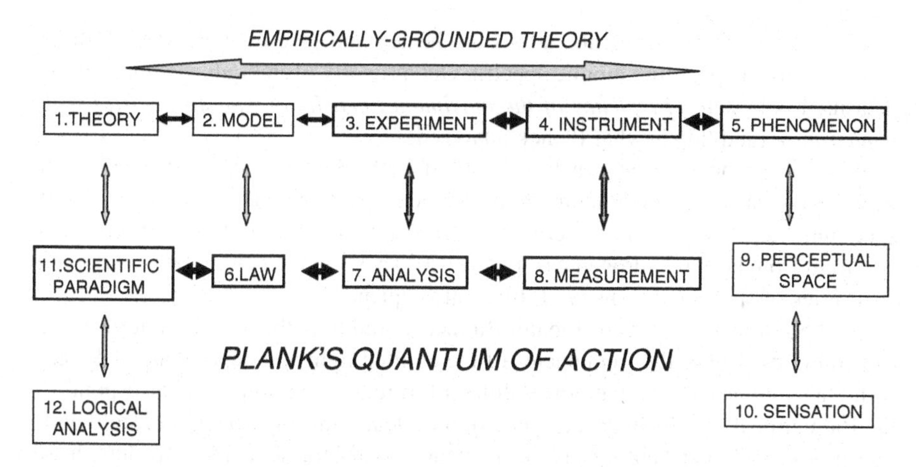


**Fig. 12.2** Scientific procedure of grounding theory on experiments

Breakthroughs in theory (paradigm shifts) have been constructed with the inductive logic of experiments (data) and deductive logic of modeling a physical phenomenon of the experiment (theory)

Significant research hypotheses (quantization of atomic energies) arise from models that fit experimental data.

Max Planck (1858–1947) was born in Kiel, Germany, and attended the University of Munich in 1874. He focused his research on the mechanical theory of heat. After his doctorate and habilitation thesis, Planck lectured as Privat Dozent at Munich, and he continued to work on the field of heat. In 1885, Planck was appointed as an associate professor at the University of Kiel . In 1889, he was appointed as a professor at the University of Berlin. In 1894, he began his studies on the physical phenomenon of "blackbody radiation." An electricity company asked him to research on how to gain the most light from new light bulbs. In 1918, Planck was awarded the Nobel Prize in Physics.

(http://en.wikipedia.org, Max Planck, 2007)

## *Illustration: Quantum Theory – Balmer and Rydberg*

Next, we review another source of the strange empirical facts (experiments) that Heisenberg cited: "Then there were the spectral laws (atomic spectra), especially the famous law of Rydberg, that the frequency of the lines in a spectrum could be written as a difference between terms, and that these terms had to be considered as characteristic properties of atoms" (Heisenberg 1983, p. 20).

In the radiation experiments, atoms are heated and give off light. This light is called a radiant spectrum, atomic spectra. One property of the radiance that can be measured is how the *frequency of the maximum intensity of radiation varies with temperature* (and this is what Planck analyzed).

A second property that can be measured is the *intensity of light at different frequencies* (and at a single temperature of the radiant gas). This is called a "light spectrum," and the first measurements were performed in the 1870s on the light spectrum from heated hydrogen gas. Figure 12.3 shows this spectrum from hydrogen as a series of spikes of intensity at different frequencies.

In 1885, a mathematician, Johann Balmer, found a mathematical pattern in the spectral lines seen in the frequencies of the emitted light and devised an analytical formula summarizing the pattern, Balmer's formula. The two peaks correspond to the Balmer peaks in the hydrogen spectrum of deuterium (hydrogen molecule with one proton and two neutrons in its nucleus). In the peak at the wavelength of 486 nm (Db peak), an electron is stimulated by heat into a higher orbit and cascades back to ground orbit, emitting light. Another electron transition from higher orbit to ground orbit occurs at 656 nm (Da peak).There are other kinds of emission mechanisms. The continual pattern from 200 to about 475 nm is due to molecular vibrations emitting light, the deuterium molecule vibrating due to heat and emitting photons as a molecular phenomenon (vibration of the molecule as a whole).

Following upon Balmer's work in 1890, Rydberg in 1890 generalized his analytical formula from hydrogen to any heated alkali metal. Rydberg showed that the spectral lines from hydrogen (Balmer's formula) was a special case of the more general alkali metal emission pattern. It was Rydberg's formula that Bohr's future quantum model of an atom would need to explain (derive).

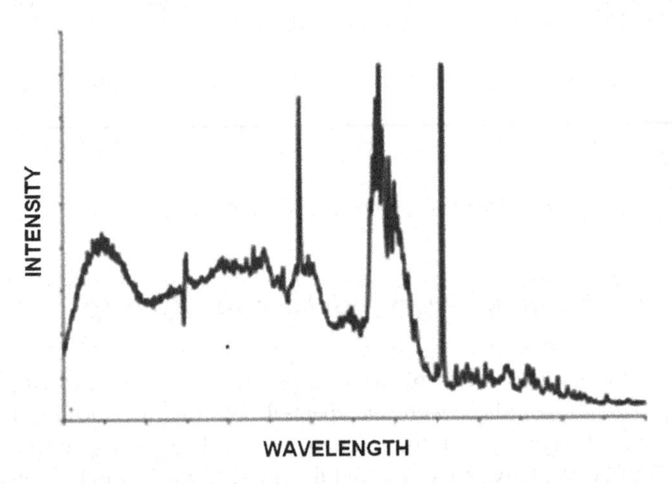


**Fig. 12.3** Hydrogen light spectrum: intensity of light radiated from heated hydrogen gas at each wavelength (http://en.wikipedia.org, Max Planck 2008)

Johann Jakob Balmer (1825–1898) was born in Lausen, Switzerland. He studied at the Universities of Karlsruhe, Berlin, and Basel. He obtained a Ph.D. in 1849 from the University of Basel. Balmer taught mathematics at a school for girls and also lectured at the University of Basel.

Johannes Rydberg (1854–1919) was born in Sweden and was educated and taught at Lund University. (http://en.wikipedia.org, Johannes Rydberg, 2007)

Johann Jakob Balmer
(http://en.wikipedia.org, Johann Jakob Balmer, 2007)

Johannes Rydberg
(http://en.wikipedia.org, Johannes Rydberg, 2007) (Fig. 12.4).

Thus, for the challenges of a new theory of an atom, which Heisenberg summarized, the physicists in the early 1900s had three experimental facts upon which to construct the new theory:

1. Rutherford's geometric atom with a positive nucleus, orbited by electrons
2. Planck's quantum of energy transfer in electrons emitting light
3. Balmer/Rydberg's formulae for a series of frequency peaks in the atomic emitted light

When there are new sets of experimental phenomena that require alteration of a scientific intellectual framework then a paradigm shift occurs.

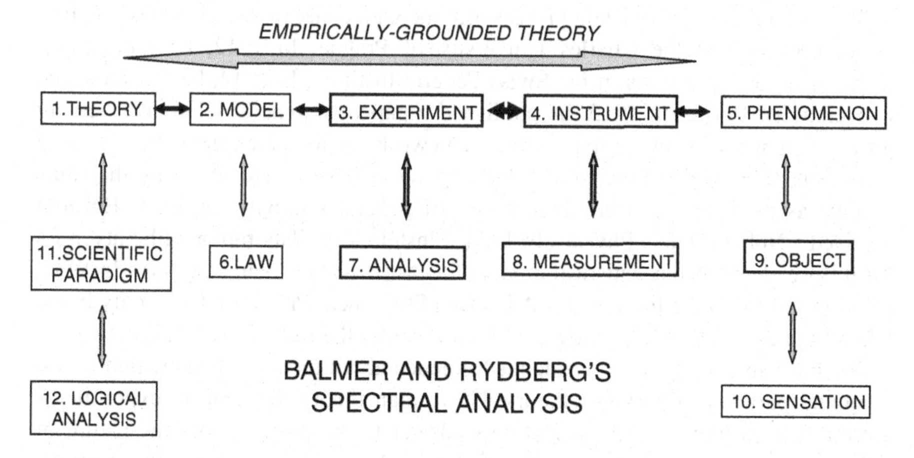


**Fig. 12.4** Scientific procedure of grounding theory on experiments: Balmer and Rydberg's spectral analysis

## *Illustration: Quantum Theory of the Atom – Einstein*

Four years later, in 1905 (after Planck's quantum idea in 1900), Albert Einstein added more physics to the idea of "quantization." Einstein analyzed the experimental data on the emission of electrons from materials exposed to incident light as evidence of quantized packets of light energy – photons. Einstein proposed that the energy of a photon ($E$) is proportional to its frequency ($\nu$) by Planck's constant ($h$): $E=hf$. He used this to derive the experimental form of electron energies emitted in the photoelectric effect.

We recall that Albert Einstein (1879–1955) was born in Wurttemberg, Germany. He graduated with a teaching diploma from the Swiss Federal Institute in Zurich, Switzerland in 1901. He looked for a teaching position, but upon not finding one, he took up a job as an assistant patent examiner in the Swiss Federal Office for Intellectual Property in 1903. And we recall that in 1905, the physics journal, *Annalen der Physik*, published four key papers by the young Einstein:

1. The photoelectric effect, which demonstrated that light interacted with electrons in discrete energy packets
2. Brownian motion, which explained the random paths of particles suspended in a liquid as direct evidence of molecules
3. Special relativity, which postulated that the speed of light was a constant in the universe, with the same value as seen by an observer, and implied that the mass of an object increased as the velocity neared the speed of light
4. Equivalence of matter and energy in that mass could be converted into energy at the quantity $E=mc^2$

We also recall that in 1908, Einstein went to teach at the University of Bern as a Privat Dozent. In 1909, he published a paper on the quantization of light as photons. In 1911, he became an assistant professor at the University of Zurich and soon after, a full professor at the Charles University of Prague. In 1912, he returned to Switzerland as a professor at the Swiss Federal Institute. In 1914, he moved to the University of Berlin. In 1915, he published a paper on general relativity, viewing gravity as a distortion in the space–time framework due to matter. In 1919, an eclipse of the sun showed the deflection of light (photons have a zero mass) by the sun's gravity, as predicted by Einstein's theory of general relativity. In 1921, Einstein received a Nobel Prize in Physics. In 1924, Einstein, in collaboration with Satyendra Nath Bose, developed a statistical model of a gas (Bose–Einstein statistics). In 1932, Einstein moved to the Institute for Advanced Study near Princeton University in the United States to avoid remaining in Germany under the rule of the Nazi party.

We have seen that the experimental evidence about the atom had created a fundamental paradox in previous theory, Newtonian theory. In science, experiments are the final arbiters of theory. Nature's answer to the puzzle about the nature of light, wave or particle, was turning out to be that light is both a wave and a particle. Light travels as an electromagnetic wave, according to Maxwell's equations. But when interacting with matter (atoms), light acts like a particle, transmitting or

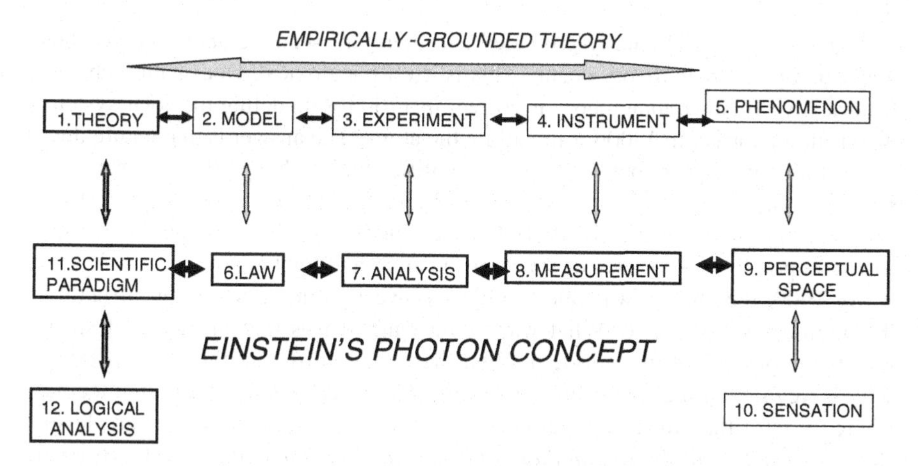


**Fig. 12.5** Scientific procedure of grounding theory on experiments

receiving energy in discrete bundles (quanta), according to $h = E/v$, so that Planck's constant $h$ is a minimum "bundle/quantum" of energy transmitted between light and atoms. We can sketch how Einstein's photon theory was grounded in experiment, in Fig. 12.5.

As shown in Fig. 12.5, Planck's new law (6) of the quantization of energy in atomic light emission ($E = hf$) occurred in a research technique of analysis (7) from experimental (3) measurements (8).

Einstein logically analyzed (12) the experiment (3) on the photoelectric effect by proposing a new theory (1) of a quantum of electromagnetism – a photon

Einstein's theoretical idea (1) of light traveling as a wave but interacting as a particle altered the scientific paradigm (11) (of motion as either particle or wave, but not both).

Einstein justified this change to the physical paradigm of mechanism based on the different perceptual spaces of matter: macro and micro. At the atomic level (micro space), the difference in scale allowed a different theory for different scales of nature. Einstein, as a theoretician, used experiment as the arbiter of theory, ground of theory.

The paradigm shift that was evolving was that, at a macro-scale in classical physics, there can be a continuous range of energy transfers between things (Newtonian mechanics), but at a micro-scale (atomic level) there are only discrete transfers of energy (quanta) between light and atoms (quantum mechanics).

Light, as both a particle and a wave, introduces a new philosophical idea into physics because the idea is paradoxical. On a macro-level, a natural object can be either a particle or a wave, but not both. Yet at an atomic level, light is both a wave and a particle.

Paradoxical ideas introduced into empirically-grounded theory require a change in the scientific paradigm in which the theory is constructed.

If theory is experimentally-grounded and therefore accurate about nature, then how can a paradox in theory occur? This is itself paradoxical. Does nature change historically? If Newtonian theory is correct in 1700 for modeling the solar system, why is it not correct in 1900 for modeling the atom? The answer is that nature does not historically change, but the foci of scientific studies change (here a change in scale focus). By 1700, Newton was focused upon the macro-level of space of the earth and the solar system. By 1900, Planck, Einstein, Rutherford, and Bohr were focused upon the micro-level of space of the atom and light emission.

By 1900, the change was due to physics investigating a new scale of nature, the atomic level of space. What was being learned was that the scale of space affects the physics of the world, and that there was a different theory at different scales. Newtonian mechanics provided a *sophisticated and adequate* intellectual framework for the physical experience of nature (experiments) at a macro-level. But it was not valid theory, *neither sophisticated nor adequate*, at a micro-level (atomic scale).

## Motion in Physical Perceptual Space

The next physical issue about scale that needed to be resolved was how electrons traveled. Did electrons travel as particles or as waves? Maxwell had analyzed that light travels as waves, but Einstein had analyzed that light interacts with electrons as particles. At the atomic scale of space, natural things appeared to show a wave/particle duality, traveling as waves and interacting as particles. Was this real?

We pause for a moment to review these two kinds of motion: particle motion and wave motion. In physical perceptual space, the *temporal* part of the transcendental aesthetic of space/time provides for the expression of *motion* of physical objects. There are two forms of physical motion: particle motion and wave motion.

### Particle Motion

As shown in Fig. 12.6, particle motion is described by the distance moved over a period of time.

The change in position (distance $=\mathrm{d}x$) of a particle in space is the difference between position $x_1$ at time $t_1$ and the position $x_2$ at a later time $t_2$: $\mathrm{d}x=(x_2-x_1)$, where the time interval ($\mathrm{d}t$) over which this distance was traveled was $\mathrm{d}t=(t_2-t_1)$. The velocity of a particle in motion is the rate-of-change ($v$) of position as the ratio of the distance traveled $\mathrm{d}x$ over the time interval $\mathrm{d}t$: $v=\mathrm{d}x/\mathrm{d}t$. The acceleration ($a$) of a particle in motion is the rate-of-change ($\mathrm{d}v$) of the velocity over the time interval $\mathrm{d}t$ of the change: $a=\mathrm{d}v/\mathrm{d}t$. (All this is in differential calculus notation, which we will briefly review in the later chapter on measurement).

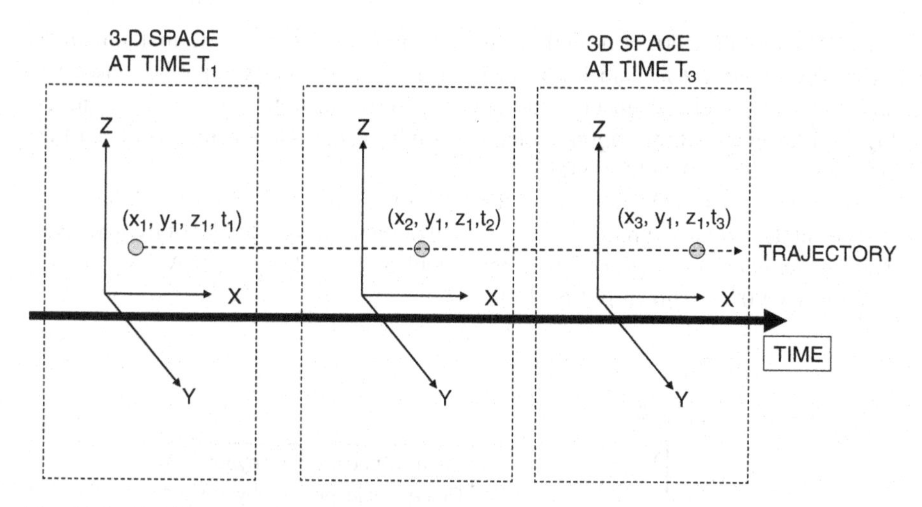


**Fig. 12.6** Classical four-dimensional space–time description of particle motion

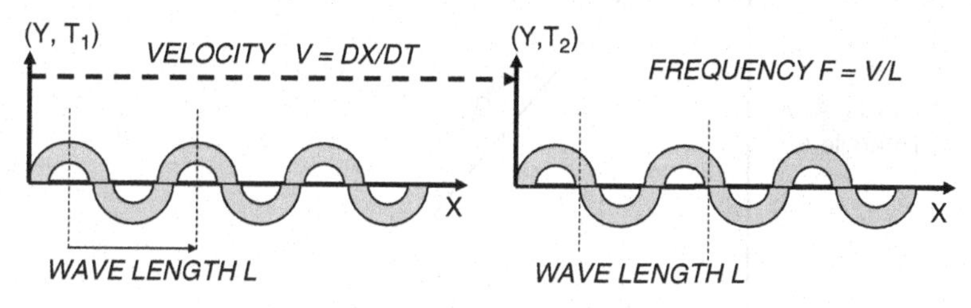


**Fig. 12.7** Wave motion

## *Wave Motion*

Wave motion must be expressed in a two-dimensional space as there is an up–down motion to the wave perpendicular to the direction of travel of the wave, as depicted in Fig. 12.7. This up–down motion occurs as it travels, so that a picture over time of the wave looks like a "sinusoidal pattern" – a wave pattern.

The term "sinusoidal" comes from the mathematical topic of trigonometry, begun in ancient Egypt and completed in European mathematics in the 1500s. Trigonometry is the mathematics of the Euclidean geometric form of the right triangle. All right triangles are similar in form with one angle a "right-angle" of 90°. The sum of all angles of any triangle adds to a total of 180° (half the degrees in a circle). So all right triangles are similar with one angle at 90° and the other two angles adding up to another 90°.

This particular Euclidean form of the 90° triangle was especially important because it could be used (in Egyptian times) to resurvey farm land along the Nile annually after the Nile flooded. It was historically called the "right triangle" probably because it was the correct triangle to use in surveying. So a whole mathematical topic was developed around the right triangle. (This is an example of where an application of science impacted the epistemology and ontology of the science.) Figure 12.8 depicts how trigonometry is a study of the relationships between the sides and angles of a right triangle.

The term "sin *x*" is a short name for "sine of *x*" and "cos *x*" for the "cosine of *x*." The mathematical expression sin $x$ = length of opposite side/length of hypotenuse. The sine and cosine functions have been calculated for all angles of A from nearly 0° to nearly 90°, as shown in Fig. 12.9.

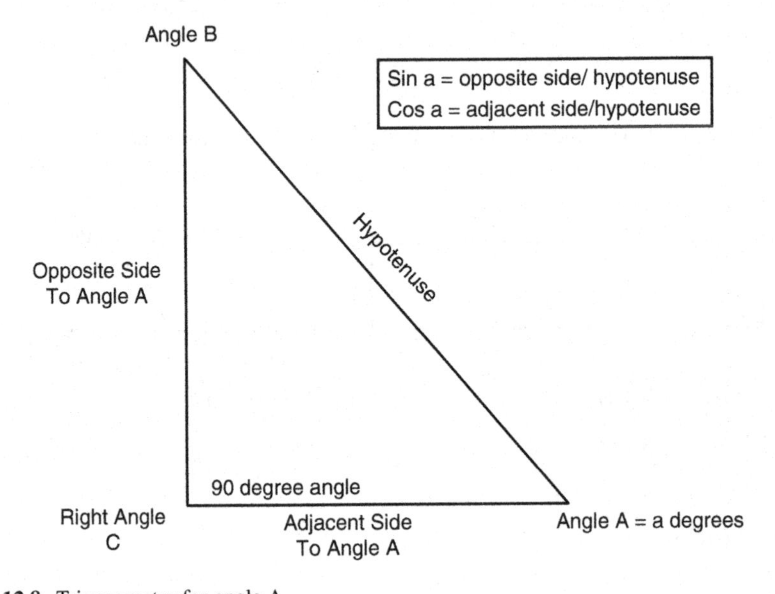


**Fig. 12.8** Trigonometry for angle A

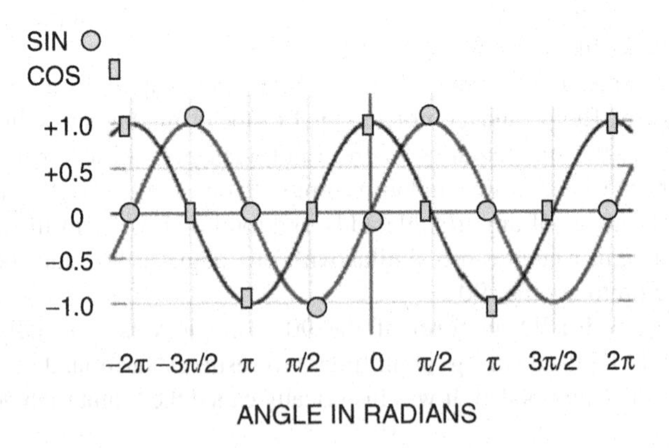


**Fig. 12.9** Sine and cosine functions (http://en.wikipedia.org, Trigonometry, 2008)

The sine function swings steadily and continuously from values of 0 to 1 and back to 0 and to a −1 and back to 0, over and over again. *The form looks just like a wave!* In Cartesian analytic geometry in two dimensions, the sine function can be used to describe the motion of a wave. Also, because the cosine function has the same form, only changed by an angle of one-half of the value of pie (90°), both the sine function and the cosine function can be used to describe a wave form in Cartesian geometry.

The form of a wave traveling with velocity $v$ in time across one dimension can be written as a sine function (or cosine function) having both place, $x$, and time, $t$, values and with $y$ being the height of the wave (on a two-dimensional $(x,y)$ Cartesian plane): $y = \sin(kx - wt)$. Here the value $k$ is called the wave number (the inverse of the wavelength) and $w$ is the frequency of the wave oscillation. Then the velocity of the traveling wave is the ratio $v = w/k$. The wave number $k$ is defined as the inverse of the wavelength. This gives rise to the famous wave property that the velocity of a wave equals the product of wavelength times the frequency.

Now the point of this background is that Newtonian mechanics in a Cartesian geometry can quantitatively describe both particle motion, $x = vt$, and wave motion, $x = \sin(wt)$. This is very powerful because it means that the two kinds of motion that humans can see in nature, particle and wave motion can both be quantitatively described by physics of mechanics.

We compare particle motion and wave motion in Fig. 12.10. In comparing particle motion to wave motion, one can see that a particle is at a single position $(x_1)$ at a time $t_1$. But at any time $t_1$, a *wave form* extends through space $(x,y)$ as a series of repeated crests and troughs. At any time $t_1$ for a wave, the wavelength $l$ is the distance between crests of the wave.

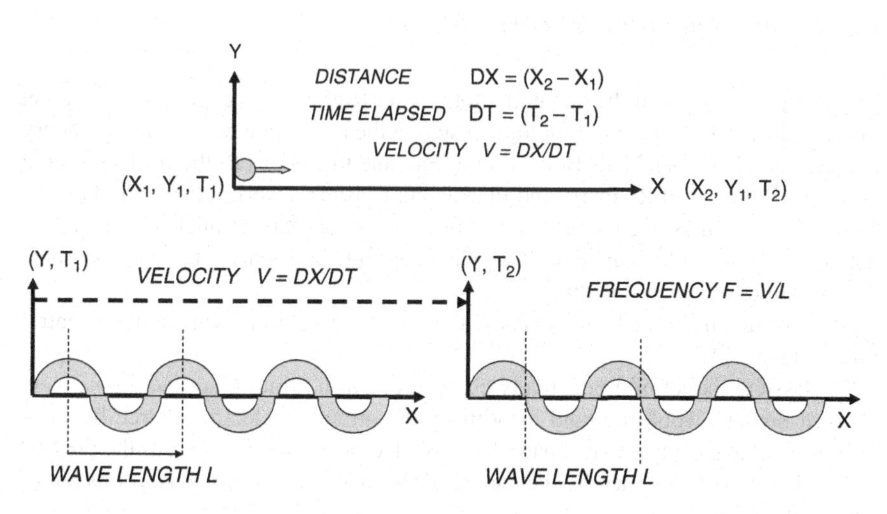


**Fig. 12.10** Particle motion and wave motion

For example, wave motion in a material medium (such as air or water) passes motion through the medium as waves. In air, changes in air pressure in recurrent sequences (waves) produce the sensation of sound. In water, changes in water pressure produce the sensation of water waves. In a vacuum (no medium), changes in electrical-magnetic sequences produce the physical waves of light. The kinematics of wave motion are described by the properties of a wave as to its velocity ($v$) and wavelength ($l$) and frequency of oscillation ($f$). The relationship between these properties of a wave form is that the wavelength $l$ divided by velocity $v$ equals frequency $f$:

$$f = l / v$$

Particle motion consists of a particle traveling a distance in space at a velocity of $v = dx / dt$

Wave motion consists of the whole form of the wave moving in space at a velocity $v = dx / dt$

The movement of a wave through space requires all the crests to move through space. This movement of the whole wave form occurs at a velocity $v$, which shows the distance traveled $dx$ of all the crests over the time $t$. In this velocity of the whole wave moving, the *number of crests* passing a given point of space $x$ in a time $t$ is called the *frequency* $f$. The number of crests passed ($f$) can be computed as the wave velocity v divided by the wavelength $l$: $f = v/l$. It is this idea of wave motion that Bohr next used to model Rutherford's atom as a quantized atom.

## *Illustration: Quantum Theory – Bohr*

We return to the story of the quantum atom. After Rutherford's experiment to model the atom in 1908, the next chronological step in the development of quantum theory occurred in 1912 when Niels Bohr went to England to work as Rutherford's research assistant (Balmer and Rydberg, and Planck and Einstein's works all occurred before 1908). Bohr would solve the problem of how electrons orbit the nucleus of an atom, but his solution would change philosophical concepts in physics – the wave–particle double nature of subatomic things.

Bohr started his model with a classical model of an orbiting body under a central force, Fig. 12.11.

A classical model of a negatively charged electron orbiting around a positively charged nucleus would be held in orbit by the attractive electrical force between positive and negatively charged particles. The electrical force $F_e$ varies as the product of the charges ($Qq$) divided by the square of the distance $r$ (where $k$ is a constant of electrical attraction):

$$F_e = kQq / r^2$$

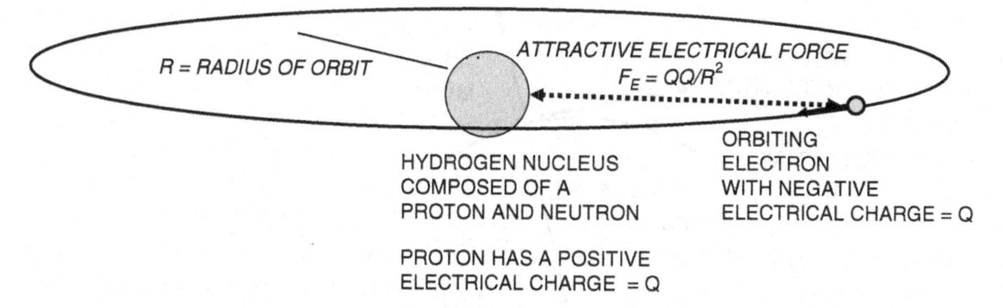


**Fig. 12.11** Classical model of a hydrogen atom (but not physically possible)

We recall that the gravitational force in Newton's model of the Copernican solar system looked just like this – with a similar form of an attractive gravitational force ($F_g$) holding the planets in orbit around the sun. The gravitational force $F_g$ also depends on the product of the mass ($M$) of the Sun and the mass ($m$) of a planet divided by the square of the distance $r$ (where g is a constant of gravitational attraction:

$$F_g = Mm / r^2$$

But Bohr had to change this classical model, since classically an electron in orbit would radiate away all its energy and fall into the nucleus. And the change Bohr made was to assume that the electron traveled not as a particle around the nucleus but as a wave! And it was the wave motion that Bohr would quantize (as earlier Planck had quantized the oscillation of electronic radiators in atomic light emission).

To understand how Bohr used Newton's mechanics first as an analogy, we need to review some of Newtonian physics. Therein two particles are related together by a force in the space around the particles. The impact of a force ($F$) on the particle is to change its motion, with the acceleration ($a$) of the motion proportional to the force, with the proportion being the mass ($m$) of the particle: $F=ma$. The dynamics of motion then can be described by the momentum (p) of a particle and the kinetic energy ($E$) of a particle. When a particle is moving in an orbit or radius ($r$), it was another angular property about momentum called its angular moment ($L$), as in Fig. 12.12.

The momentum ($p$) of a particle is equal to the velocity of a particle times its mass: $p = mv$

The kinetic energy ($E$) of a particle is equal to one-half the square of the velocity times its mass: $E = \frac{1}{2}mv^2$

The angular moment is a product of mass times velocity times radius: $L=mvr$.

Also in Newtonian mechanics, when a body is swung in a circle then its direction of velocity is continually changing and so its inertial reaction (Newton's First Law) is to pull the body away from the swing as a centrifugal force. For example, when one ties a rock to a string and swings the rock about in a circle, one feels the pull on the string as the centrifugal force on the rock, and this is called an inertial centrifugal force $F_c$.

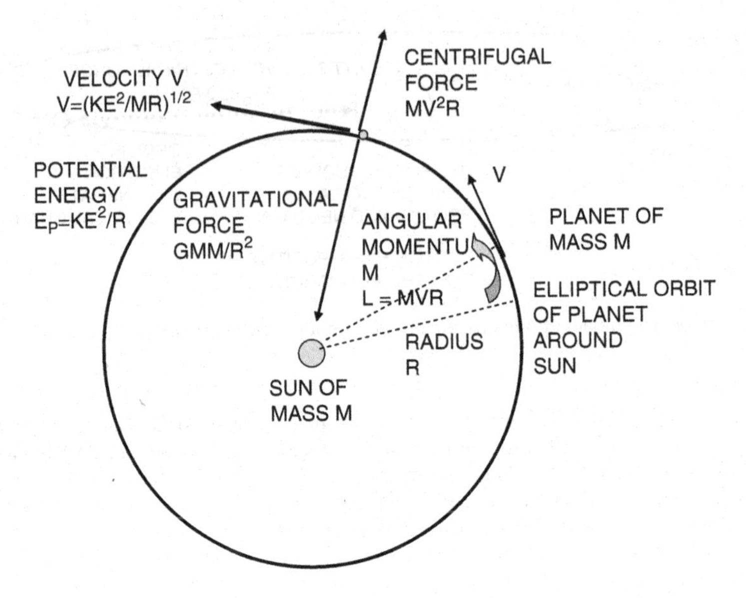


**Fig. 12.12** Newton's gravitational model of the solar system

When a body of mass m circles in a radius r with a velocity of v, the centrifugal force $F_c$ is: $F_c = mv^2 / r$

First, Bohr used a Newtonian model of the atom. In the natural phenomenon of electricity, there is a static electrical force $F_e$ between charged particles ($Q$ and $q$), which is attractive when their charges are opposite and repulsive when their charges are similar. $F_e = ke^2/r^2$, where $r$ is the radius (distance) between them and $k$ is an electrical constant and $e$ is the same sized charge of each particle. (The charges of an electron and of a proton are opposite but equal in strength.) (Fig. 12.13).

In writing out the equations for this model, Bohr had to express how the electrical attractive force ($F_e$) of the orbit balances out the centrifugal force ($F_c$) (which an object in orbit feels as a result of its constantly changing direction in the orbit). This in balance: $F_e = F_c$.

What Bohr did was to balance the centrifugal Force ($F_c = mv^2r$) against the attractive electrical force ($F_e = ke^2/r^2$).

When Bohr set the attractive electrical force ($F_e = ke^2/r^2$) equal to the centrifugal force due to the inertia of the electron mass ($F_c = mv^2r$), he could solve for the velocity ($v$) an electron must have in a stable orbit or radius ($r$) for under the attractive electrical force:

$$F_e = F_c$$

$$ke^2 / r^2 = mv^2 r$$

$$v = \left(ke^2 / mr\right)^{1/2}$$

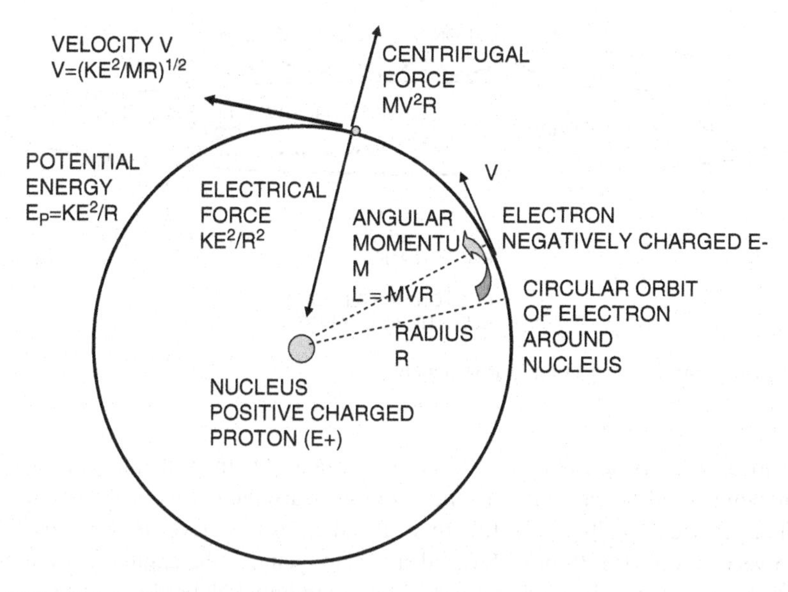


**Fig. 12.13** Bohr's electrical model of an atom

(The sign $()^{1/2}$ denotes the square root of the expression in the parenthesis.) This reads that for an electron to be in a stable orbit around the nucleus of an atom, its velocity in orbit must be proportional to the square root of (the electrical charge of the electron squared divided by the product of the electron mass times the square of the radius of the orbit).

Next, Bohr changed the model from a classical Newtonian model to a new quantum model of the atom.

Bohr quantized the angular momentum of the electron orbit.

The way he did this was as follows: Bohr solved for the velocity of the orbit in order to express the angular momentum ($L$) of an electron in an atomic orbit by multiplying the mass times the velocity times the radius of the orbit:

$$L = mvr$$

$$L = m\left(ke^2/mr\right)^{1/2} r$$

$$L = \left(m^2ke^2r^2/mr\right)^{1/2}$$

$$L = \left(mke^2r\right)^{1/2}$$

Thus the angular momentum ($L$) of an electron whirling around in an orbit could be calculated from the mass ($m$) of the electron times the square of the charge ($e$) of the electron times the radius ($r$) of the orbit – all taken to the square root $()^{1/2}$. This calculation was useful because the mass ($m$) and charge ($e$) of the electron had

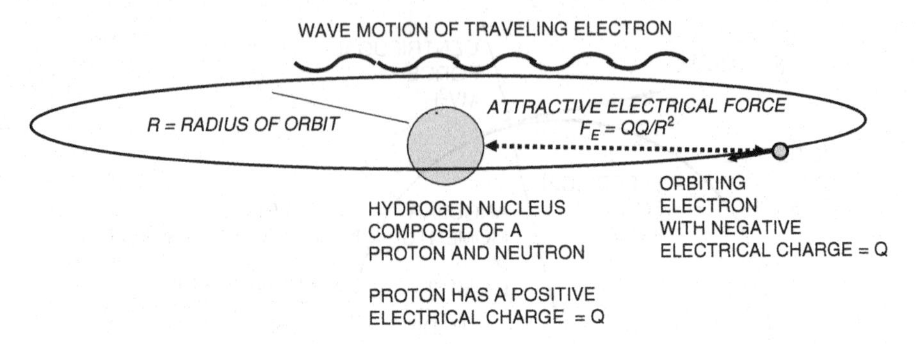


**Fig. 12.14** Quantized model of a hydrogen atom

been measured, as well as the electrical constant ($k$). In principle, the angular momentum could be calculated if one could measure the radius of the orbit. This had not yet been possible in Bohr's time. But there was a clever trick yet by Bohr and a very clever idea. Bohr's clever idea was to quantize the angular momentum, and it was such a good idea that it would later win him a Nobel Prize in Physics.

His reasoning behind this idea was that a quantized angular momentum would give some stable orbits for electrons as they would then move in the form of a wave, instead of a particle. One can envision this in Fig. 12.14.

If the electron moved as a wave, then the number of crests in the circumference of an orbit must divide into that circumference length exactly by some integer $n$. Then as the electron wave form goes around and around in the orbit, the wave form reinforces itself and does not cancel itself out. One of the distinct properties of wave forms is that they are additive, waves add to themselves. Thus Bohr's conjecture about quantizing the angular momentum was a way of asserting that an electron must travel in orbit as a wave and not as a particle. Thus, a physical consequence of quantization was that subatomic particles (such as electrons) must have characteristics both as particles and as waves. For classical mechanics, this was a paradox; but for the new atomic theory, this was merely a puzzle. The idea of quantization came to Bohr from the prior work of Max Planck and Albert Einstein.

And Bohr's model worked! Bohr next derived the Balmer/Rydberg empirical spectral law (phenomenological law) from his quantized-atom model. The emission of a light particle, photon, occurs in the transition of an electron from higher to lower orbit. This is the set of empirical measurements that Bohr could use to judge whether or not his atomic model was real. Bohr had derived a formula in which the *radius* ($r$) of stable orbits of electrons varied by the square of an *integer ($n$)* times a physical constant product ($h^2/mke^2$). But *the Balmer spectral series also varied by an integer ($n$)!*

> There were integers ($n$) in Bohr's theory and integers ($n$) in experimental reality (Balmer's series).

This was an amazing coincidence of the appearance of integers in both experiment and theory. Or was this real theory? It was both amazing and real theory!

Bohr's derivation of the Rydberg formula from his quantized-angular-momentum model is shown in the appendix to this chapter. It is quantitative. But whether or not one chooses to read the math, the important philosophical point here is that Bohr needed to change the *description of physical reality* to get his quantum model to work – to match theory with experiment, to ground a theoretical model in empiricism (the empiricism of Rydberg's formula).

> The change in the description of physical reality at the atomic level meant that electrons traveled as waves (although they interacted with matter as particles, a wave/particle duality of atomic matter).

Thus, quantum mechanics began with (1) the idea of "discreteness" in spectral light emission from atoms (Planck), (2) the idea of "discreteness" in the interaction of light with matter (Einstein), and (3) the stability of atomic orbits from the idea of discreteness of the angular momentum of electron in orbits in the atom (Bohr). For the quantitative reader, Bohr's derivation of the Rydberg formula follows (otherwise, please skip to the next section).

## *Bohr's Model*

In quantitative expression, Bohr's derivation was in the following steps:

1. Bohr set the attractive centripetal force on the electron equal to the inertial outward centrifugal force to keep the electron balanced in orbit.
2. Bohr then quantized the angular momentum of the electron in orbit in integer multiples of Planck's constant. This means that only electron orbits with quanta of Planck's constant were stable orbits in which the electron could exist without radiating away energy.
3. Bohr then inserted the quantized angular momentum expression for the stable orbit radii into the potential energy of an electron in orbit due to the electric attractive force between positive nucleus and negative electron. This gave an expression for the energy of the stable orbits in terms of integer multiples of Planck's quantum constant, with energy proportionate to $1/h^2n^2$.
4. Next, Bohr would derive the Rydberg formula for the spectral lines of hydrogen as the emission of photons with energy $E_{photon}$ equal to quantized frequency $f_{photon}: E_{photon} = hf_{photon}$.
5. Bohr defined the Rydberg constant $R$ as $R = mk^2m^4/2h^2$. Then the formula for the energy of the electron in a stable $n$-th orbit simplifies as $E_{electron} = R/n^2$. This says that each stable orbit of a quantized electron orbit is proportional to the inverse square of the integer $n$ of the orbit.
6. Bohr then could compute the energy difference of two stable electron orbits, say in the $n$-th orbit and in the $n+1$ orbit as $E_n = R/n^2$ and as $E_{n+1} = R/(n+1)^2$.
7. Bohr then assumed that when an electron fell from a higher $(n+1)$ orbit to a lower $n$ orbit, the energy difference $(E_{n+1} - E_n)$ would be given up to the energy $(E_{photon} = hf_{photon})$ of the emitted photon in the spectral emission. So Bohr set the

two energy expressions as equal to the energy loss of the electron transition from higher to lower orbit and the energy of the emitted photon:

$$\left(E_{n+1} - E_n\right) = E_{\text{photon}}$$

$$R / (n+1)^2 - R / n^2 = hf_{\text{photon}}$$

And this derived the Rydberg formula for the frequency f of spectral lines in the hydrogen spectrum: $f_{\text{photon}} = R/h\left((n+1)^2 - n^2\right)$

And in the experiments, the pattern of the frequency lines in the hydrogen spectrum do differ by the inverse difference of squared integers. Remarkable! Bohr was able to quantitatively model the spectral emission phenomena of real hydrogen atoms – using Newtonian mechanics and two odd assumptions:

1. Quantization of photon frequencies.
2. Quantization of angular momentum of electrons in stable atomic orbits.

Bohr's model of the atom consisted of a quantization of electron orbits as discrete orbits. A transition by an electron from a higher energy orbit to a lower energy orbit resulted in the emission of a photon at a light frequency, *f*, just equal to the energy difference (d*E*) of the orbits divided by Planck's constant: $f = dE/h$. Bohr's atomic theory just matched experiment when he derived the Rydberg formula from the model.

Niels Bohr (1885–1962) was born in Denmark. As a young man he went to England as an undergraduate at Trinity College, Cambridge. He returned to Denmark and received a doctorate from Copenhagen University in 1911. He returned to England, did postdoctoral research under Ernest Rutherford at the University of Manchester. There Bohr learned of Rutherford's experiments and devoted himself to theoretically modeling the structure of the atom. In 1913, Bohr published his model of the atom, for which he received the Noble Prize in Physics in 1922.[2]

## Quantum Mechanical Theory

Bohr had successfully modeled Rutherford's atom. Later, other physicists would begin to explain these quantum assumptions with a new physical assumption. At the atomic level, all matter travels as waves and interacts as particles – the wave–particle duality of matter at the atomic level. Bohr's model was presented in an incomplete theory. It was correct in representing phenomena, a phenomenological model. But it was not able to predict all the observable phenomena of atomic spectra. It worked for the hydrogen spectral emission pattern but not for the helium spectral emission. More theory development in quantum mechanics was needed. And this continued between 1925 and 1930. Then there was a completely new theory for atomic-scale phenomena, quantum mechanics. And quantum mechanics was not only a new complementary theory to classical mechanics, but it was also a paradigm shift in mechanism.

Paradigm shifts are rare in science. Most science is performed within a paradigm, which Kuhn had called "normal science." Normal science provides the details of nature when the correct way, the paradigm, of perceiving and understanding nature has been constructed, for example, Newtonian theory, quantum theory, Special relativity, etc. But paradigm shifts may be necessary when new realms of nature are discovered and explored by science and create "discontinuities" in the progress of science.

In 1927 at Bell Labs, Clinton Davisson and Lester Germer experimentally demonstrated that electrons did travel as waves, shooting electrons into a crystalline nickel target and observing diffraction patterns typical of wave motion. Louis de Broglie extended the Einstein–Bohr conjectures about light and electrons to all atomic-level matter, for which he received the Nobel Prize in 1929. In 1925, Heisenberg, Born, and Jordan would create quantum theory in a matrix formulation. In 1926, Erwin Schrödinger would create quantum theory in a wave differential formulation.[3]

Davison and Germer (http://en.wikipedia.org) 2008

Erwin Schrödinger

Louis de Broglie

## Scientific Objects and Models

We recall that the logical positivists assumed that theory was directly induced from experiments on objects. But we have seen instances in the history of science that contradict this assumption. We have just seen in this history of quantum theory that Bohr's model of the quantized atom stimulated the development of quantum theory, while the experiments of blackbody radiance and spectra of atoms stimulated phenomenological laws, not theory.

And earlier we saw that it was Newton's gravitational model of the Copernican solar system that stimulated Newton's mechanics theory, whereas Brahe's astronomical measurements stimulated the phenomenological law of Kepler (elliptical orbits). And we also saw that Watson and Crick's chemical and geometrical model of DNA stimulated molecular biology theory.

> Significant events in the history of science about theory construction have indicated that *modeling* is the direct linkage between the *empirical studies* of a natural object and *construction and validation of a scientific theory* about the object.

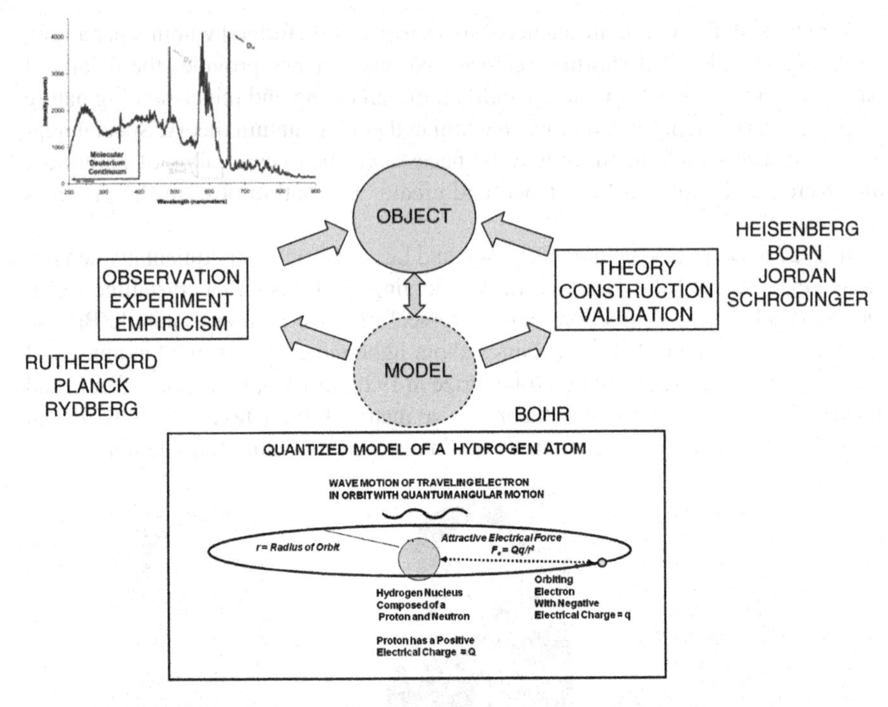


**Fig. 12.15** Modeling as the intermediate technique between experiment and theory

We picture the relationship between objects and their model with regard to empiricism and theory, in Fig. 12.15. The empiricism of the object interacts with the construction of its theoretical model as an intermediary in connecting empiricism to theory. In that figure, we show again how the atomic spectrum of the hydrogen atom (Rydberg's phenomenological law) was important to Bohr in his construction and validation of his quantum model of Rutherford's atom. Next, building upon Bohr's model, Heisenberg, Born, and Jordan constructed a matrix version of quantum theory, which was soon followed by Schrodinger's wave version of quantum theory.

Models construct a formal abstraction and generalization (theoretical model) of the object to derive the observed properties and relationships and phenomenological laws of natural objects

## Summary

1. Paradigm shifts may be necessary when new realms of nature are discovered and explored by science and create "discontinuities" in the progress of science.
2. In science, one can directly measure the properties of a *model* in *experiments*.

3. Theory is indirectly validated through the predictions/prescriptions from a model of a natural object.
4. The difference between theory and models is that a model describes a specific object within a theory; whereas a theory provides a semantically specialized language for expressing a model.

## Notes

[1]A biography of Heisenberg is Cassidy (1993).
[2]There are several biographies of Bohr, including Pais (1991).
[3]A detailed history of the development of quantum theory is Mehar and Helmut (1982). The classic and authoritative exposition of quantum theory is Dirac (1981).

# Chapter 13
# Models in the Social Sciences

## Introduction

We ended the last chapter with a strong methodological statement: in science, one can directly measure the properties of a model in experiments. Models rather than theory are directly experimentally validated. Predictions/prescriptions from a model of a natural object can be compared to experiments. Theory is indirectly validated by experiments, through the models which are constructed within the theory. We saw that quantum theory would be indirectly validated through the Bohr model of the atom (from which the empirical formula of Rydberg's phenomenological law could be directly derived).

> Theory is grounded in empiricism, grounded and indirectly validated by its theoretical models of natural objects.

Now let us apply this methodological point to the social sciences. As in all other science, social science models provide the direct way to empirically and normatively ground theory, intermediate to the grounding of social theory. Let us next examine the kinds of models that are used in the social sciences to explore how they can be used to ground social theory, to create *empirically and normatively grounded social science theory*. Particularly important types of social science models are: (1) topological connective, (2) topological flow, (3) dynamic systems, and (4) optimization models.

1. Topological models qualitatively formalize the relationships between properties of natural objects.
2. Flow-topological models quantitatively formalize the relationships between properties of natural objects.
3. Dynamic models quantitatively formalize the temporal relationships between properties of natural objects.
4. Optimization models quantitatively formalize the control over relationships between properties of natural objects.

F. Betz, *Managing Science*, Innovation, Technology, and Knowledge Management 9,
DOI 10.1007/978-1-4419-7488-4_13, 

## Use of Social Science Models

Social science models find practical use in business for decisions and in government for policy analysis and policy.

In businesses, models are used to plan and control the operations of production and distribution. For example, a manufacturer may use sales models to control the production of a proper mix of its product brands and models and to schedule production, in order to minimize product inventory costs. A business may also use marketing models to plan where to build manufacturing facilities regionally in order to minimize transportation costs of product deliveries nationally and globally.

In government, models are used for policy analysis and implementation. For example, public health organizations can use models of infectious disease outbreaks to control epidemics. One can sketch the policy application of social science models in Fig. 13.1

When policies are formulated to address a societal problem, a societal model can assist in anticipating intended and unintended consequences of a proposed policy. Such policy analysis can assist in choosing an effective policy. After policies are implemented, monitoring the effectiveness of the policy can be assisted by proper measurements on how the policy is working, so that a societal model can identify the factors creating the intended or unintended consequences of the policy. A policy then may be revised to improve the policy.

As an illustration, Jessica Twentyman reported in 2010 upon the impact of information technology data and models on public health in Rwanada: "Huge stacks of paper-based medical reports from local health centers are being input into a computer at Nyamata Hospital, miles south of Kigali, Rwanda's capital.... The reports contain a wealth of valuable information that is helping medical staff tackle some of the district's most pressing health challenges, from monitoring the health of expectant mothers to ensuring that HIV patients take their antiretroviral drugs correctly.... It is part of an ambitious, \$32 m e-health initiative launched in September 2009 and supported in part by the Rockefeller Foundation" (Twentyman 2010).

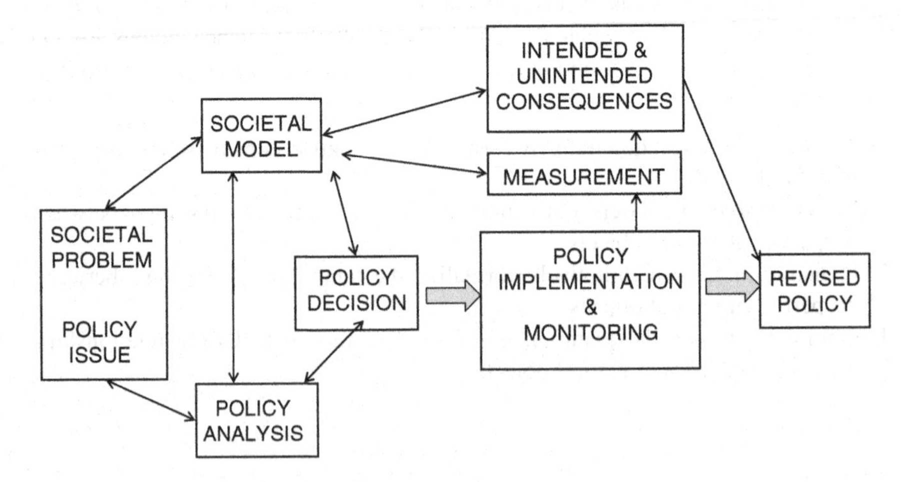


**Fig. 13.1** Role of "ideal-type" societal models in policy

The coordinator of e-health at Rwanda's Ministry of Health explained the government use: "By converting information into digital formats and implementing new systems, we can get reports and monitoring tools into the hands of frontline health workers, enabling them to respond more efficiently and effectively to issues as they arise" (Twentyman 2010, p. 3).

Data needs to be input into appropriate models, in order to provide tools for monitoring and implementation of effective public policies.

Social science models are used within a context of applications.

## Topological and Flow Models

Topological models qualitatively formalize the relationships between properties of natural objects. The term "topology" comes from the mathematical topic. A topological graph is a technique for connecting parts of a system, and consists of "nodes" which depict a component part of a system and the line connections between nodes, which interact in the system.

As an example, we return to an earlier illustration about energy systems, shown again in Fig. 13.2.

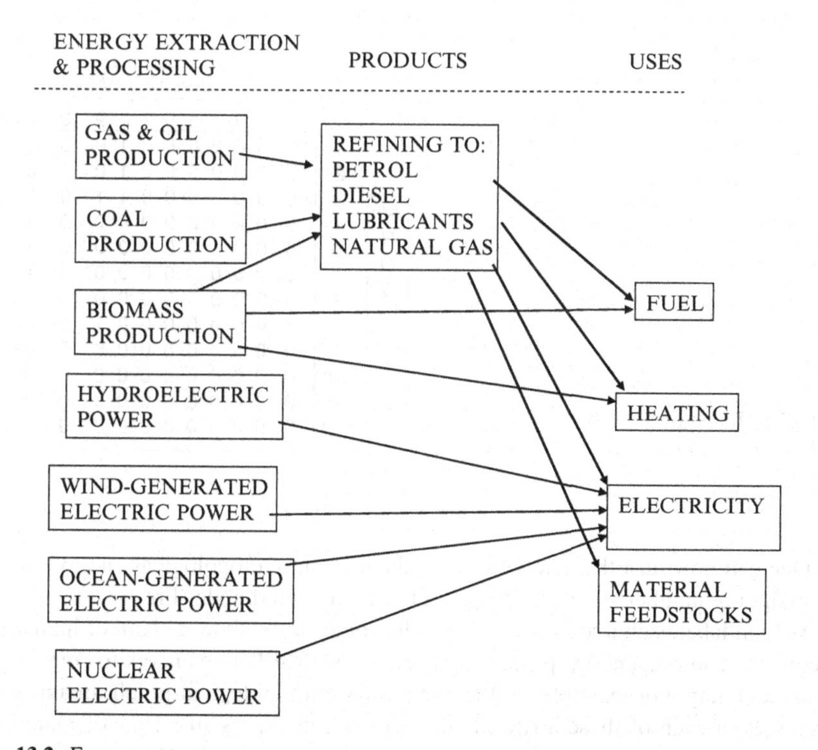


**Fig. 13.2** Energy systems

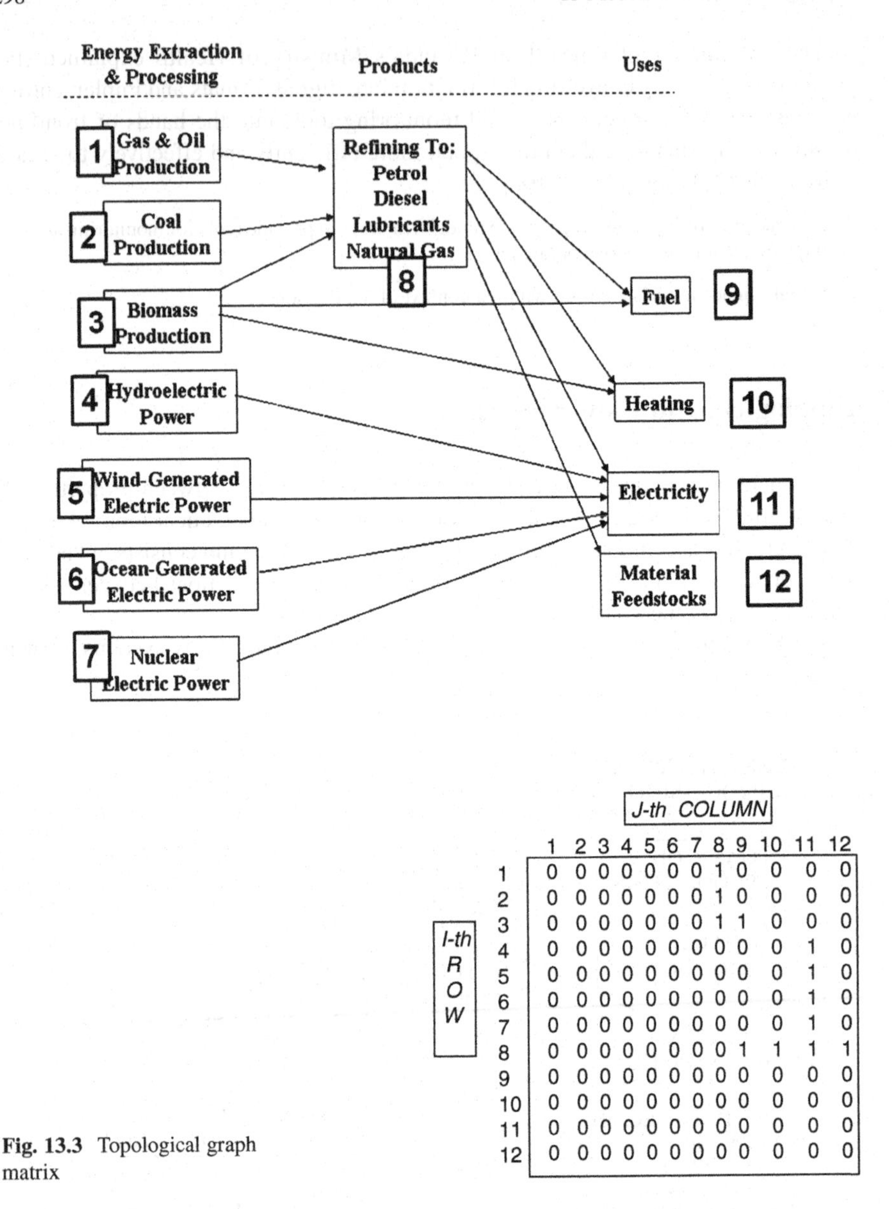


| | 1 | 2 | 3 | 4 | 5 | 6 | 7 | 8 | 9 | 10 | 11 | 12 |
|---|---|---|---|---|---|---|---|---|---|---|---|---|
| 1 | 0 | 0 | 0 | 0 | 0 | 0 | 0 | 1 | 0 | 0 | 0 | 0 |
| 2 | 0 | 0 | 0 | 0 | 0 | 0 | 0 | 1 | 0 | 0 | 0 | 0 |
| 3 | 0 | 0 | 0 | 0 | 0 | 0 | 0 | 1 | 1 | 0 | 0 | 0 |
| 4 | 0 | 0 | 0 | 0 | 0 | 0 | 0 | 0 | 0 | 0 | 1 | 0 |
| 5 | 0 | 0 | 0 | 0 | 0 | 0 | 0 | 0 | 0 | 0 | 1 | 0 |
| 6 | 0 | 0 | 0 | 0 | 0 | 0 | 0 | 0 | 0 | 0 | 1 | 0 |
| 7 | 0 | 0 | 0 | 0 | 0 | 0 | 0 | 0 | 0 | 0 | 1 | 0 |
| 8 | 0 | 0 | 0 | 0 | 0 | 0 | 0 | 0 | 1 | 1 | 1 | 1 |
| 9 | 0 | 0 | 0 | 0 | 0 | 0 | 0 | 0 | 0 | 0 | 0 | 0 |
| 10 | 0 | 0 | 0 | 0 | 0 | 0 | 0 | 0 | 0 | 0 | 0 | 0 |
| 11 | 0 | 0 | 0 | 0 | 0 | 0 | 0 | 0 | 0 | 0 | 0 | 0 |
| 12 | 0 | 0 | 0 | 0 | 0 | 0 | 0 | 0 | 0 | 0 | 0 | 0 |

**Fig. 13.3** Topological graph matrix

One can now turn this energy system diagram into a topological energy model by assigning numbers to each component as a numbered node (Fig. 13.3).

We can label each node (1–12). Then the connections show a chain of industrial sectors in acquiring energy, processing energy, and distributing energy to consumers in an economy. For example, in the energy system topological graph, quantity of resources in each of these arrowed connections can express the flow of materials and energy in the economic system.

## Matrices and Topological Graphs

Mathematically, any topological graph can be translated into a matrix model. This is done by constructing a two-by-two matrix ($M_{ij}$) with each node labeled as *j*th column along the top of the matrix and also each node labeled as an *i*th row down the side of the matrix (where *i* and *j* run from number of the nodes in the topological graph). Then nonzero entries into the matrix $M_{ij}$ occur only where there is a connection between two nodes *i* and *j*.

For example, the topological graph of Fig. 13.3 can be converted into a 12 × 12 element matrix Fig. 13.4. The 12 nodes of the graph form 12 columns (*j*) along the top of the matrix and also 12 rows (*i*) down the side of the matrix. Where there is a connection between two nodes *i* and *j*, then an entry of "one" is listed in box labeled as (*i,j*). Where there is no connection between nodes *i* and *j*, then an entry of "zero" is listed in that box labeled as (*i,j*).

We read this matrix ($M_{ij}$) as:

When the *i*th column indexed node connects to a *j*th row indexed row then $m_{ij} = 1$
When there is no connection between an *i*th and *j*th nodes, then $m_{ij}$ equals 0

This topological matrix of the energy topological graph then shows:

Node 1 and node 2 each only connect into node 8
Node 3 connects into node 8 and into node 9
Node 4, node 5, node 6, and node 7 all only connect into node11
Node 8 connects into nodes 9 and 10 and 11 and 12
Nodes 9–12 do not connect into further nodes

> In the mathematical topic of matrices, a matrix is an array of numbers. The elements *m* of a matrix *M* are indexed (named) in rows and columns of the matrix. A matrix can be of *n*-dimensions. A two-dimensional matrix ($n = 2$) is a two-dimensional array of numbers ($m_{ij}$), which each number ($m_{ij}$) in the matrix is named (indexed) by a row index (*i*) and a column index (*j*) (Fig. 13.4).

In a similar way, topological flow graphs can be converted into matrix flow models by entering the quantity ($q_{ij}$) of a flow between the *i*th and *j*th nodes. The usefulness of matrix flow models is that such matrices ($m_{ij}$) can be mathematically manipulated as matrices, such as matrix addition or subtraction or matrix algebraic equations. For example, let $M + N = P$, then the matrix elements ($p_{ij}$) of *P* can be calculated from the matrix elements $m_{ij}$ of *M* and $n_{ij}$ of *N* by adding these elements: $p_{ij} = m_{ij} + n_{ij}$.

J-TH COLUMN

I-th ROW

| | | | |
|---|---|---|---|
| M11 | M12 | M13 | M14 |
| M21 | M22 | M23 | M24 |
| M31 | M32 | M33 | M34 |
| M41 | M42 | M43 | M44 |

**Fig. 13.4** Two-dimensional matrix $M_{ij}$

So algebraic models can be constructed from topological graphs and expressed as matrix algebraic equations. Matrices can also be labeled as time dependent to obtain temporal matrix algebraic equations.

## *Illustration: Input–Output Economic Model*

As we emphasized, the difference between theory and models is that a model describes a specific object within a theory; whereas a theory provides semantically-specialized language for expressing a model. We can illustrate a social science matrix-algebraic flow-model of a national economy. As we noted, one of the sub-systems in the theoretical taxonomy of societal systems is the economic system. A specific societal object that can exist within an economic system is a "national economy."

Wassily Leontief formulated a model of any national economy as a topological flow model of all the products from industrial sectors produced or consumed in the economy. He described the total production ($P_i$) from an economic sector (such as manufacturing or agriculture) and traces that quantity of production $P_i$ as it is distributed into the economy for consumers ($C_i$) or for other industrial sectors ($X_j$) or exported to other countries ($E_j$). Then a Leontief input–output matrix equation describing the economy as sectors can be written as:

$$P_i = \Sigma_J \left(C_i + X_{ij} + E_{ij}\right).$$

This is read as the quantity of production $P_i$ in the *i*th economic sector is distributed to a summation of all (a) the consumers of the *i*th products and (b) the industrial consumption and (c) the exports. The summation ($\Sigma_J$) taken over all other *j*th economic sectors and all the other *j*th countries. (In mathematical notation, the quantities of *P* and *C* are vectors and the quantities of *X* and *E* are matrices.)

What is interesting about this model is that it is a *system model of an economy* – with *inputs* to the economy by production ($P_i$) in *i*th sectors of the economy and *outputs* from the production into the economic sectors of consumer consumption ($C_i$) and the other *j*th sectors of industrial consumption ($X_{ij}$) and exports ($E_{ij}$) to other *j*th nations.

Leontief's input–output model of a national economy is a *topological flow model.*

For example, one can apply Leontief's model to measure direct economic impacts of technological innovation upon an industrial sector (*i*) in the following way (Fig. 13.5).

We recall that national innovation systems invest in science and technology funding in Research and Development (R&D) budgets. Successful investments from R&D budgets eventually fund the innovation of high-tech products and services sold nationally and internationally. Very successful innovations (such as the Internet) stimulate over time dramatic economic development. The long-term economic impacts of such innovation can be traced in the changes of product

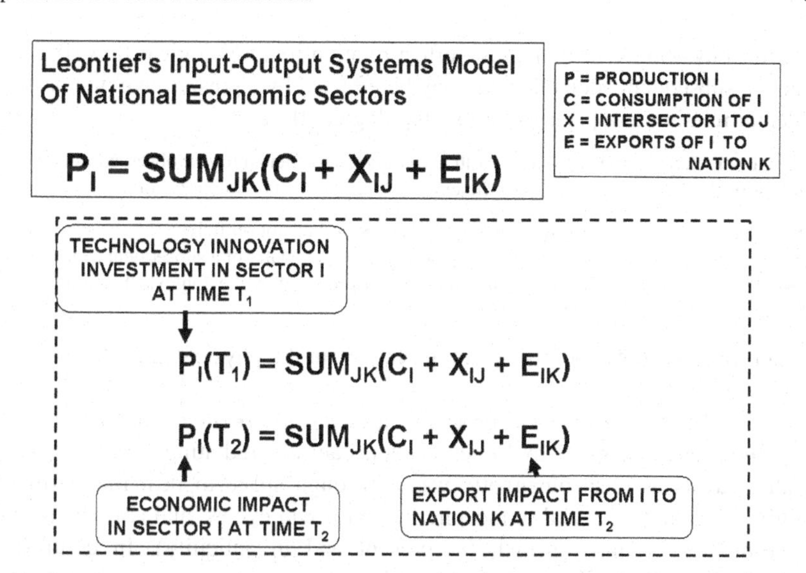


**Fig. 13.5** Measuring the direct economic impact of R&D funding

volumes of an impacted *I*th industrial sector and in the changes due to such innovation in exports ($E_{IK}$).

Wassily Leontief (1905–1999) was born in St. Petersberg, Russia. He entered the University of Leingrad in 1921 and earned a master's degree in economics. In 1925, he left Russia for political reasons. In Germany, he entered the University of Berlin and obtained a doctorate in economics in 1925. In 1931, he went to the United States, and began teaching at Harvard University in1932. In 1949, Leontieff modeled data on the US economy, which divided the economy into 500 sectors. In 1973, he won a Nobel Prize in Economics for his model. In 1975, he joined New York University.

Wassily Leontief (http://en.wikipedia.org, Wassily Leontief, 2008)

## Explanation in Social Science Models

Earlier we emphasized that one of the important methodological differences between the physical and social sciences was in the modality of their phenomenological laws. In physical science, a subsequent physical event is causally affected by a prior physical event in a cause–effect relation in a topological flow. In social science, a subsequent societal event may be affected by a prescriptive, thematic, or accidental relation to a prior event in a topological flow.

The causal relationship in physics is both necessary and sufficient for the observed existence of a prior physical event to explain the existence later of a subsequent event.

A prescriptive relationship in sociology is necessary but not sufficient for the observed existence of the prior societal event to explain the existence later of a subsequent event; and sufficiency in a social science explanation must be provided by the context (the setting) of the observed societal events.

One can illustrate these points by comparing again the physical and social models of Figs. 10.1 and 10.2.

In Fig. 10.1, all the physical processes are *causally related.* The sun burns by fusing hydrogen nuclei to hydrogen nuclei, causally radiating light: (1) which causally travels as electromagnetic waves through space to heat the Earth and causally drives hydrological, atmospheric, and ocean cycles, and (2) which causally stimulates biomass growth (at present and in antiquity). In contrast in Fig. 10.2, all the flows of energy from production to consumption are *prescriptive relations*, wherein production of energy is necessary to explain the energy available for consumption, but not sufficient. The contexts of production and distribution are required to fully explain whether a source of energy delivered energy to a consumer and how much.

In *causal relationships* (causal explanation), the explanation is *independent of context*

In *prescriptive relationships* (prescriptive explanation) the explanation is *dependent upon context.*

For all social science explanation using social science models, the *context of the model* must be added for *sufficiency of explanation.*

## *Illustration: Failure in the Context of Oil Drilling*

An example of the importance of the *context of a prescriptive model in its application* can be seen in oil-well accident of British Petroleum in the Caribbean Sea in April 2010. Although oil is created by ancient natural physical process, oil reservoirs are found by empirical search and then drilled through social processes, as engineering technology used by commercial firms in economic systems. The seismographic model BP used for prescriptively located the drilling site was accurate (necessary but insufficient to find oil). But the *context* of the model's application, *the drilling operation*, was a *flawed management operation.*

This contextual failure created a major oil spill, even with an accurate, prescriptive oil-locating model

BP management failed to have properly prepared safety engineering for deep sea ocean drilling. It was a failure, from a methodology perspective, to properly integrate engineering knowledge and management knowledge – a multidisciplinary failure.

Deepwater Horizon drilling rig explosion (http://en.wikipedia.org, BP Oil Spill, 2010)

The drilling technology was a floating platform tethered to the ocean floor down at 1,500 m (5,000 ft). The drill reached the natural petroleum reservoir filled with gas and oil. Then a cement plug was poured to cap the well, but the technology of the cement capping was flawed. On April 20 at 9:45 p.m. central time in the US methane gas in an undersea reservoir under high pressure broke through the cement plug, shot up the well pipe and exploded at the surface in a fire ball. Eleven workers on the platform died. One hundred fifteen survived and were rescued. Following the fire, oil began gushing from the well head at the bottom of the sea in millions of barrels of oil. The drilling rig was built by Hyundai Heavy Industries, owned by Transocean, and leased to British Petroleum. The cement capping was done by Halliburton Energy Services. (Jervis 2010)

The societal event was the oil spill. The well-head was covered by the technology of a "blow-out preventer" which didn't work. Individual actions in the spill consisted of BP's leasing of a rig which had insufficient technology to prevent a methane gas eruption, and Halliburton's ineffective cement cap. The explanation of the societal event requires a relation of before drilling and after drilling sequence of an "accident." With the "accident" as not necessary but sufficient to explain the spill. The "context" of the accident was poor technology and technique in the safety of the drilling operations. The drilling-for-oil even was itself a social relation of "prescription" – if global society wants additional oil as an energy resources, it must (necessarily) drill in deep sea depths. The tragic societal event continued, as the BP engineers were unable to stop the gushing of oil from the sea bottom.

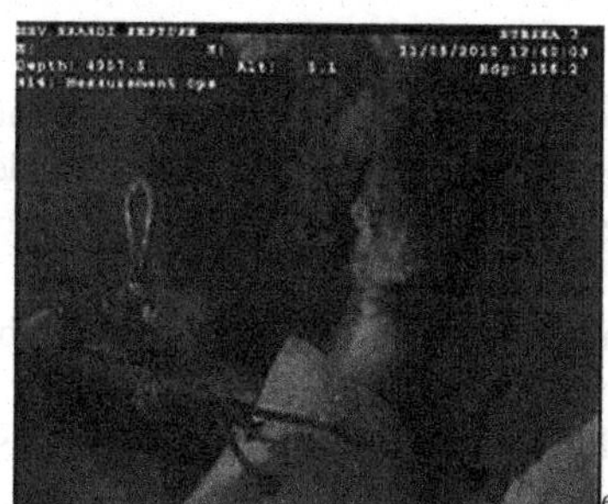

Photo of oil spill live video, May 11, 2010 (http://en.wikipedia.org, BP Oil Spill, 2010)

Two days later, a slick of oil floated up to spread around the damaged rig. BP tried to use technologies of remote operated vehicles (ROVs) to close valves on the spill containment barrel over the well head, but the containment apparatus didn't work (technical failure). Oil continued to gush from the well head. Two more engineering attempts were made to stop the flow of oil, dropping a bigger cap over the well head and then pumping mud down the pipe to plug the well. But these did not work and oil continued to well up all of May and into June. Finally in June, part of the oil flow was captured, when the broken pipe was sawn off and a new pipe was clamped upon the well head. However, not until two new relief wells were finished drilled in August was the spill expected to be fully contained.

Meanwhile environmental catastrophe and economic damage was experienced on the beaches of Louisiana and by the fishing industry in the Gulf of Mexico.

The oil slick as seen from space by NASA's Terra satellite on May 24, 2010. (http://en.wikipedia.org, BP Oil Spill, 2010)

On June 16, British Petroleum established an escrow fund, as Jackie Calmes and Helen Cooper reported: "Four days of intense negotiations between the White House and BP Lawyes allowed President Obama to announce Wednesday that the oil giant would create a $20 billion fund to pay damage claims to thousands of fishermen and others along the Gulf Coast" (Calmes and Cooper).

One sees in this illustration that social science models, such an economic model for energy production can only use "prescription" (necessary and not-sufficient) in the explanatory relations among the topological connections in its model.

All social science models must also be contextually understood to explain how such prescriptive models worked in real societies and their histories.

## Modeling Organizational Systems

In addition to the use of topological flow models in the economics discipline, another important application is in the management science discipline to model organizations. Organizations are complicated enough to require not one but three planes of topological flows, which we will next review.

Earlier we had noted a systems view of a business organization as depicted by Michael Porter, shown again now as Fig. 13.6.

But earlier than Porter, Jay Forrester had proposed a "systems dynamics" approach to modeling manufacturing organizations. Forrester proposed two planes of modeling: (1) a transformation plane for producing a product and also (2) a control plane for information controlling the production processes. This was because the materials flow in a production system (transformation plane) needs to be controlled

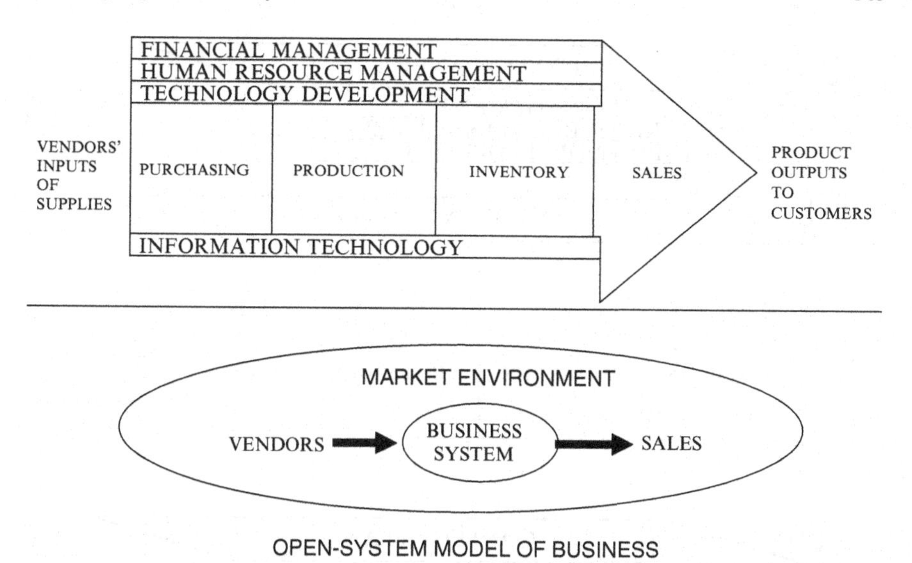


**Fig. 13.6** Michaelporter's value-added (open-system) model of a business enterprise

by information about the production. Jay Wright Forrester (1918) was born in Nebraska, USA. He attended the Massachusetts Institute of Technology, obtaining a bachelor's degreeand a doctorate in electrical engineering. He did research at MIT and in the early 1950s he developed the first electronic computer with a ferroelectric stored memory technology. In 1956, he moved from the engineering school to the business school at MIT. He developed a systems approach to studying the dynamics of manufacturing, called "systems dynamics" (Forrester 1961).

Jay Forrester (http://en.wikipedia.org, Jay Forrester, 2008)

The author combined both Porter's ideas and Forrester's ideas to construct a general form for modeling the operations of a business, as shown in Fig.13.7 (Betz 2001).

Accordingly, any managed system (manufacturing or services) can be modeled on three planes, as shown in Fig. 13.8:

1. Activities of *support* for the transforming activities (overhead activities)
2. Activities of the *functional transformation* (production activities)
3. Activities of *control* of the transforming activities (controlling operations)

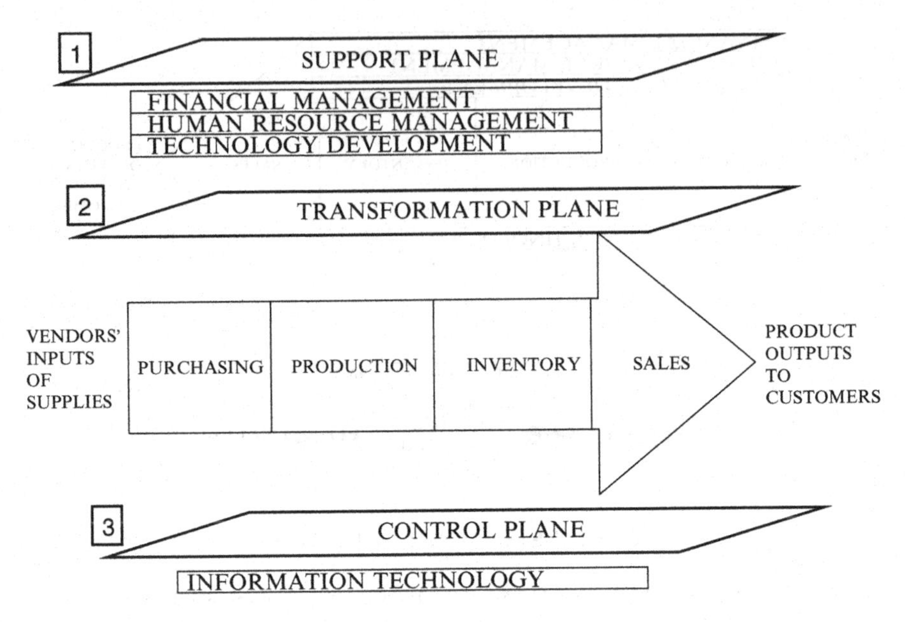


**Fig. 13.7** Three-plane model of organizational systems

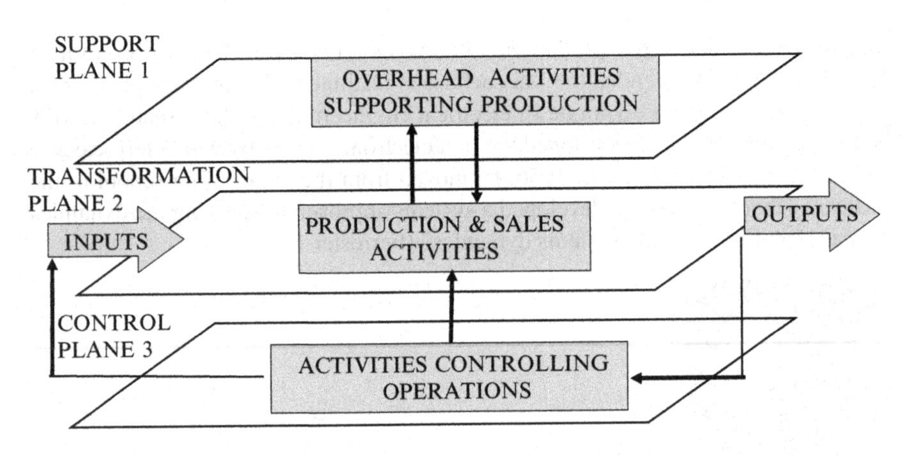


**Fig. 13.8** Model of a managed system

## *Illustration: Modeling a Manufacturing Organization*

As an illustration of manufacturing organization, we can show the production processes in making an automobile in Toyota in 1983 in Fig. 13.9.

Toyota purchased both manufactured components and materials for Toyota's own manufacturing processes from various supplying vendors. Toyota purchased components such as electrical parts, bearings, glass, radiators, batteries, and so on. From other suppliers, Toyota also purchased processed materials, such as steel sheets and rolled steel, nonferrous metal products, oils, paint, and so on. Purchased

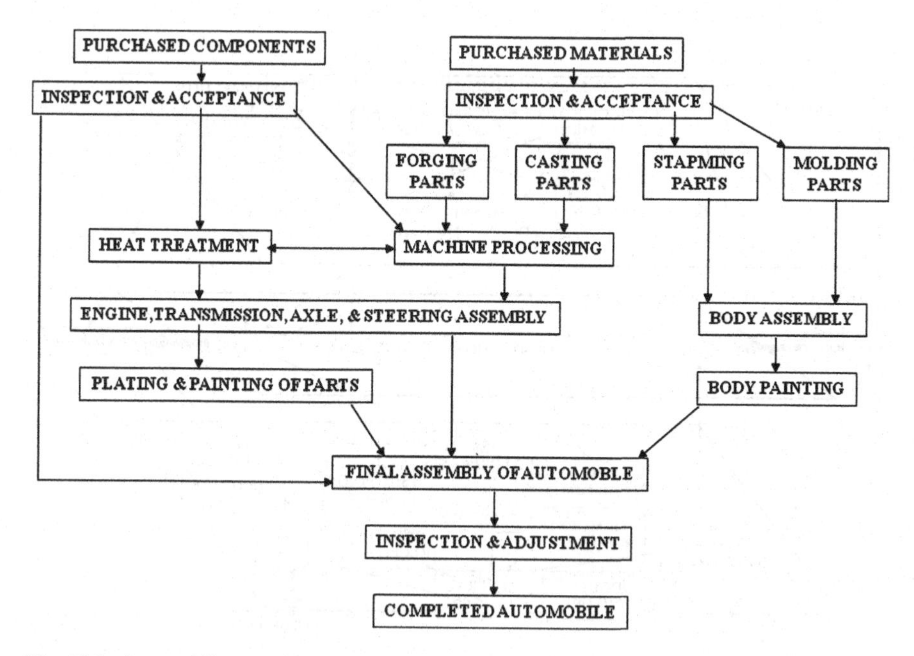


**Fig. 13.9** Automobile manufacturing system flow

components and materials were subjected to acceptance inspection. Next, materials went through various production processes to be formed into parts (such as forging, casting, machining, stamping, plastic molding). In addition, some of the purchased components also went through further processing to be finished as components (such as heat treatment or additional machining).

Materials, components, and parts eventually were all used for three subassembly systems in fabricating the automobile:

1. Power subsystems – engine, axles, transmission, steering assembly, etc.
2. Chassis subsystems – frame, suspension, brakes, etc.
3. Body subsystem – body, seating, windows, doors, etc.

Various plating and painting processes prepared the power and chassis subsystems for final assembly, and the body was painted for final assembly. Then finally the three major fabrication subsystems were attached together as an automobile. After adjustments and inspection, the product emerged as a completed automobile.

All this flow of production, processes can be pasted upon the Transformation Plane of Toyota's manufacturing operations, as indicated in Fig. 13.10.

## Transformation Plane Activities

The continual operations that produce and sell the products (services) of the enterprise are performed as a sequence of value-adding activities, as depicted in the

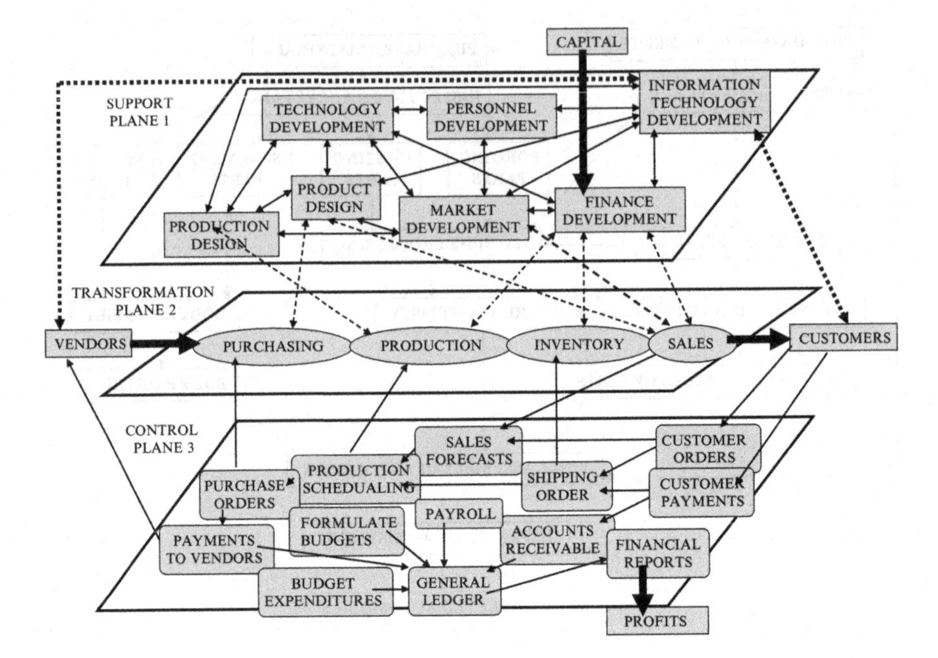


**Fig. 13.10** Three-plane flow model of a manufacturing organization

transformation plane. Products and services are produced and delivered by acquiring appropriate supplies, materials, resources from vendor and then forming and assembling and using these to create the product or service sold to the customer. The activities in the transformation are organized in this sequence; and one can depict the direct value-adding activities in the enterprise value chain as:

- Purchasing
- Production
- Product inventory
- Product sales

Purchasing activities are usually organized in a purchasing department; production in a production department, and sales in a sales and marketing department. Product inventory is stored in product inventory warehouses until shipped to dealers and customers.

## *Support Plane Activities*

Projects in the support functions of a business are necessary to change operations for improvement. The kinds of improvements that are useful are in product, production, personnel, markets, finances, information technology, and other technologies. Accordingly it is useful to have and depict the explicit project activities attending to these areas for improvement as listed in on the Support Plane:

- Product design program
- Production design program
- Personnel development program
- Market development program
- Financial development program
- Information technology development program
- Technology development program

Strategic change in operations is planned and implemented in the form of specific projects – design projects, training projects (and programs), product-lines and brand projects, financial analysis projects, information technology projects, technology research and development projects. Since all businesses need some kind of change annually to continue to adapt to the future, some projects in some of these areas will be occurring each year. Long term programs for change are usually organized within engineering departments, marketing departments, and research units and performed as discrete projects by multifunctional teams. New product designs are performed by multifunctional design teams, led by engineers in the engineering department. Production improvement projects are performed by multifunctional teams led by manufacturing engineers in the engineering department. Product-line and brand market analysis projects are performed by multifunctional teams are led by sales personnel in the marketing department. Cost analysis projects for product-lines and new product-lines are performed by multifunctional teams led by financial personnel in the finance department. Personnel development projects are performed by multifunctional teams led by personnel people in human resources department. Technology development programs are performed in multidisciplinary teams led by scientists and engineers in the research & development (R&D) laboratories.

The arrows within the Support Plane connecting these different areas of development indicate the kinds of informational interactions that occur between multifunctional projects in carrying out the projects for strategic change.

For example new technology developed in R&D projects might be used either in new product design or in new production design and also in market and financial analyses and projections. Projects to improve information technology in the business can impact the processes of technology development, product design, production design, change in markets and financial performance of the business. Information technology development projects can also improve communication and interactions with customers and with vendors.

Financial analyses can look at the strategic implementations for operations change as to the requirements for capital.

### *Control Plane Activities*

The three-plane organization systems model uses both Porter's ideas of value-chain description of a business and also Forrester's ideas of dynamics of business operations. In a Forrester kind of "systems dynamics" model, one needs to relate the flow of information of business operations to the activities that produce products (services) of the business. Within the Control Plane, information systems process information on the operations and report performance of the operations. These systems must include the ability to:

- Receive customer orders
- Receive customer payments
- Create product shipping (or services) orders
- Forecast sales
- Schedule product production (or service delivery)
- Purchase supplies
- Pay vendors
- Formulate budget plans
- Control budge expenditures
- Pay personnel
- Control customer payables
- Account for finances
- Produce financial reports

The directional arrows depict where the information system on the Control Plane directly impacts the control of operational activities on the Transformation Plane. Customer orders control shipping orders and sales forecasts. Shipping orders control inventory depletion and co-control production scheduling. Actual sales control the sales forecasts, which co-control production scheduling, along with inventory depletion. Production scheduling controls the production kind and rate and build-up of inventory. Purchasing orders control the kinds and rates of purchases from vendors and payments to vendors.

The General Ledger records customer accounts receipts and receivables, payments to vendors, payroll and budget expenditures. Financial reports (daily, quarterly, and yearly) summarize financial performance of operations and calculate profits. Information systems record and assist in the control of the performance of operations.

## Control Models: Dynamics and Optimization

As we saw, topological flow models can be converted algebraic matrix equations. In such form, then, control of modeled systems can be mathematically assisted by dynamical calculations (dynamic models) and by optimizing decisions based upon the model (optimization models). Dynamic models quantitatively formalize the temporal relationships between properties of natural objects. Optimization models quantitatively formalize the control over relationships between properties of natural objects.

There are many kinds of these models used for control of processes and procedures; and there are several social science areas in which these kinds of models are constructed, such as in: operations research, industrial engineering, communications, computer science, econometrics, etc. All the different kinds and uses of dynamic and optimization models is too large a topic to review for this book. The interested reader must be referred to the many books in each area.

The point to be appreciated is that most systems and control models are constructed on top of a topological-flow model architecture.

## Summary

1. Topological graphs can be used to depict the components and connection of a social science object.
2. Topological graphs can become flow graphs when activities flow in the connections between system components.
3. Matrix models of topological graphs can be constructed by listing the components along the top and down the side of a matrix and entering the flows between components within the matrix.
4. Organizational systems require further connections between three planes of matrix flow-models, as support, transformation, and control planes.

# Chapter 14
# Multidisciplinary Research

## Introduction

We have focused upon disciplinary science, since disciplines are the primary organizations of scientific communities. But not all research is disciplinary in boundary. When science is connected to technology, then research often crosses disciplinary boundaries, becoming interdisciplinary research or multidisciplinary research. The reason for this is that disciplines are intellectual boundaries but not nature's boundary. Also, technologies have more than one science base and the need by industry for simultaneous progress in both science and technology, cutting-edge research now is often multidisciplinary. Multidisciplinary research involves researchers from different disciplines working together in multidisciplinary research centers.

For this reason, a modern research university now has both academic departments and multidisciplinary research centers. A research center (1) encourages an integrated strategy for conceiving research projects, (2) facilitates the obtaining of research funding to support research projects, and (3) targets progress in science and technology. To handle the challenges of multidisciplinary research, one needs to understand three issues: (1) how to organize multidisciplinary research into research centers, (2) how to plan multidisciplinary research strategy, and (3) how to manage multidisciplinary research centers.

## *Illustration: MIT Biotechnology Process Engineering Center*

We recall that the biotechnology industry began in 1971 on the basis of the scientific technique of recombinant DNA, invented by Boyer and Cohen. Boyer and Cohen had used the technique on *Escherichia coli* bacteria. This was then used by the new biotech industry to produce their products, making genetic modifications of *E. coli* bacteria. From a scientific perspective, *E. coli* bacteria were fine, easy to culture, and study. But from an industrial technology perspective, *E. coli* bacteria

F. Betz, *Managing Science*, Innovation, Technology, and Knowledge Management 9,
DOI 10.1007/978-1-4419-7488-4_14, © Springer Science+Business Media, LLC 2011

were awkward and difficult producers of therapeutic proteins, the products of the new biotech industry. Therapeutic proteins are used in medicine, and their forms are as a protein assembled by DNA in a cell. The problem with using *E. coli* as protein-producing cells is that they do not secrete out the protein from the cell. They exist as individual animals and use only internally all the proteins they construct. So to get a produced protein from an *E. coli* cell, the biotech industry needed to smash the cell and extract the desired protein from all the debris of the cell – highly inefficient.

Consequently by 1985, the biotech industry wanted to produce therapeutic proteins from a connected-tissue-type cell, a mammalian cell which nature had evolved to secrete out some of its proteins through pores in its cell wall. This was the scientific challenge the biotech industry wanted university scientists to undertake – to enable the efficient production of therapeutic proteins from mammalian cell lines from scientific understanding of the metabolism of mammalian cells in their complex, obscure and untiring operations. For efficient production, a biotech firm had to grow thousands of these cells in the same vat and close together. But when mammalian cells began to grow together densely, they sickened and died. Why?

To address this scientific issue, Massachusetts Institute of Technology (MIT) professors along with biotech scientists proposed a multidisciplinary research center in biotech processing, the Biotechnology Process Engineering Center (BPEC) at the Massachusetts Institute of Technology (MIT). This university research center for biotechnology processing technology was established in 1985 under an NSF/ERC grant. It has successfully operated over 2 decades. (http://web.mit.BPECresearch.html) It had been founded and was then directed by Professor Daniel Wang. In this center, university and industrial researchers had planned basic research for the technologies of producing biotechnology products grown in mammalian cell cultures (Betz 1996a, p. 445).

The scientific understanding of biotech protein production from mammalian cells required researchers from two different disciplines, biology and chemical engineering. Biology would describe and explain the biological nature of mammalian cell metabolism. Chemical engineering would describe and explain the physical processes in the containers growing the cells, and these containers were called "bioreactors." After the cells produced proteins and secreted them outside the cells into the bioreactor brew, then the proteins had to be obtained, separated out. This process was called the "separation processes" in biotech production. And because technology provided clear technical goals for the scientific understanding, the scientific projects and issues could be clearly and explicitly planned as a group of research projects – a center research plan.

The research vision of the BPEC director was to view the production process for biotechnology products as a technology system: a *bioreactor* (which grows the cell cultures) and a *recovery system* (which recovers the desired proteins produced by the cells). This vision was articulated in a research plan which laid out the scientific issues to be studied by the center in the following way.

### Bioreactor Technology Subsystem

Within the bioreactor, the technological performance variables for protein production are

- Number of cells
- Functioning of each cell
- Quantity of the protein produced by the cells
- Quality of the protein produced by the cells

The products of biotechnology cell production are particular proteins that have commercial use in therapeutic medicine. The quantity of protein is a measure of the productivity. The quality of protein depends on being properly constructed and properly folded. Proteins are long chains of organic molecules that fold back in particular configurations. The usefulness of the protein depends not only on having the right molecules link together but also on the ability to fold up into the right pattern. One of the technical bottlenecks in protein production was to get the proper folding.

For the design of these bioreactors (for the production of proteins from mammalian cells), the biotech engineer needed to design reactors that

- Grow a high density of cells
- Formulate the proper medium for their growth
- Provide proper surfaces for mammalian cells to attach to
- Control the operation of the reactor
- Provide for recovery of the proteins that are grown in the cells

### Recovery Technology Subsystem

For the design of the recovery processes of the protein product from the bioreactors, the biotech engineer needed to control the physics and chemistry of protein structures, including

- Protein–protein interactions that avoided aggregation of proteins or mutations of proteins
- Protein–surface interactions that affect aggregation of proteins or absorption of proteins to separation materials
- Protein stability that affected the durability of the protein through the separation processes

## Multidisciplinary Research Strategy

Creative and accurate strategic vision is essential for exploratory activities. Research is an exploratory activity. How does one plan an activity of research exploration? To plan an activity of research exploration, one needs to have a methodological vision of an important research issue. Vision is a cognitive act of intuition, and the cognitive act of intuition is fed by prior experiences. One intuits an idea based upon synthesizing prior ideas and experiences. Successful research vision is based upon an experiential base of prior research. The past basis of experimental experience of nature provides the grounds for and the foundation for research intuition about future research directions. Accordingly, a center director must have a methodological capability of envisioning a research issue and the proper methodological direction of research for scientific and technological progress. A research director must next transform this research vision into a concrete research plan by assembling a team of researchers who can formulate research projects along the methodological direction of the research issue – a strategic research plan for the center.

Methodology in center research lies in envisioning a multi-disciplinary research issue, and organization in center research lies in assembling a proper multi-disciplinary research team with projects about the research issue.

Although a multidisciplinary research center has an overriding vision of its strategic research issue, this is implemented as a portfolio of research projects. As we have noted, all university research is performed as individual research projects – organized around a graduate student's PhD dissertation research. Thus in universities, scientific research proceeds in projects as single scientific events – piecemeal. So too the projects in a university research center are pieces of a larger research vision of the center. The management role of the faculty research-group leaders in the center is to formulate research projects and guide doctoral candidates in performing the research projects. The role of the director is to ensure that all the research projects add up to a "critical mass" of research to advance science and technology, toward the strategic research issue of the center.

The research skills for a center director thus differ from the research skills for a research group leader or research project leader in a center. The research skills of a research project leader and manager in a center is capable in formulating and performing "pieces" of research. The research skills for a center director is envisioning what research "pieces" should be done and how they will fit together as a "critical mass" of scientific and technological progress. Then the center director must tie all the research projects together into a coherent and coordinated research plan of exploration.

A center-director's research vision must assemble a "critical mass" of research projects which lay out and assemble all the necessary pieces of research to integrate and accomplish *simultaneous scientific and technological progress*.

A center director needs to recruit participation of other professors in the university as research project managers. And, the center director may need to supplement existing research expertise in the university by hiring research scientists from outside the university for pieces of the critical mass of research effort.

Also a center director needs to hire an administrative staff for operations of the center. The minimum staff for a small university research center will consist of an administrative assistant to the director, a secretary, and an industrial liaison person.

The administrative assistant must keep records and administer the research funds in the center. The secretary must maintain communications of the center.

Also, a center director needs to recruit industrial sponsors for the center. Industrial researchers need to meet with the center at least twice a year, in a center industrial meeting. The meeting reviews research progress and recommends on the selection of research projects in the center. The industrial group of researchers know in detail what are the real technological problems and challenges. An industrial liaison person hired by the center director can help in maintaining communications with industry between meetings and facilitate the continuing recruitment of industrial sponsorship.

The difference between disciplinary research and multidisciplinary research is this. A purely disciplinary research is defined solely within the research issues of a specialty in a scientific discipline. In contrast, multidisciplinary research is performed both within the issues of a specialty but also within the context of a set of coordinated projects – all together defined by the context of a practical application – strategic science/technology progress.

## *Illustration (Continued): MIT Biotechnology Process Engineering Center*

With the system analysis of biotech industrial production process into a bioreactor and a recovery system, the center could identify the important technological variables for the process. Then, the MIT Biotechnology Process Engineering Center identified the research areas necessary to provide improved knowledge for the technology. For the bioreactor portion of the system, the center had listed the following scientific phenomena as requiring better understanding:

1. Extracellular biological events
   - Nutrition and growth factors of the cells
   - Differentiation factors of the cells
   - Reduction–oxidation conditions in the cellular processes
   - Secreted products of the cells
2. Extracellular physical events
   - Transport phenomena of materials in the medium
   - Hydrodynamics of fluid–gas interactions of the medium
   - Cell–surface interactions
   - Cell–cell interactions
3. Intracellular events and states
   - Genetic expression of the proteins in the cells
   - Folding of the proteins and secretion from the cells

- Glycosylation of the proteins
- Cellular regulation of secretion
- Metabolic flows in the cells and their energetics

We see this is a list of the biological and physical phenomena underlying the cellular activities in the bioreactor.

For the recovery process of the system, the center identified the scientific phenomena which required better understanding:

1. Protein–protein interactions
    - Aggregation of proteins into clumps
    - Mutations of protein structure
2. Protein–surface interactions
    - Aggregation of proteins through denaturation
    - Adsorption of the proteins to surfaces
3. Protein stability
    - Surface interaction
    - Chemical reaction
    - Aggregation in the solvent
    - Stabilization

Accordingly, in 1992, the Center had organized its research into two areas:

- Engineering and scientific principles in therapeutic protein production
- Process engineering and science in therapeutic protein purification

In the area of therapeutic protein production, the research projects were:

- Protein trafficking and posttranslational modifications: glycosylation and folding
- Redox potential
- Pathway analysis
- Intercellular energetics
- Regulation of secretion
- Hydrodynamics: gas sparging
- High-density bioreactor designs
- Substrata morphology for cell attachment
- Expression: transcription factors

The research projects in the second research area of protein separation were:

- Protein adsorption: chromatography and membrane
- Protein aggregation
- In vivo protein folding
- Protein stability in processing, storage, and delivery

In summary, the research at the Center created a critical mass of projects, all focused upon improving scientific understanding of the complete set of technologies which formed the production system for producing therapeutic proteins from mammalian cells.

The first director of the center who wrote the proposal to the center and directed it for the first decade of its research was Daniel Wang, a professor in the Chemical Engineering Department at MIT. He obtained bachelor of science in biology from MIT in 1959 and a master's degree in chemical engineering in 1961. In 1963, he obtained Ph.D. in chemical engineering from the University of Pennsylvania. His research interests are in bioreactor engineering and the production of recombinant proteins. He has been a member of the U.S. National Academy of Engineering since 1986.

Daniel Wang (http://web.mit.edu/cheme/people/faculty/wang.html)

## Organizing and Planning Multidisciplinary Research

In the science–technology interaction, research for technology cannot be planned until after a basic technological invention has occurred. After the basic technology invention, research then can be planned by focusing on the generic technology system, production processes, or underlying physical phenomena. Technology-focused-and-targeted basic research can be planned for:

1. Generic technology systems and subsystems for product systems
2. Generic technology systems and subsystems for production systems
3. Physical phenomena underlying the technology systems and subsystems for product systems and for production systems

Research can be targeted to improve a technology system through improving any aspect of the system:

1. Improved components
2. Improved connections
3. Improved materials
4. Improved power and energy
5. Improved system control
6. Improved system boundary

The physical phenomena underlying a system can be focused on any of the system aspects:

1. Phenomena involved in the system boundary
2. Phenomena underlying components
3. Phenomena underlying connections
4. Phenomena underlying materials
5. Phenomena underlying power and energy
6. Phenomena underlying system control

## University/Industry Cooperation in Multidisciplinary Research

Industrial researchers are very sophisticated about current technology and, in particular, about its problems and locating and identifying the roadblocks to technical progress. However, because of the applied and developmental demands on industrial research, they have limited time and resources to explore ways to leapfrog current technical limitations. On the other hand, academic researchers have time, resources, and students to explore fundamentally new approaches and alternatives that leapfrog technologies.

Together, industry and university researchers can see how to effectively bind a technological system in order to envision a next generation technology (NGT). This boundary is an important judgment, combining judgments (1) on technical progress and research directions which together might produce a major advance and (2) over the domain of industrial organization that such an advance might produce a significant competitive advantage. (Betz 1996b)

To effect industry and university research cooperation on next-generation-technology innovation, a bridging institution is necessary – because industries and universities live in almost completely different universes. The industrial universe is a world of technology, short-term focus, profitability, and markets. In contrast, the university universe is a world of science, long-term view, philanthropy, and students. This is the role of a university research center to bridge the two views, creating a balance between (1) technologically pulled research and scientifically pushed research, (2) short-term and long-term research focus, (3) proprietary and nonproprietary research information, and (4) vocationally relevant education and professionally skilled education. These are the issues inherent in industry and university cooperation. Properly handled, these provide creative tension:

1. Linking technology and science in real-time operation
2. Creating progress in knowledge and developing the technological competitiveness of nations

For an industry to effectively support and use university research centers, several requirements must be met:

1. A diversified firm must have a corporate research lab as well as divisional labs.
2. The corporate research lab must be tightly linked into divisional labs.
3. The company should work with multidisciplinary strategically focused university research centers.
4. Such centers must be large enough to perform a critical mass of research useful to industry.
5. Such centers should link with both the corporate research lab and the divisional labs of a company.
6. Company personnel should participate in both the governance and research of the university center and provide mentorship to students on their research projects.

7. The corporate research lab and divisional labs should be performing joint applied and development projects parallel to the research projects of the university center.
8. The firm should be hiring university graduates from the university center.
9. The firm should join with several other firms and with government in financially supporting the university center.
10. The firm should participate in and support several university centers sufficient to provide it with a long-term competitive edge in strategic technologies.

For an integrated research vision of simultaneous progress in both science and technology, the research vision of a research center should focus upon a next-generation of scientific technology. Both a system analysis of a technology and identification of the important technological variables in the commercialization process are necessary to identify the research issues necessary to provide progress in knowledge for a next-generation technology. Research projects should be formulated for improved understanding of the underlying scientific phenomena that could be used for inventions, design aids, and standard control procedures of the technology system.

The following procedure facilitates the appropriate kinds of cooperation for technology transfer of next generation technology (NGT):

1. Corporate research should strategically plan next-generation technology, and this is best done within an industrial–university–government research consortium.
2. Corporate research should plan NGT products jointly with product development groups in the business divisions.
3. Marketing experiments should be set up and conducted jointly with corporate research and business divisions, with trial products using ideas tested in the consortium experimental prototype testbeds.
4. The CEO team should encourage long-term financial planning focused on NGT.
5. Personnel planning and personnel development are required to transition knowledge bases and skill mixes for NGT.

Consensus on vision and testing technical feasibility in a program of technology transfer between corporate research and strategic business units should pose and address the following questions:

1. What will be the boundaries of the next generation of a technology system?
2. What NGT ideas are technically demonstrable and should now be planned into product strategy, and what NGT ideas must still be technically demonstrated and should not yet be planned into product strategy?
3. What is the pace of technical change, and when should the introduction of products based on NGT be planned?
4. What professional developments and training should be planned for product development groups in order to prepare for NGT products?

## University Science to Industrial Technology

The NSF Engineering Research Centers Program was established in 1984 to fund multidisciplinary research centers at universities which performed both scientific and engineering research – science and technology progress. (http://www.nsf/erc.org) As a ERC program officer, the author observed some lessons about connecting university science to industrial technology, some lessons about what didn't work and some about what worked. (Betz 1996).

And a clear lesson about governmental technology transfer policy was: "what doesn't work." The traditional way of handing off basic research in a *sequential* manner from universities and government laboratories into industry simply does not work quickly or effectively. It had not provided the USA with a timely competitive edge from its huge investment in sponsoring basic research since the Second World War. The reasons for the sequential model of technology transfer not working were: (1) a failure of proper industrial focus to the project, (2) an incompleteness of research results from the project as a base for a new technology, and (3) lack of interest and/or proper incentives for technology innovators to use the research for subsequent innovation. This is because of the disciplinary focus of university research and the technology focus of industrial research. These do not exactly matchup but must be integrated.

The lack of an industrial focus on a project means that the experimental data cannot be immediately used in technology design, but must be performed again in another experiment with a proper focus. For example, if an experiment measures the properties of a material similar to a material used in a technology but not the same material, the experiment must be redone with the right material. Incompleteness of research results means that the research data cannot be immediately used for technology and engineering design projects in industry, without additional experiments to complete the range and set of experimental information required for a design effort. A lack of industrial interest in university research results means that industry will not pay any attention to the university research effort, nor even consider sponsoring it. Universities must consider industry as one of their clients, as well as government research funding agencies, and society in general.

And there is the problem of locating high-tech industries near a university to facilitate interaction between university and industrial researchers. This is why many "research parks" and "technology parks" have been constructed near research universities. For example, the first successful example of a science park occurred in the 1950s when Stanford University established its "Science Park." This research park facilitated the growth of "Silicon Valley" in Palo Alto, CA, USA, which was the famous industrial region for the growth of the early semiconductor industry.

Research universities are the principal producers of new science; while high-tech industries are the principal producers of new technology. Coupling simultaneous progress in science and technology requires coupling university and industrial research. The importance of university research for technology transfer to industry for a next-generation-technology is to have research projects in (1) science that connects to technology, (2) science that connects to engineering, (3) science that connects to commercialization. In this way, the scientific research in a multidisciplinary university research center can correctly focus and complete a critical mass of research projects of direct relevance to an industry's future.

## *Illustration: Center for Biofilm Engineering at Montana State University*

In addition to organizing and planning multidisciplinary research, one must manage the center (lead/direct). As a further illustration of the leadership challenges in proposing and running a multidisciplinary research center, we next look at another of the NSF ERCs, Center for Biofilm Engineering (CBE). This was funded in 1990 at Montana State University; and that Center would put Montana on the high-tech map for biofilm technology. An NSF award a provided a $7.5 million grant for 5 years to the Montana State University to study the science of biofilms and the technology of controlling biofilm growth. It was to be a multidisciplinary research center because (1) the research was to contribute to Chemical Engineering technical capability of controlling biofilm pollution in industry and (2) for this purpose there would also be the need for scientific progress in biology to better understand the phenomenon of biofilms.

The professor who wrote the proposal for the Center for Biofilm Engineering (CBE) and would direct the new center was William Characklis. He won one of the three awards in NSF/ERC Program made that year (out of 100 proposals that had been submitted that year). Competition for a multidisciplinary research center funded by NSF was keen and a tough research competition, involving (1) sophisticated methodological issues of relating science to technology and (2) difficult organizational issues of relating university research centers to industrial firms' research laboratories.

> Scientific research and engineering research and policy research pursued simultaneously requires multi-disciplinary research, coupling scientific progress with technological progress.

And, this turned out to be a dramatic case of science progress created by a focus for technology progress. New science was discovered from multidisciplinary biology and engineering research that had been motivated by, focused upon, and planned for as progress in industrial technology. The issue was that biofilm fouling was a big problem for industry and the environment, and the existing technologies of dumping harsh chemicals into water-based technologies did not really kill biofilms but only temporarily harmed them, while also harming the environment.

Biofilms are one of the oldest forms of life on earth, appearing in biological taxonomy halfway between single-celled bacteria living independently and multi-celled organisms in which cells were attached to each other. Biofilms are ubiquitous, living anywhere there were running channels of water across surfaces to which they could attach their colonies. Biofilms are in water pipes, irrigation systems, cooling pipes, and so on. For any industrial use of flowing water in pipes, biofilms clogged the pipes. Biofilms in oil wells cause the formation of sulfuric acid, rusting well pipes and polluting the atmosphere. Also biofilms live in human lungs, such as pneumonia which is a biofilm. Many of the diseases in the developed world involve biofilms (as much as 80%).

Traditionally, chemical engineers tried poisoning the biofilms when they wanted to get rid of them. But after poisoning, biofilm colonies revive – since the poisons often never reach the core of the bacteria living in the center of their biofilm tower complexes. In medicine, this is the same problem about cystic fibrosis, in which the biofilm colonies in the lung survive a doctor's penicillin treatments.

Was there no better way to control biofilm damage? Perhaps, but only if one really understood biofilms in their natural complexity and obscure details. At the time of funding the Center in 1990, scientists did not understand the details of biofilms, their structures, and how the structures came to be. They had not yet even seen what biofilms really looked like, at the microscopic level. But then the timing of the Center was fortuitous. Research on biofilms then took a dramatic step forward due to new instrumentation to observe biofilms. Biofilm structures are very small and then previously had only been viewed by using scanning electron microscopes. But this required dried samples of biofilms, and biofilms lived in water. It could now be seen just how completely the drying of a biofilm sample destroyed the proper observation of the structure. Viewing the dried biofilms under the electron microscopes, scientists and engineers saw just a tangle of strings of the polymer strands bacteria excreted to tie themselves into a biofilm community.

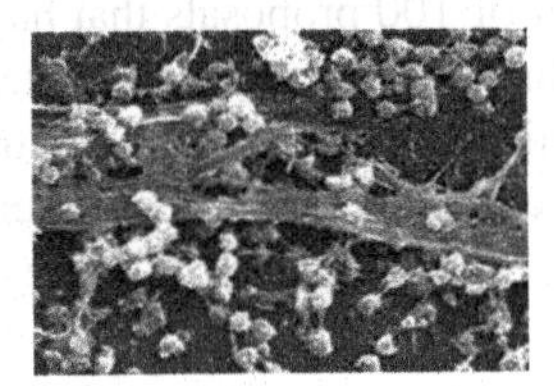

Dried sample of a *Staphylococcus aureus* biofilm (http://en.wikipedia.org, Biofilms, 2008)

William Characklis was a professor in chemical engineering, and after being notified on his award, he held an industrial sponsors meeting for the new Center in the summer of 1990.

The industrial sponsors of the new Center had supported the previous research activities of Characklis and were enthusiastic about the NSF support – which would add $2 million dollars a year research support to their annual half-a-million dollars support. Research activities could be dramatically expanded, with the additional NSF funding adding new research staff and many more graduate students.

Characklis had invited a biology researcher from another laboratory, Dr. Brenda Fraser, to the first center meeting of the new ERC. She showed new pictures, providing startling new evidence about the structures of biofilms. Her government laboratory research in Atlanta had acquired a new research microscope instrument using laser light, confocal scanning laser microscope. She used this to take the first pictures of biofilms in water. And in the watery world in which biofilms lived, biofilms could now be seen, not as a tangle, but as having built orderly and elaborate structures of polymer strands in which bacteria lived (much like bacterial-occupied apartment towers with flows of water around and through each tower). These were amazing pictures which changed the direction of biofilm research. As shown in the following sketch, the life cycle of biofilm colonies consist of (1) bacteria floating onto a surface, (2) bacteria attaching themselves to the surface with a polymer secretion, (3) bacteria multiplying within secreted polymers, (4) bacterial colonies constructing polymer apartment towers up into the water flow and separated one tower from another to prevent blocking of nutrient flow, and (5) some bacteria detaching to drift downstream to find new material surfaces to colonize.

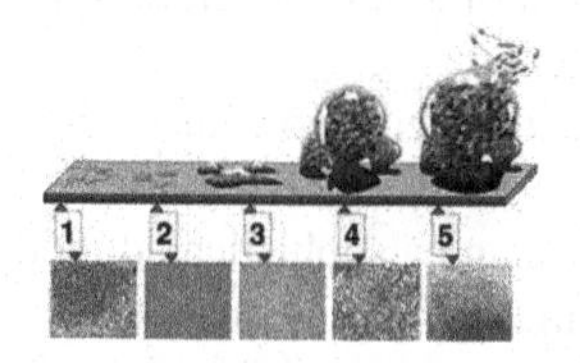


(http://en.wikipedia.org, Biofilms 2008)

The center research then focused how and why such biofilm colony architecture was formed.

William G. Characklis (1942–1992) was born in Maryland. He attended Johns Hopkins University, graduating in 1964 with a bachelor of science degree in chemical engineering. He worked for a year at Olin Mathesen Chemical in New Haven, Connecticut. There he leaned about the problem of biofilm fouling in industry. He then entered the University of Toledo and obtained a masters degree in chemical engineering. He returned Johns Hopkins University for a doctorate in environmental engineering. He began teaching at Rice University, and in 1979 moved to Montana State University. There in 1980 he established a research program in biological and chemical process analysis at Montana State, sponsored by twelve industrial firms. He co-authored a text on biofilms. And in 1989, he submitted a research center proposal about biofilms to NSF. The new Center for Biofilms added Characklis's previous industrial sponsors – who now jointly sponsored the new Center with NSF. But after staffing and running the center for less than two years, Characklis died of a malignant lymphoma cancer.

William Characklis (http://www.biofilms.montana.edu, 2008)

## Multidisciplinary Research Strategy

As we have emphasized, research grants for modern university research centers are multidisciplinary research grants. They are necessary to advance both scientific and technological progress, when a strategic (managed) multidisciplinary research approach is necessary. A research center proposal is justified when there is a clear *scientific-technological research goal* that requires several research projects to be performed simultaneously and coordinated over a period of time (5–10 years). The goal may be:

1. The advancement of an area of science, focused upon a clearly envisioned and stated and significant direction of advance, organized in a research center.
2. The development of technology focused upon a vision of a next generation of a current technology, also organized as research in a center.
3. The advancement of the engineering science base focused upon an existing industrial technology.
4. The development and maintenance of a large database and a sophisticated modeling effort in the social sciences to provide capabilities for policy analysis.

We can see that Characklis's reason for establishing a research center for biofilms (CBE) was the need for a *multidisciplinary approach* to study biofilms and to improve technological control of biofilm fouling. The scientific study of biofilms required research by different disciplinary perspectives, biology and chemical engineering. Biofilms are biological phenomena, bacteria living in colonies, held together by secreted polymers attached to surfaces immersed in flowing water. The interaction of the water flow and nutrients accessibility is a kind of chemical transfer phenomenon studied in chemical engineering. The study of the chemical and physical forces involved in the attachment of polymers to material surfaces is also within the perspective of chemical engineering. So, the science of biofilms crossed over two science and engineering disciplines – biology and chemical engineering.

> When the whole phenomenon of nature can be wholly studied only by more than one science or engineering discipline, then multidisciplinary research centers are necessary for effective research about the phenomenon. And this is true not only for the physical and biological sciences, but also for the social sciences.

The first focus of the research was to do both physical studies of the biofilm colonies to establish their physical structures and processes. The second focus of the research was to do biological studies to learn how bacteria functionally constructed the structures. The third focus of the research was to develop new technology for controlling unwanted biofilm growth.

> Development of new scientific technologies to dramatically improve an existing technology requires multi-disciplinary research in centers which span both science and engineering disciplines in universities.

The management of multidisciplinary research requires a research group to both understand and communicate between different scientific paradigms. Different paradigms are used by the different disciplinary researchers, all of which together are needed to address both the scientific and technological issues in the research focus of the center.

## *Illustration (Continued): Center for Biofilm Engineering at Montana State University*

Making a large research award that runs for several years requires management procedures not only to select a proposal for a grant award – but also to monitor and oversee research progress on the grant. First, the NSF/ERC Program ran an award selection process, called a request for proposals (RFP). Then after making an award, the NSF/ERC program also ran an annual grant review process of the awards. This procedure required a center, each year during its award, to write a progress report and request the next year of funding under the grant. Then, NSF assembled a review team – a group of external peer reviewers and led by an NSF/ERC program officer, to conduct an annual site visit to the center – to judge research progress and identify problems.

It was at the first of these site visits, in the fall of 1991 to Montana State, that the NSF review team uncovered two methodological–organizational problems about research progress in the center. The methodological problem was that Montana State did not have the appropriate biofilm expertise in their professors in the Department of Biology and had not proceeded to hire new professors with that research focus. The organizational problem was that Montana State did not have an appropriate accounting system in its university administration to properly control the research expenditures on the large NSF grant.

The biology department of Montana State had not contributed any biology professors to participate in the center. In fact, there were no biology professors in the university who had scientific publications in the biological specialty of biofilms. Characklis had hired two biological research scientists into the center staff to provide technical expertise, but these had not yet been given faculty appointments by the biology department. To the NSF review team, it was apparent that the Biology Department faculty had not been cooperating with the Chemical Engineering faculty to seize the opportunity to build research strength in biofilms. And additional flaw was that the university financial system had not set up appropriate separate budget accounts for the NSF grant. One million dollars of NSF research funds had been expended, without the university being able to properly report to NSF about the expenditures.

But after the NSF site visit in November of 1941, William Characklis died in April 1992. Informed of his death, NSF ERC program officers were shocked and saddened at the news.

NSF/ERC administrators were faced with a decision of whether (1) to terminate the grant and close the center or (2) to temporarily continue the grant to allow the

Montana State University administration time to find a new center director, fix its accounting problem, and hire new biology faculty with expertise on biofilms. The administration of NSF ERC Division decided to send an ERC program director again out to Montana, to recommend whether or not to continue the NSF grant.

Meanwhile, even after the Director's death, research activities in the Biofilm Engineering Center were continuing. Researchers held their next scheduled biannual industrial meeting in July 1992. Going out to that meeting, the NSF program officer observed that there was still an active center. Center researchers and graduate students presented their continuing progress. Eight industrial sponsors were present, interested in the research and continuing their funding to the center. The industrial sponsors listened to and commented upon the student research projects and their project plans; they were still providing individual mentoring on each of the student research projects, in addition to the faculty supervision. This combination of industrial mentoring and faculty supervision of student research was one of the goals of the NSF/ERC program. The NSF program officer concluded that there could still be an active center for biofilms at Montana State, if the problem of replacing the center director could be solved.

Next, the NSF/ERC program officer met with the President of Montana State University. Three issues were discussed: (1) recruitment of a new qualified director from outside Montana State, (2) improved involvement of the biology department in the research and educational program of the center, and (3) accounting for the NSF funds expended at the university. The President emphasized the importance of the NSF grant toward building future research capability at Montana State University. He would commit to NSF (1) to provide new faculty positions in biology and in chemical engineering for the center and (2) to hire new research leadership for the center directorship. He would also encourage the biology department to build an educational program specifically for biofilms. Also, the President committed Montana State to reimburse NSF for the million dollars of research funding not properly accounted for by the previous financial system in Montana State. This would end the need for an NSF audit of previously expended grant funds. For the future, the President committed to an immediate improvement in the university financial accounting system to properly account in the future for all research expenditures in the university. These two actions would solve both the current and the future problems for proper university accounting for the expenditure of funds on NSF grants.

With these commitments by the President of Montana State, the NSF program officer recommended to NSF that the center grant be continued for another year, giving time for Montana State to recruit for a new center director. For the rest of the year of 1992, research continued at the center conducted by the research staff and graduate students, with continuing industrial research mentoring. During that time, the university conducted a wide search for a new center director and deputy director, one in biology and the other in chemical engineering. By December of that year, the chair informed NSF that they had selected as director a Canadian scientist, J. William Costerton, who was a recognized expert in biofilms.

A year after Costerton became director of the Biofilm Engineering Center, NSF sent a peer-review team to review the center's progress. These outside experts recommended NSF to renew the center grant for another 5-year period. And in each following year, NSF continued to send review teams to monitor research progress. Research depicted the details of the physical structure of a biofilm and the flows of water and nutrients through the biofilm colonies.

Then significant research progress was made in connecting biological function to physical structure in biofilms. The biological community had discovered that individual bacteria could communicate with chemical signals. Living in a fluid medium, a bacterium could read the level of a chemical excreted by fellow bacteria; and when that level was high enough, it could trigger an action by the bacterium. This method of communication by the level of a commonly excreted chemical by bacteria was called "quorum sensing." M.B. Miller and B. L. Bassler summarized this scientific phenomenon: "Quorum sensing is the regulation of gene expression in response to fluctuations in cell-population density. Quorum sensing bacteria produce and release chemical signal molecules called auto-inducers that increase in concentration as a function of cell density. The detection of a minimal threshold stimulatory concentration of an auto-inducer leads to an alteration in gene expression. Gram-positive and Gram-negative bacteria use quorum sensing communication circuits to regulate a diverse array of physiological activities. These processes include symbiosis, virulence, competence, conjugation, antibiotic production, motility, sporulation, and biofilm formation" (Miller and Bassler 2001).

Costerton used the new concept for research on the architecture of biofilms. His group demonstrated that when biofilm colonies began to grow to close together the concentration of signaling molecules told bacteria in adjacent colonies to stop growing toward each other (Davies et al. 1999). The center experimentally demonstrated that chemical signals (1) limited overall bacterial colony size in the biofilms and (2) created distances between colonies. The quorum sensing in the bacterial colonies of a biofilm created an "architecture" of the colonies, a structure. These structures allowed both clumps of bacterial colonies and water channels between them so that the water could continue to flow between colonies, bringing vital nutrients to them.

The multidisciplinary research vision – of the original director and sustained by the second director – had succeeded! New science and possibly new technology about biofilms had begun to emerge as research results from the Center. The Center discovered the chemicals that were the quorum-sensing signals. And, the Center also learned that some signals could also tell a colony to migrate, to release their sticky hold on a surface and drift downstream to colonize a new space. This signal might provide a new kind of technology to rid pipes of unwanted biofilms (instead of the older type of technology of trying to poison the biofilms with noxious chemicals).

> Multidisciplinary scientific research, focused upon a technological problem, had produced a major scientific discovery in biofilms, a fundamental form of the nature in bacterial existence.

J. William Costerton was born in Canada. In 1955, he obtained a bachelors degree in biology from the University of British Columbia and a masters degree in 1956. In 1960, he obtained a doctorate in biology from the University of Western Ontario. In 1978, he held a position of a postdoctoral student at the University of Calgary, studying how bacteria anchored in cattle stomachs with a slimy and protective excretion. He called these colonies "biofilms." "The researchers found that biofilm bacteria were often team players, different species working in physiological co-operation. Costerton and his team have applied their knowledge to developing new technologies in areas ranging from oil production to bacteria-resistant medical devices. He holds several biological patents. He became president of the Microbios company. He is an expert in microbiology and electron microscopy." (http://www.science.ca/scientists, 2008) In 2008, Costerton moved to University of Southern California to establish a new center there in biofilm applications in medicine.

J. William Costerton (http://usc.edu, 2008)

## Multidisciplinary-Research Center Grants

Since funding a multidisciplinary research center is expensive, research funding agencies need to oversee the performance in a research center grant on two points. First, the quality of research being performed toward achieving scientific progress should be evaluated. Second, the expenditure of the research funds on items appropriate to the research should be audited. The first is an oversight question of research quality; and the second is an oversight question of administrative responsibility.

In university research, the question of research quality is a responsibility of the professor, who is the research grant recipient (the principal investigator on the research grant). To judge the quality of research performance, a funding-agency program officer needs to use scientific peers of the researcher to judge quality. As noted earlier, the quality of research in disciplinary scientific research can be effectively judged only by scientific peers who are (1) familiar with the frontiers of scientific research from reading the scientific literature and (2) trained in the research methodologies used in the research. For example, in the annual review of an NSF ERC center (such as the Montana Biofilm Engineering Center) during the period of the NSF grant, NSF sent out a group of scientific peer-reviewers (along with an NSF ERC program officer) yearly to review progress of research in an ERC center.

For the fiscal responsibility of expenditure of research funds in a university research grant, a government funding agency (such as NSF) depends upon the university administration (such as Montana State University) to guarantee fiscal responsibility. Accordingly, a government research grant is formally made to an institution (to the university) and not to an individual. The professor wins the grant, but the grant is given to the university. This makes it easier for a government agency to audit its research grants by performing periodic audits of universities. For example, in the case of the Biofilm Engineering Center in 1991, after the first year of the

NSF ERC award to Montana State University, the university administration had not yet set up accounts for properly tracking the expenditure of NSF funds.

Universities are responsible to government research-funding agencies for the expenditure of research funds won by its professors in research grants.

## Role of a Multidisciplinary Research Center Director

The justification for a university research center is to perform multidisciplinary research – when a "critical mass" of research projects are needed to advance together both scientific and technological progress. As we have noted, the connection between progress in both science and technology is due to the concept of a *scientific technology* – improving technology by means of a deeper understanding of the science base of knowledge beneath the technology. Technology is so sophisticated and knowledge of science is so deep that extreme specialism is necessary at the frontiers of technology and science. Different specialists need to work together to advance those frontiers.

But all technologies are knowledge-based upon more than one science discipline!

For example, in the case of biofilms, two academic disciplines were necessary for scientific progress in understanding the biology of biofilms and the physical processes in biofilm communities:

1. Biology, with a science base of biochemistry and quorum sensing
2. Chemical engineering, with a science base of hydrodynamics and mass transfer

Accordingly, the research vision of the director needs to envision how a multidisciplinary approach could simultaneously advance both scientific and technology (engineering) progress – a science/technology research vision. A professor who desires to become a center director then needs to express such a vision in a research proposal and win external research funding for the center. The proposal should describe the major research issues needed to be addressed for an integrated science/technology progress. It also should describe what kinds of disciplinary methodologies could provide a research approach in a "critical mass" or research efforts.

Upon winning a research grant, the new center director then needs to complete the recruitment of a full multidisciplinary research staff for the center and direct the research projects into a coherent and strategic research plan.

This role of the director in conceiving an exciting multidisciplinary research vision and in translating that vision into a methodology of sets of research projects is the critical factor for successful center management. If, as happened in the Biofilm center, a founding director is lost to a center, then another such visionary director must be found or the center grant be terminated. The search for a new director for that center required finding a researcher in biology, because by then biological nature of biofilms structure had become the leading scientific issue. Also, the subsequent director, Costerton, needed to have visionary capability about biofilms, as had the original director, Characklis.

## Critical Mass of Research

We used the metaphor of a "critical mass of research" to indicate how a university research center needs to assemble a portfolio of research projects targeted for and sufficient to advance a scientific technology. We ground this metaphor in an idea basic to the dramatic progress in scientific technologies – a next generation of a technology. We will describe why and how industry needs university science to make for industrial innovation.

An industry directly uses technology in its production of goods and services and not science. Accordingly, industry needs new science only (1) when technological progress in an existing technology cannot be made without a deeper understanding of the science underlying the technology and (2) when new basic technologies need to be created from new science.

For technological innovation in industrial firms, industry knows how to incrementally develop new technology for its needs. Industry can identify technology bottlenecks in current technology systems and organize research programs in industrial laboratories to address these bottlenecks. It is when industry needs new science that major problems in the R&D infrastructure arise and when industry–university research partnerships prove valuable.

As we earlier emphasized, science is expensive and seldom contributes to near-term profitability in a business firm, in a direct economic sense. The average time from scientific discovery to technological innovation has varied historically, from as long as 70 years to as short as 10 years. In the twentieth century in the best cases, it usually took at least 10 years from laboratory to product. Also, science research is primarily performed in research universities. And, the scientific research is integrated with graduate education. Research projects are performed by doctoral candidates for their thesis requirements. Therefore, university research projects are divided into a size of a doctoral thesis. Moreover, such doctoral research projects are rarely on a scale, breadth, or timeliness of research focus – suitable for the industrial needs for new science or technology.

This is the problem. When industry needs science: (1) it can do it itself at great expense, high risk, and trouble, or (2) it can turn to universities to do the science – but then performed in perspective, size, and timing unsuitable to industrial needs. Proprietary issues about intellectual property are important to industry, and industry ultimately must put knowledge to use in making profits. While there is a natural basis for cooperation between industries and universities, the problem is to effect their cooperation, despite their major differences in perspectives, goals, and timing.

In planning basic research in a university center for progress in a scientific technology useful to industry, center research planning should:

1. Describe the current technology system and identify current technical limitations in the system (technology "bottlenecks")
2. Identify technology performance variables for a next-generation technology system, which are dramatically better than the performance of the current system

3. Identify the underlying scientific phenomena of a new technology that needs better scientific understanding
4. Formulate a "critical mass" of sets of scientific research projects to investigate the scientific phenomena
5. Use the scientific results as the basis for new inventions or design aides or control procedures for a next-generation technology
6. Demonstrate to industry the technical feasibility of a next generation of a scientific technology in a research "test bed"

## Summary

1. University research centers coordinate methodology and organization of research to provide an integrated process for posing and answering questions of nature – scientific inquiry.
2. Scientific research and engineering research pursued simultaneously requires multidisciplinary research performed in a research center – to couple scientific progress with technological progress.
3. The "critical mass" of a center director's research vision is to lay out and assemble pieces of research as center projects – that together will coherently yield integrated scientific and technological progress.
4. Universities are the loci for science, while industry is the loci for technology. While there is a natural basis for cooperation between industries and universities, the problem is to effect the cooperation – despite great differences in research perspectives and goals and in project sizes and timing.
5. In planning basic research in a university center for development of new technology for industry, research planning should (1) describe the current technology system, (2) identify technology performance variables for a new technology system, (3) identify the underlying scientific phenomena of a new technology that needs better scientific understanding, (4) formulate sets of research projects to investigate the scientific phenomena, and (5) use the scientific results as the basis for new inventions or design aides or control procedures for a new technology.
6. The reasons for university research to fail in technology transfer to industry are improper focus or incompleteness or irrelevance.

# Chapter 15
# Measurement

## Introduction

Research methodology uses techniques for finding: (1) the distinctive quality of a natural thing (what it is), (2) the quantity of these things existing in nature (how many), (3) from some properties (quality), deducing other properties of the natural thing (what else), and (4) from observation, inducing the properties of natural objects (measurement).

We will look at the following topics in measurement: expressions of quality, expressions of quantity, probability, populations, sampling, statistical distributions, statistical correlations. And to discuss these ideas for measurement, we need to review some mathematical topics, including: set theory, algebra, analytical geometry, and statistics. Measurement requires numbers and sampling, and so this chapter necessarily is quantitative.

This is a difficult chapter and can only be well understood if the reader has previously studied the mathematical topics of modern algebra, calculus, probability, and statistics. A lot of math is reviewed here.[1]It can't be helped. To understand measurement as a research technique, one needs a lot of math. But the point of this review is to show what mathematics is needed and how it is used for measurement – for quantitative expression in science. (If the reader does not have the math background, please look then at *how* math is used in science, ignoring technical math details.)

This is a basic fact of the scientific life. Mathematics is a literacy barrier to understanding the quantitative expressions in science. Math gives science great power; but it gives great trouble to the understanding of science in technical depth. If you haven't had all the math reviewed here (and it is a lot), don't worry. It is only math. It is only a syntactically specialized language of science. It is only what is at the bottom of it all: quantity, counting. The important point is to understand the difference between quality and quantity. These two ideas are at the heart of all mathematics – the quantitative expression of quality.

> The philosophical point of this chapter is to see how the mathematical topics fit together, all of them being essential to the task of measurement in empirical research.

F. Betz, *Managing Science*, Innovation, Technology, and Knowledge Management 9,
DOI 10.1007/978-1-4419-7488-4_15, © Springer Science+Business Media, LLC 2011

## Quality and Quantity

As paradigms and perceptual spaces of the physical sciences (geometric) and the social sciences (functional) differ, so too do their methods of observing nature. Physics uses experiments, whereas the social sciences use case studies and surveys. In the physical sciences, experiments on physical nature observe physical nature. In the social sciences, case studies and surveys about the history of social/cultural events and individuals are used in observing social nature.

For example, in the physical sciences, we reviewed how Rutherford, Marsden and Geiger bombarded metal foils with radioactively generated alpha particles to experimentally observe how the alpha particles were deflected by the atoms of the metal foil. When rare but actual back-scattered events of the alpha particles were observed, Rutherford could infer that the geometric structure of the atom consisted of a tiny positively charged atomic nucleus surrounded by orbiting electrons.

But within the social sciences, it was not experiments but cases of observing social history and culture behavior that became the empirical approach. As an example in psychology, Sigmund Freud created his theory of unconscious behavior from the cases of psychologically disturbed patients that he treated. Similarly, Carl Jung developed his theories of personality types from the cases of patients he observed. In sociology, for example, Max Weber's early theory of the relationship of religion to capitalism rose from his history of protestant ethics and business practices. So also, Weber's theory of bureaucracy arose from the observation of German government organizations of the early twentieth century. In anthropology, many anthropologists, such as Franz Boas, Margaret Mead, and so on, all visited technically primitive societies to observe cases of different cultures. In economics, Adam Smith's observation of economic trading in England in the 1700s provided the basis of modern economic theory.

Case studies in the social sciences have been central to the creation of social theory and to its application in practice.

For example, in management practice, theory is sometimes called "management principles." These principles are expressed as a way to communicate appropriate lessons of successful practice observed in one case which can be applied to other cases of management practice. In the second half of the twentieth century, the Harvard Business School popularized the use of case studies for business education. In practice, the use of empirically based theory or principles is important to improving performance of organizations. For example, David Besanko, David Dranove, and Mark Shanley once nicely expressed the importance of theoretical principles in management: "There is a keen interest among serious observers of business to understand the reason for profitability and market success.... However, observers of business often uncritically leap to the conclusion that the keys to success can be identified by watching and imitating the behaviors of successful firms. (And this is often called 'benchmarking' or 'best practices').... However, uncritically using currently successful firms as a standard for action *assumes* that the successful outcomes are associated with identifiable key success factors, and by *imitating* these factors, other firms can achieve similar successful results" (Besanko et al. 2000, p 4).

The important methodological idea is the *critical analysis* of what are the key success factors. Uncritical imitation of a prior success by a different company may not prove successful in a new situation. In any action, no two situations (nor two actors) are ever absolutely identical. Action is always a particular set of factors and activities – all of which together explain a particular success (or failure). Teasing out of a "benchmarking" case to see just what is really general and transferable is what critical analysis is intended to accomplish. Theoretical principles are the result of critical analysis over a range of particular and unique "benchmarked" cases in order to identify, abstract, and understand the key success factors of practice which can be generalized.

Case studies identify and provide the fundamental objects to be observed and their properties and their relationships. However, because each case study is a particular study of an individual or a group or a social pattern in a society, such research results also need to be generalized across all individuals, groups, or societies. This generalization is accomplished by surveys which examine the quantitative extent of the results across populations (individuals, groups, societies). We focus in this chapter on the generalization across populations of particular research results. This is the topic of measurement and statistics to estimate the validity of measurements.

Generalization of quality over populations requires surveys (social science) or experiments (physical science).

## Quality: Mathematical Set Theory

The idea of quality distinguishes one object in nature from another. The idea of quantity compares the amount of quality an object has or the amount of quality comparisons between objects. The *quality* of a thing in nature is defined by the *property* of an *object*. Expressing this in mathematical language, an object can be defined as a *set of things* sharing a *common property*. Set theory is the mathematical topic which was constructed to express the idea of quality. A set can be defined as (1) a list of members or (2) a common property which all members share.

*Set Membership* of $b$ in a set $B$, ($b \ni B$).

Let $B$ denote a set whose members $b$ share a common *property*. In set notation each member $b$ is said to belong ($\ni$) to the set $B$ ($b \ni B$), when $b$ has the common property.

*Set Equality*, ($A = B$).

Two set $A$ and $B$ are equal ($A = B$) when all members are the same, $a = b$.

*Set Inclusion*, ($A \subset B$).

One set $B$ is said to include ($\subset$) a set $A$, when all members $a$ of $A$ also belong to $B$. $A$ is called a subset of $B(A \subset B)$.

*Union of Sets*, ($C = A \cup B$).

Another set $C$ can be formed as the union ($\cup$) of $A$ and $B$ when all members $a$ and $b$ (and only $a$ and $b$) also belong to $C$.

*Intersection of Sets*, ($D = A \cap B$).

Set $D$ is said to be the intersection ($\cap$) of two sets $A$ and $B$, when the only members $d$ of $D$ are members also of both sets $A$ and $B$.

Function Between Two Sets.

If one can match the elements of one set $A$ to the elements of another set $B$, this is called a mapping between the two sets. A rule to create such a mapping is called a functional relation $F$ between the two sets. Two sets $A$ and $B$ are then said to have a functional relationship $b = f(a)$, when there exists a rule-of-combination, function $f(\ )$, so that for each member $a$ of $A$, there is thereby uniquely identified a member $b$ of $B$, so that $b = f(a)$.

The property shared by things defines membership of the things in a set as an object.

The number of things belonging to the object-set is the quantity of the things observed in the object-set.

## Illustration: Set Notation in Defining Natural Objects

Natural objects can be described as having common properties, so that a set whose members share a common property can be assigned to the object class. For example, an atom as a chemical element can have properties of atomic number and atomic weight. The chemical element table thus classifies atoms by the atomic properties of number and weight. For example, any atom having the property of atomic number 1 (one orbiting valence electron) and atomic weight one (a proton as its nucleus) is classified as the set of hydrogen atoms. All atoms with atomic number one and atomic weight two (one proton and one neutron as its nucleus) are deuterium atoms. Thus the classification ($A$) of a given atom ($a$) by atomic number ($z$) and atomic weight ($n$) is an example using sets to classify kinds of matter, so that:

$$a \ni A, \text{ if and only if } a = z \cup n.$$

An illustration in the social sciences, we saw that Weber classified all social actions as to their properties of social anticipations in interactions, expectation properties: utility ($u$) or identity ($i$) and reciprocity ($r$) or authority ($a$). With these one could then define four categories of social objects (societal systems), as societal systems ($E$, $C$, $P$, $T$).

1. With an Economic System denoted as ($E$), and its member as ($e$);
   $e \ni E$, if and only if $e = u \cup r$ (union of utility and reciprocity)
2. With a Cultural System denoted as ($C$), and its member as ($c$):
   $c \ni C$, if and only if $c = i \cup r$ (union of identity and reciprocity)

3. With a Political System denoted as ($P$), and its member as ($p$):
   $p \ni P$, if and only if $p = i \cup a$ (union of identity and authority)
4. With a Technology System denoted as ($T$) and its member as ($t$):
   $t \ni T$, if and only if $t = u \cup a$ (union of utility and authority)

A natural object is classified as a member of a set of objects, each set defined by properties, observed and measured within a perceptional space of a scientific paradigm

## Quantity: Counting

In the concept of quantity, the first idea is that of counting, or numbers. The simplest set of numbers is called the "integer set." A special set, the Integer Set, can be constructed as lists of members, using the set relationship of Union ($\cup$).

First define the set $Z_0$ as having no members: $Z_0$ is an empty set. We name the empty set $Z_0$ as zero: 0.

Next introduce the idea of the cardinality of a set. The cardinality of the null set $Z_0$ is 0. Cardinality is just a technical name for the intention to label sets with a "number." The point of this exercise is to construct the idea of "numbers" using the mathematical language of "set theory." The null set has no members. Next construct a One Set $Z_1$ as having $Z_0$ as its only member: $Z_0 \cup Z_1$. We will call this set $Z_1 = 1$, with the zero set $Z_0$ as its only member. The cardinality of the One set $Z_1$ is 1, since it has only one member ($Z_0$) in its list of members.

Next construct a Two set $Z_2$ as the union of the One Set $Z_1$ with itself: $Z_2 = Z_1 \cup Z_1$. Then Two Set $Z_2$ set has two members ($Z_0$ and another $Z_0$) as its list of members, and the cardinality of the Two Set $Z_2$ is termed as 2.

Next construct a Three Set $Z_3$ as the union of the Two Set with the One Set: $Z_3 = Z_2 \cup Z_1$. The cardinality of the Three Set $Z_3$ is termed as 3, since the members of $Z_3$ are the list of three elements ($Z_0$ and another $Z_0$ and another $Z_0$).

The idea of "cardinality" is a counting idea, counting the number of members of a set.We can see by repeating the pattern of constructing successive sets $Z_{n+1}$ as unions of the One Set $Z_1$ with the preceding set $Z_n$, one can gain a set of sets with cardinalities (number of members) from 0, 1, 2, 3, … $n$, $n$ + 1, etc. This set of sets of cardinalities is called the Integer Set (0, 1, 2, 3, … $n$, $n$ + 1). The number of elements in each integer number set is one larger than the previous set.

This is a systematic way to construct a set ($I$) of integer numbers, using only the ideas of a set and union of sets. The idea of cardinality then defines as the idea of the number of elements in a set, or quantity. If one matches the cardinality of an Integer in a one-to-one correspondence with the list of members of any set, one can count the numbers of the set (cardinality of any set). This is how one defines the basic quantitative operation of counting using the concepts of Set Theory. There is a little logical circularity here in definitions, as the term "cardinality" means the same as the term "number." But the procedure is a systematic and unambiguous way to construct the integer number set, using the ideas of sets and union and membership by list. (Set theory is used as a meta-language to construct the language of integers.)

With this idea of a set of integer numbers ($I$), one can then count the cardinality (quantity) of elements in any set $S$ by matching the elements of the integer number set with the elements of $S$ in a one-to-one match. When all the elements in $S$ have been matched to successive integers, the largest integer in the match can then be said to the cardinality (number of elements) of the set $S$.

Naming the integers is done by means of a *base* of an integer set. The base of ten uses the ten symbols of (0, 1, 2, 3, 4, 5, 6, 7, 8, 9) in repeated manner to *name* all the integers in columns of numbers (units, tens, hundreds, thousands, millions, billions, etc.). Humans having ten fingers, found it convenient to count in a decimal system; whereas computers have two transistor states (on and off) found it convenient to count in a binary system. The binary number base of two uses the two symbols (0, 1) in repeated columns to name the integers:

| Base 10 | Base 2 |
|---|---|
| 0 | 0 |
| 1 | 1 |
| 2 | 10 |
| 3 | 11 |
| 4 | 110 |
| 5 | 111 |
| 6 | 1110 |

For example, in the physical science, counting is used to measure the quantity of a property. A count of the number of electrons ($z$) determines the atomic number of a particular atom ($a$) and a count of the number of nucleons ($n$) determines the atomic weight of a particular atom ($a$).

For example, in computer science, counting is used to sequence logical operations in a computer. For example in the idea of a stored program computer, Von Neumann used Turning's idea that any mathematical calculation could be expressed as a numbered sequence of operations. Hence a program can be written as numbered sequential logical/mathematical operations.

## Quantity: Mathematical Operations

However, numbers (counting) are not sufficient for quantitative expression. One must operate on numbers and combine two to get another number. Two kinds of combining (numerical operations) are fundamentally important: addition and multiplication. A set of numbers operating under one operation (addition or multiplication) is called a "mathematical group." A set of numbers operating under two operations (addition and multiplication) is called a "mathematical algebra." We review these topics.

## *Mathematical Group*

The set of integer numbers ($I$) can be extended into a general set of numbers by next adding the idea of an *operation* within a set. The set of integers ($I$), or the set of numbers in general

### Operations Within a Set ($c = a * b$)

If an operation (*) is defined between any two members $a$, $b$ of a set $S$, so that another member $c$ belonging to the set $S$ is uniquely identified by the operational combination ($a * b$), then the set $S$ is said to be complete under the operation (*) and $c = a * b$.

### Group (G) as a Set Complete Under an Operation (*)

A set $S$ is called a group G under an operation (*), when:

1. There is a rule-of-combination (operation *), so that $c = a * b$, and $a$, $b$, $c$ belong to $S$.
2. The operation (*) is associative, so that $ab * c = a * bc$.
3. There exists an identity element (e) in the set S, so that $e * a = a$.
4. For every element a in the set $S$, there is also an inverse element ($a^{-1}$), so that $a^{-1} * a = e$. (For example, the inverse of addition is subtraction, and the inverse of multiplication is division.)

## *Addition Group*

The first important operation that can be defined on the integer set ($I$) is called addition (+).

1. There exists a rule-of-combination, *addition*, so that $c = a + b$, where $a$, $b$, $c$ belong to $I$.
2. The operation of addition (+) is associative, so that $(a + b) + c = a + (b + c)$.
3. There exists and identity element (0) in the set $I$, so that $0 + a = a$.
4. For every element $a$ in the set $I$, there is also an inverse element ($-a$), so that $(-a) + a = 0$.

The rule-of-combination of addition is defined by adding the unit integer to each integer to calculate a successive integer (e.g., $1 + 0 = 1$, $1 + 1 = 2$, $1 + 2 = 3$, etc.) This rule is consistent with the set definition of constructing the integer set as successive unions of a prior set with the null set. Thus the set definition of integers and the group definition of addition are mutually consistent. The operation of addition on the Integer set ($I$) extends the set $I$ from positive integers ($a$) to also include in $I$ all the negative integers ($-a$). The operation of *subtraction* is then defined as the addition of an inverse element: $a - b = a + (-b)$.

## Multiplication Group

A second basic operation that can be defined on the integer set ($I$) is called *multiplication*.

1. There is a rule-of-combination, *multiplication* (*), so that $c = a * b$, where $a$, $b$, $c$ belong to $I$.
2. The operation of multiplication is associative, so that $(a * b)* c = a *(b * c)$.
3. There exists and identity element (1) in the set $I$, so that $1 * a = a$.
4. For every element $a$ in the set $I$, there is also an inverse element ($a^{-1}$), so that $a * a^{-1} = 1$.

The rule-of-combination of multiplication can be constructed as a repeated operation of addition, where $a * b$ is the instruction to add $b$ to itself a times. Thus the unambiguity in addition (as defining integers with successive additions of one to a preceding integer) prescribes an unambiguous way of constructing the multiplication table for integers (repeated addition).

Division is the inverse operation of multiplication. The multiplication operation on the integer set require expanding the set of integers into the larger set of scalar numbers – integers, ratio numbers, irrational numbers, and complex numbers. The division ($a/b$) of an integer $a$ by an integer $b$, and these ratios ($a/b$) are called the rational number set. The inverse square (square root) of many integers, such as square root (2) is not expressible as a finite string of integers (1.414141414…). These square root numbers are called not-ratio, or irrational number set. Finally, the square roots of negative numbers turn out to be expressible only with two numbers, or a complex number ($a + ib$), the complex number set. So for the number group under division, the number set must include integers, rational, irrational, and complex numbers.

> One sees the existence (ideal existence) of numbers is due to *definitions of logic* and *completion of an idea in logic.*

## Algebra

A *group* is a number set operated on by one rule-of-combination (either addition or multiplication), and an *algebra* is a number set operated upon by two rules-of-combination (both addition and multiplication). An algebra consists of a number set and the two operations of addition and multiplication upon the set with association and commutation rules between the two operations.

### Addition Group

1. There exists a rule-of-combination, addition, so that $c = a + b$, where $a$, $b$, $c$ belong to $I$.
2. The operation of addition (+) is associative, so that $(a + b) + c = a + (b + c)$.
3. There exists and identity element (0) in the set $I$, so that $0 + a = a$.
4. For every element a in the set $I$, there is also an inverse element $(-a)$, so that $(-a) + a = 0$.

### Multiplication Group

1. There exists a rule-of-combination, multiplication, so that $c = a * b$, where $a$, $b$, $c$ belong to the group.
2. The operation of multiplication is associative, so that $(a * b) * c = a * (b * c)$.
3. There exists an identity element ($l$) in the set $I$, so that $1 * a = a$.
4. For every element a in the set $I$, there is also an inverse element $(a^{-1})$, so that $a * a^{-1} = 1$.

And a rule of distribution between the operations of addition and multiplication and two rules of commutation.

1. Distributive Law: $a * (b + c) = a * b + a * c$.
2. Commutative Law of Addition: $a + b = b + a$.
3. Commutative Law of Multiplication: $a * b = b * a$.

An addition group and a multiplication group with distribution and commutation rules – all together define an *algebra* of numbers.

## *Vectors*

A quantity ($q$) expresses information about the property of a natural object, how much ($q$) of the property. Some quantitative information can be expressed by a single number, and then such a $q$ is called a scalar quantity (single number). For example, the mass ($m$) of a physical object is a scalar; so too is the electrical charge ($q$) of an electron ($-q$) or a proton ($+q$). A quantity expressed in one bit of information is called a scalar quantity.

When information about a property must have more than one number (more than one bit of information), then a vector quantity is required for description. A vector is a row or column of numbers required provide a complete description of a property, $\underline{v} = (a_1, a_2, a_3, \ldots)$. Vector notation is marked by an underline ($\underline{v}$), and the component numbers of the vector are marked by numbers $a_1$, $a_2$, $a_3$, etc.

For example, quantitatively describing the position ($\underline{q}$) of a physical object in nature requires a vector of four numbers, three of space ($x$, $y$, $z$) and one of time ($t$), or $\underline{q} = (x, y, z, t)$ Physical space is a three-dimensional frame, requiring three dimensions (X, Y, Z), as information to describe a position. And time is a scalar. In mathematical terms, the position of a physical object thus requires three bits of information, three numbers $(x, y, z)$, and so is expressed by a three-dimensional vector, $\underline{v} = (x, y, z)$. Adding time to the description, space/time, then a position vector of a physical object requires expression by a four-dimensional vector $\underline{v} = (x, y, z, t)$.

Vectors are related one to another by (1) addition, (2) multiplication, or (3) transformations. For example, addition of $\underline{v} = \underline{w} + \underline{z}$, or multiplication of $v = \underline{w} * \underline{z}$, or transformation of $\underline{v}=T\ \underline{w}$, (where $\underline{v}$, $\underline{w}$, $\underline{z}$, are all vectors and $T$ is a matrix, and we have defined a matrix in the earlier chapter on social science models). Also, there are two kinds of vector multiplication: scalar multiplication and cross-product multiplication.

Scalar numbers and vector numbers can be used together to form a vector algebra by multiplying a vector ($\underline{v}$) by a scalar (a) which gives another vector($\underline{w}$), so that the components of $\underline{w}$ are the components of $\underline{v}$, each multiplied by $a$. For example:

$$\text{If } \underline{v} = (a_1, a_2, a_3), \text{ then } b * \underline{v} = \underline{w} = (b * a_1, b * a_2, b * a_3)$$

It all gets further complicated. But the point to remember is that:

1. Quantitative expression of a property uses a scalar number (single number) when only one bit of information describes a property.
2. Quantitative expression of a property uses a vector number (several vector component numbers) when more than one bit of information (several bits of information) describes a property.

## *Differential Algebra (Calculus)*

In addition to the mathematical operations of addition and multiplication in constructing two groups for algebra, there is a third important operation in modern algebra, called differentiation (differential calculus). We can introduce this mathematical idea of a differential through the example of physical motion (with distance $r$ and time $t$ and velocity $v$ and acceleration $a$).

Let the symbol ($t$) denote time and the symbol ($r$) denote the distance over which a physical object moves through space (transcendental esthetic space/time in the scientific paradigm of Mechanism). One of the interesting quantitative expressions one can express in algebra is the idea of "velocity" ($v$), how fast a physical object moves. Suppose one measures an interval of time from $t_1$ to $t_2$ as the time difference $(t_2-t_1)$, which we symbolize as "delta $t$" ($\Delta t = t_2 - t_1$); and suppose the physical object during that time interval of $\Delta t$ has moved the distance of $(r_2 - r_1 = \Delta r)$. Then average velocity ($\underline{v}$) during this transit of the object is the ratio: $\underline{v} = \Delta r/\Delta t = (r_2 - r_1)/(t_2 - t_1)$.

Next suppose one measures the time interval $\Delta t$ smaller and smaller, and thus one approaches the measurement of the velocity of the object at time nearly $t_1$, From the idea of an "average velocity" (v), one gets to the idea of an instantaneous velocity $v(t_1)$ as a number toward which the ratio $\Delta r/\Delta t$ approach as $\Delta t$ gets very small, which is called the mathematical limit of the series of ratios $\Delta r/\Delta t$ as the time $t_2$ is near to $t_1$. This was Isaac Newton's idea for differential calculus.

What Newton did was to add to the ideas of operations of addition and multiplication in an algebra of numbers, a third operation, which he called a "fluxion" and is now called a "differential." The differential operation upon any algebraic number ($a$) by a time interval $\Delta t$ is called the differential ($\mathrm{d}a/\mathrm{d}t$).

> The differential operation a ($\mathrm{d}a/\mathrm{d}t$) on an algebraic value ($a$) is defined as the mathematical limit of the ratio $(a_2-a_1)/(t_2-t_1)$, as $\Delta t$ gets very, very small.
>
> When the algebraic number is a measure of distance ($r$), then $\mathrm{d}r(t)/\mathrm{d}t = v(t)$ is the instantaneous velocity at time $t$; and $\mathrm{d}v(t)/\mathrm{d}t = a(t)$ is the instantaneous acceleration of the motion of the physical object at time $t$.

The mathematical topic of differential algebra is more commonly called "differential calculus." To this differential is added an inverse operation of integration, and together these make up the modern topic of calculus. Differentiation takes tiny slices of algebraic variables, and integration integrates these back into a whole. One sees that in the history of science, a "physicist" Isaac Newton invented new mathematics, differential calculus, in order to describe the physical reality of planets moving around the sun in space. This is an example of how the disciplines of physics and of mathematics have interacted in modern science.

> New mathematics has been sometimes needed to quantitatively express new physical reality; and newly observed physical reality has sometimes stimulated new mathematical developments.

One can see that many of the mathematical topics get constructed by adding logical complexity to the initially simple idea of a number, counting. And all this logical (and very useful) complexity is "progress" in the scientific discipline of mathematics. As we have noted earlier, historically, mathematics has provided syntactically specialized languages for quantitative expressions in the semantically specialized languages of physics, chemistry, biology, and social science.

## Analytic Geometry

The next general idea is how properties are described mathematically requires a quantitative construction of a perceptual space, the space in which objects are observed. In the history of the physical sciences such a quantitative perceptual space was invented by Renes Descartes, a topic which mathematicians now call "analytic geometry." We recall that Descartes combination of geometry and algebra was essential for Newton to construct his gravitational model of the Copernican solar system.

The set of all material objects that are coexistent and close to each other can be described as occupying different points in a geometrical space. We recall that a three-dimensional geometric space has three orthogonal axes (*X*, *Y*, *Z*). The point of geometrical space in which a material object resides can be described by the three coordinate projections upon these axes as (*x*, *y*, *z*). Two different objects will be described at two different points $(x_1, y_1, z_1)$ and $(x_2, y_2, z_2)$ as the two vector positions of $q_1$ and $q_2$ (Fig. 15.1).

The distance along each axis is denoted by lengths, as a scalar algebra. For example, length along the *X* axis is measured as inches, or feet or miles of meters or kilometers, etc. Analytic Geometry consists of mapping to the length of a dimension an algebra of scalar numbers. This usefulness of this quantitative expression of space and distance is to be able to chart (graph, see) all the possible solutions to an algebraic sentence.

For example, the equation (algebraic sentence) of a straight line is $y = m*x + b$. Here, ($y$) is the noun and (=) the verb and ($m*x + b$) is the prepositional object. To find a solution, one can use the rules of algebra to manipulate the sentence into another form wherein ($x$) becomes the noun. For example, in a straight line, the sentence with ($x$) as the noun reads: $x = (y - b)/m$. This is called an algebraic solution for $x$ in terms of $y$. All the numbers $y$ as a function of $x$ that satisfy this sentence lie upon a straight line with slope $m$ and intersect $b$ on the vertical axis (Figs. 15.2 and 15.3).

An algebraic formula can be expressed in analytical geometry as the set of points in the two-dimensional space that satisfy the functional mapping of $y = f(x)$. For example, the equation $y = mx + b$ graphs in the space of scalar algebra as a straight line with slope $m$ and $y$-intersect $b$. The equation $x^2 + y^2 = r^2$ graphs as a circle.

A use of this geometrical space technique in the social sciences is called "cluster analysis." When a social object can be described by two quantities ($x$ and $y$), one

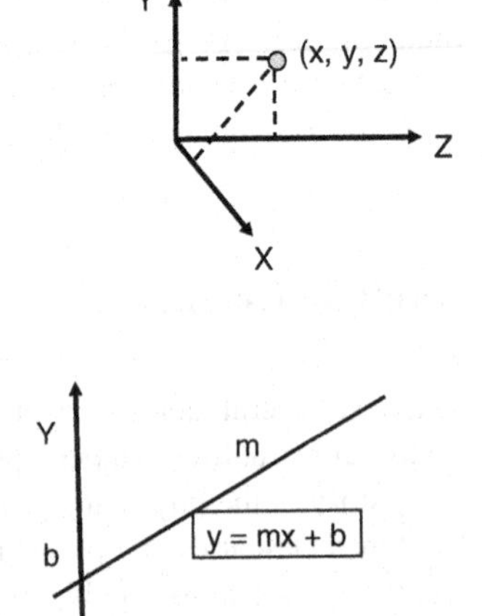


Fig. 15.1 Three-dimensional geometric perceptual space

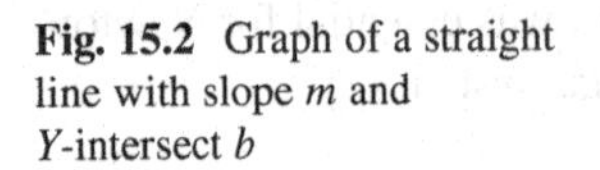

Fig. 15.2 Graph of a straight line with slope *m* and *Y*-intersect *b*

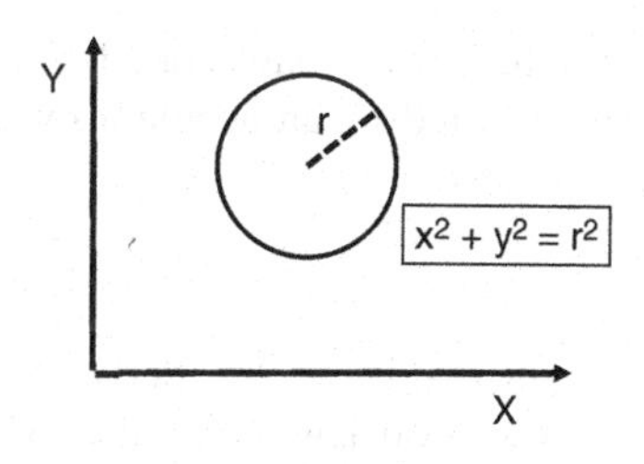


**Fig. 15.3** Graph of a circle with radius $r$

can plot the sampled population upon a two-dimensional geometric space to see if some measured objects lie closer together than other objects, which thus can results in grouping together of some of the objects, "object clusters."

## *Illustration: Quantitative Sentences and Deductive Inference*

We have reviewed numbers (counting) and algebras (operations) in order to construct a syntactic language for quantitative expression. The importance of the idea of an algebra of numbers is that one can form quantitative sentences and reason with the quantitative numbers. We next introduce the idea of a *quantitative sentence*, a statement about a natural thing in numbers.

We recall that in a language, a sentence is constructed from a subject and a predication about the subject. The subject of a sentence is denoted by a noun, and predication by a clause that asserts some property of the subject. A predication clause consists of a verb and an object of the verb. For example, take the sentence: The dog barks at a stranger. The subject is a canine, termed a dog. The dog is a *noun*, a label for the thing-in-nature observed as a canine. The *predication clause* is: barks-at-a-stranger. The *verb* bark is an action asserted about the subject, and the stranger is a noun as the *object* of the action. The terms of *the* and *a* are called *propositions*, or terms to clarify the subject or propositional clause in the sentence. The point is that, in any language, a sentence consists of a subject and something asserted (predicated) about the subject.

So too, one can construct an algebraic quantitative sentence of a subject and verb and object, such as: $x = a + b*c$. Here $x$ is the *subject* and the *propositional clause* asserted about the subject $x$ is equal to $(a + b*c)$. Here the equal sign = as the verb and the object of the sentence is the algebraic phrase $(a + b*c)$. In algebraic sentences, the verb is usually an equality (=) or a greater than (>) or a less than (<).

An example of a physical law expressed in an algebraic equation were Galileo's laws of motion:

1. The first law stated that a physical object in motion with a velocity ($\underline{v}$) will remain in that motion with a constant value ($\underline{c}$), unchanging until acted upon by an external force: $\underline{v} = \underline{c}$, as long as no external force occurs.

2. The second law stated that when an external force ($\underline{F}$) occurs, motion is changed by an acceleration ($\underline{a}$), which is proportional to the mass ($m$) of the object:

$$\underline{F} = m * \underline{a},$$

where $a$ is the time derivative of $\underline{v}$, or $a = d\underline{v}/dt$, so $\underline{F} = m * d\underline{v}/dt$.
3. The third law stated that when an external force ($\underline{F}_e$) occurs upon a physical body, the inertial ($m$) of the physical body creates an equal and opposite reactive force ($\underline{F}_r$) upon the source of the external force: $\underline{F}_e = \underline{F}_r = -\underline{F}_e$

*The power of quantitative expression, algebraic equations, is that solutions to the equations (deductive inference) can reveal important behavior of the observed natural object.*

Newton used Galileo's second law, $\underline{F} = m * \underline{a}$, to quantitatively express the astronomical behavior of the solar system held together by a gravitational force, earth orbiting the sun, quantitatively expressed as a differential equation:

$$\underline{F}_g = \underline{F}$$

$$g * m * M/\underline{r}^2 = m * \underline{a}$$

$$g * m * M/\underline{r}^2 = m * d\underline{v}/dt$$

$$g * m * M/\underline{r}^2 = m * d^2\underline{r}/dt^2;$$

(where we noted that velocity $\underline{v}$ is a differential of the distance $\underline{r}$. and the acceleration ($\underline{a}$) is a second-order differential ($d^2\underline{r}/dt^2$) of the distance, A second-order differential is a rate-of-change-of-a-rate-of-change.)

The point of writing the orbital behavior of planets as an algebraic equation (using differentials) is that Newton could then mathematically derive all orbits which were solutions to the differential algebraic equations: circles ($x^2 + y^2 = r^2$) and ellipses ($ax^2 + by^2 = c$).

An algebraic equation is a quantitative sentence.

With quantitative algebraic sentences involved a verb of equality (=), one can perform kind of deductive reasoning on the sentence; and this is called a solution to the equation.

A deductive inference (deductive reasoning) in algebra thus occurs in finding a solution of an algebraic equation to obtain an expression with variables $y$ and $x$.

Quantitative sentences are important because deductive and inductive reasoning are very powerful logical techniques in mathematics, solutions to equations (deduction) and data samples (induction).

As an example of quantitative expression in the social sciences, a topic of operations research has evolved in the discipline of management science. This topic formulates decisions as quantitative expressions so that a solution to an operations equation identifies an optimal decision.

In the economics discipline, quantitative expressions of demand for a product and supplies of a product intersect when demand equals supply. At this equilibrium point, the price of that product is optimal for the economy in which the product is produced and marketed.

## Measurement: Random Variables

When measuring properties of an object, each act of measurement produces a slightly different value from other measurements. This variation of measurements by small amounts is called a "measurement error." All measurements have errors. For example, even measuring a simple property, such as the length of a pencil, has an error of measurement. But some measurements involve more complex phenomena than, e.g., a simple length measurement; a complex phenomenon that is influenced by more than one (several) phenomenal forces appears as a *random variable* in a measurement.

In physical measurements, one way to understand the origin of measurement errors in physical measurements is the following. In measuring a physical phenomenon, there is a primary, dominant force, and secondary, much smaller forces, influencing the measurement, then the measurement can be said to be a variable influenced by random smaller forces – a random variable.

As an example, the firing of an artillery gun lobs a shell over an arc from the place of firing to the place of impact. The dominant primary force in this firing phenomenon is the chemical-explosive force aimed by the barrel. But the chemical compound may vary slightly form shell to shell; and the recoil of the gun may alter its aiming very slightly. So repeated firings of shells aimed at the same distant target will place the impacts nearly together around the same spot, *but not exactly on the same spot*. Repeated shelling will not strike exactly the same spot because of the influence of secondary forces, such as (1) variation in the strength of the chemical explosion from shell to shell, (2) aerodynamic variation in the aim of a shell due to variation in the spin placed upon the shell by the rifling of the barrel, and (3) variation in the pointing of the barrel from firing to firing due to the inertial reaction of kickback on the artillery piece from the firing. These secondary forces will influence the target placement of each shell, so that n firings will be scattered in nearly n positions around the center of impacts.

> In physical action, secondary forces in any action will make the principle value of the action into a random variable, introducing randomness into the measurement.

Thus when measured, any variable will show variations in the measurement, which are due to random forces in (characteristics of) the measurement. In the mathematical topic of statistics, any measured variable is called a random variable. A random variable describes a property of a phenomenon influenced by a primary force and by much smaller secondary forces. The much smaller secondary forces introduce randomness into the measurement, which is included in the measurement error.

In the artillery example, the scattering of the impacts is called the *distribution* of the variable. The center of the distribution of impacts is called the *mean-value* of the variable.

The form of the distribution impacts is called the *shape* of the distribution. And the half-width of the shape of impact-distribution is called the *dispersion* of the random variable.

## *Illustration: Measurement Histogram*

Quantitative measurements must be repeated because there is always some error in any measurement. The instrument used to observe and measure the property of a natural object interacts with the object, thus introducing small changes in the measurement. For example, the repeated measurement of the length of a physical object using a measurement stick will yield a sample of the length, varying by small differences. The distribution of measurement results can be quantitatively expressed, plotted. And this is called a histogram plot. Some histograms will plot symmetrically around a central (mean) value; and some asymmetrically.

As an example of an asymmetric histogram plot, we look at a nineteenth-century research project by a Swedish physicist, Theodore Svedberg. Svedberg observed a solution of microscopic gold particles suspended in water. Because of the kinetic collisions between gold particles and water particles, the gold particles danced around in the solution. This phenomenon was called Brownian motion, after first being observed by Robert Brown in Scotland. As shown in Fig. 15.4, Svedberg

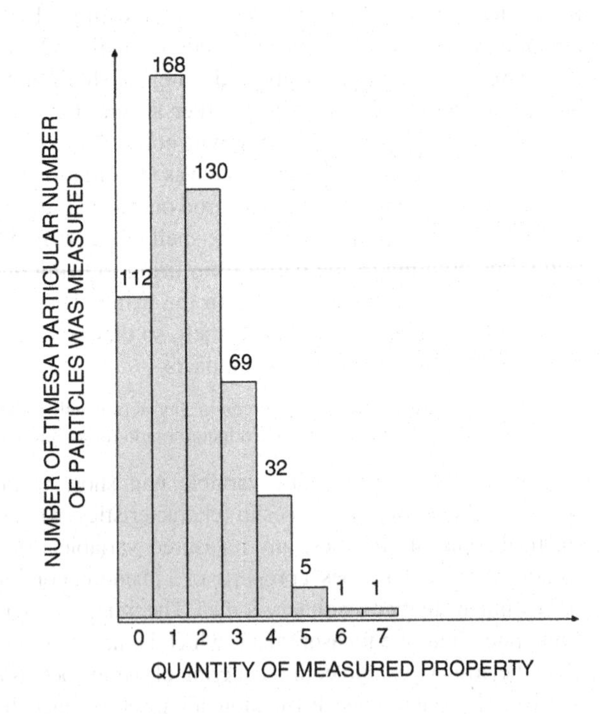


**Fig. 15.4** Histogram of number of gold particles in a volume of Brownian motion in a solution

focused upon a small region in the water and repeatedly counted the number of gold particles that danced into and out of the small region.

He counted 518 observations. In 112 of these, he saw 0 gold particles in the region; and in 168 observations, he saw one gold particle in the region. In 130 observations, he saw 2 gold particles; and in 69 observations, he saw 3 particles; with 32 observations of 4 particles, and 5 observations of 5 particles and 1 observation of 6 particles and 1 observation of 7 particles. One can see that the distribution of measurements is asymmetric because there are no measurements with less than zero gold particles dancing into a volume, and the average number of gold particles in a small volume varies mostly between 0 and 5.

Theodore Svedberg (1884–1971) was born in Sweden and graduated from Uppsala University with a Ph.D. in chemistry. He remained there as a professor and studied colloids (materials in solution) to empirically ground Einstein's theory of Brownian motion. He developed the technique of analytical ultra-centrifugation and used it to distinguish between pure proteins. He received the Nobel Prize in Chemistry in 1926.

(http:en.wikipedia.org. Theodore Svedberg, 2010)

## Probability

To quantitatively express a random variable and its histogram of measurements, one uses the mathematical topic called probability theory.

Probabilities denote the frequency of occurrence of events, but the events that occur should measure the properties of the observed object. Just denoting probabilities alone does not insure that a measurement is meaningful. Measurements should observe real properties, and probabilities estimate the frequency of occurrence of real events.

The next question one can ask about this kind of measurement is one of *prediction*, or expectation. If one makes a new observation, what is the likely number of gold particles one will observe in a single observation? This prediction (expectation) can be calculated as the probability of an occurrence of the state of number of gold particles in a small region of the solution, while in Brownian motion.

One can quantitatively express this idea of the variation in a sequence of measurement through the mathematical idea of probability. Suppose a set $S$ contains members $b$ described by a common property that defines the set $S$. Let $N$ be the cardinality of the set $S$ (the number of members $b$ in the set). Further suppose that the set $S$ can be partitioned into $k$ subsets $S_i$ so that $S$ can be expressed as the union of the sets $S_i$. $S = (S_1 \cup S_2 \cup S_3 \ldots \cup S_k)$. Let $n_i$ denote the cardinality of $S_i$. Then for each set $S_i$ a probability can be defined as $p_i = n_i/N$.

The probability $p_i$ is the ratio of the cardinality $n_i$ of $S_i$ to the cardinality $N$ of $S$.

If one randomly selects a member from the $S$, then the interpretation of the probability $p_i$ is the frequency that a member of set $S_i$ will appear in any pick.

Probability $p_i$ is the frequency of appearance of a member of set $S_i$ in random selections from the total set $S = \Sigma S_i$.

In the example of Svedberg, focusing upon a small region in the water and repeatedly counting the number of gold particles that danced into and out of the small region, Swedberg observed the following probabilities. The probability of finding 0 gold particles in the region was (112/518). The probability of finding one gold particle in the region was (168/518). The probability of finding two gold particles in the region was (130/518). The probability of observing 3 gold particles in the region was (69/518) observations. The probability of observing 4 gold particles in the region was (32/518). The probability of observing 5 gold particle in the region was (5/518); and the probabilities of observing 6 or 7 gold particles in the region were (1/518) and (1/518).

From a histogram, one can calculate the probabilities (frequencies of occurrence) of a measurement for an observation as the number in a column (of the measurement histogram) divided by the total number of observations. $p_i = n_i/N$.

$p_i$ = ($n$ gold particles in an $i$-th area) divided by (total number $N$ of gold particles)

## *Sum of Probabilities: Either-Or*

What is the probability of *either* a member $b$ of $S_i$ occurring *or* a member $c$ of $S_j$ occurring? The probability of an either-or event $P(i \text{ or } j)$ can be calculated from the probabilities of each event ($p_i$ or $p_j$) as:

$$P(i \text{ or } j) = (n_i + n_j)/N = n_i/N + n_j/N = p_i + p_j.$$

The probability of events occurring as either-or events is the sum of the individual probabilities.

## *Product of Probabilities: Both-And*

What is the probability of *both* an event of A occurring *and* then following an event of B?

Let $P(\text{A})$ = Probability of A occurring.

Let $P_{\text{A}}(\text{B})$ = Probability of B occurring, after A has occurred.

$$P(\text{A and B}) = P(\text{A}) * P_{\text{A}}(\text{B})$$

The joint probability is two independent events occurring is equal to the product of the two probabilities.

To see this result, suppose that the probability $P_A(B)$ is calculated from the occurrence $k$ events in the total of $n$ observations, given that A has previously occurred, so that $P_A(B) = k/n$.

Then $P(A \& B) = k/n$. But also $P(A) * P_A(B) = (m/n) * (k/m) = k/n$
Thus $P(A \& B) = P(A) * P_A(B)$

So the probability for the joint occurrence of mutually independent events is equal to the product of the probabilities

## Complete Set of Events

The probability of some event occurring in a complete set of events sums to one. This is so because one of the events in the complete set of events will occur. For example, let $S$ be partitioned into two sets $S_1$ and $S_2$, with cardinalities of membership $n_1$ and $n_2$, so that $n_1 + n_2 = N$.

Then $P_1 = n_1/N$ and $P_2 = n_2/N$.
Thus $(P_1 + P_2) = n_1/N + n_2/N = (n_1 + n_2)/N = (n_1 + n_2)/(n_1 + n_2) = 1$.

*For a complete set of events,* $\Sigma P_i = 1$

## Sampling Theory

If $x$ and $y$ are two independent random variables with each respective mean $\chi$ and $\mu$ and, then the mean of the product $x * z$ equals the product of the two means $\chi * \mu$.

If $x$ and $y$ are two independent random variables with each respective means $\chi$ and $\mu$, then the mean of the sum $x + y$ equals the sum of the two means $\chi + \mu$.

## Bernoulli's Formula

We saw in the illustration of the Brownian motion of gold particles, which Svedberg measured, that the histogram was asymmetric because there could be no appearance of negative events. However, most probable distributions turn out to be symmetric around a central value, a mean value. We next explain why symmetric distributions happen most often in observations of nature. But to explain this, we must first review a central formula in statistics called Bernoulli's Formula

His formula arises from taking several samples of properties of a population. Each time one samples to measure a property, one calls this a statistical "trial," so that in $n$ trials, there will be $n$ samples measured . In the repeated $n$ trials, an event

K occurs each time in a trial $n_i$; then, over the $n$ trials, the event K can occur $k$ times. So $n$ denotes the number of trials and $k$ the number of times an event K occurred in the $n$ trials.

Example: In a cotton fiber combing operation, long cotton fibers are viewed as advantages to spinning stronger thread. In this example, suppose that cotton fibers are found to be longer than 45 mm 75% of the time. We call finding a long fiber event A: $P(\text{A}) = 3/4$. And 25% of the time, a shorter fiber will be found; and we call finding a shorter fiber event B: $P(\text{B}) = 1/4$.

One can next calculate the probability that in $n$ trial, $k$ events will occur. Suppose we ask what is the probability that in three trials ($n = 3$), two of the three fibers will be long? (event A and $k = 2$) This can occur in three sequences of picking fibers in three trials:

$$\text{AAB},\quad \text{ABA},\quad \text{BAA}.$$

The probability for the combination then adds, since all three are independent trials:

$$P(\text{C}) = P(\text{AAB}) + P(\text{ABA}) + P(\text{BAA}).$$

As the events are also independent, the probability for any combination is multiplicative:

$$P(\text{AAB}) = P(\text{ABA}) + P(\text{BAA}) = P(\text{A}) * P(\text{A}) * P(\text{B}) = (3/4)(3/4)(1/4) = 9/64$$

The combination probability $P(\text{C})$ is calculated by adding the product of the probabilities of events but the number of possible combinations must be calculated for finding $k$ events in $n$ trials.

The formula for calculating this is called Bernoulli's formula.

Let $p$ be the probability of any K event occurring. If the K event occurs $k$ times in $n$ trials and not occurring $n - k$ times, the probability for this combination will be the product of the individual probabilities: $p^k(1 - p)^{n-k}$

The notation for number of $k$ ways a combination can occur in $n$ trials without regard to the order in a combination is: $C^n_k$.

$$C^n_k = \frac{n(n-1)\ldots[n-(k-1)]}{k*(k-1)*\ldots 2*1}$$

Where $n(n - 1) \ldots [n - (k - 1)]$ is number of ways $n - k$ things can be ordered. And $k * (k - 1)* \ldots 2 * 1$ is the number of ways $k$ things can be ordered. The assumption is that one does not care about the ordering of things in a combination, only the number in a combination.

The notation $n!$ is called "$n$-factorial" and means

$$n! = n*(n-1)*(n-2)*\cdots*1$$

$$C^n_k = \frac{n(n-1)\cdots[n-(k-1)]}{k(k-1)\cdots 2*1} = \frac{n(n-1)\cdots[n-(k-1)]}{k!}$$

$$C^n_k = \frac{n(n-1)\cdots(n-(k-1))}{k!} = \frac{\{n-k!\}*\{n(n-1)\cdots[n-(k-1)]\}}{\{n-k!\}*\{k!\}}$$

$$C^n_k = \frac{n(n-1)\cdots[n-(k-1)]\times(n-k!)}{k!(n-k)!} = \frac{n!}{k!(n-k)!}$$

$$C^n_k = \frac{n!}{k!(n-k)!}$$

The probability of an event K occurring $k$ times $n$ trials when the probability of a single event $P(\mathrm{K}) = p$ can be thus calculated as:

$$P_n(k) = (C^n_k)(p^k(1-p)^{n-k}) = \frac{n!(p^k(1-p)^{n-k}}{k!(n-k)!}$$

*This is called Bernoulli's formula.*

Bernoulli's formula will yield histograms (distributions) of a bell-shaped form. To see this, one next calculates the ratio:

$$\frac{P_n(k+1)}{P_n(k)}$$

$$\frac{P_n(k+1)}{P_n(k)} = \frac{\frac{n!}{(k+1)!(n-k-1)!}p^{k+1}(1-p)^{n-k-1}}{\frac{n!}{(k)!(n-k)!}p^k(1-p)^{n-k}}$$

$$= \frac{k!(n-k)!p}{(k+1)!(n-k-1)!(1-p)} = \frac{(n-k)p}{(k+1)(1-p)}$$

Next consider the ratio of $P_n(k+1)$ to $P_n(k)$:

$$\frac{P_n(k+1)}{P_n(k)} = \frac{(n-k)}{(k+1)}\frac{p}{(1-p)} > 1$$

when $(n-k)p > (k+1)\ (1-p)$ or when $np-(1-p) > k$

As long as $k$ does not exceed the value $np - (1-p)$, the probability of $P_n(k+1)$ keeps increasing.

$$\frac{P_n(k+1)}{P_n(k)} = \frac{(n-k)p}{(k+1)(1-p)} = 1 \quad \text{when } k = np - (1-p)$$

$$\frac{P_n(k+1)}{P_n(k)} = \frac{(n-k)p}{(k+1)(1-p)} < 1 \quad \text{when } k > np - (1-p)$$

As long as k exceeds the value $np - (1 - p)$, the probability of $P_n(k + 1)$ keeps decreasing.

Thus Bernoulli's formula produces a bell-shaped curve of a probability distribution with a maximum peak when $k = np - (1 - p)$

The most likely value of $k$ ($k_0$) must satisfy:

$$np - (1-p) \le k_0 \ge np - (1-p+1)$$

$$np - (1-p) \le k_0 \ge np + p$$

Dividing the inequalities by $n$ next yields

$$\frac{p-(1-p)}{n} \le \frac{k_0 > p}{n} + \frac{p}{n}$$

When $n$ is large $(1 - p)/n$ is very small, as is $p/n$.

So that nearly $p = k_0/n = p$, for large $n$.

*This indicates that when the number of trials n gets very large then the most probable frequency of the occurrence of an event approaches p – which is practically equal to the probability of the occurrence of an event K in a single trial.*

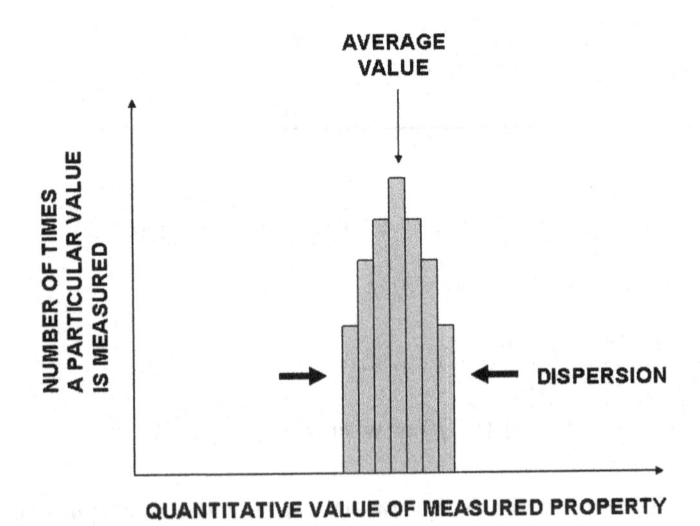


**Fig. 15.5** Histogram normal distribution

For a symmetric distribution (normal distribution) the histogram is symmetric around the average value and the dispersion is the width of the histogram, Fig. 15.5.

## *Mean Value and Squared-Dispersion of a Random Variable*

Let $x_i$ denote a measurement of a random variable $x$ in the $i$-th trial, occurring with a probability of $p_i$ for $n$ trials.

The *mean value* $\chi$ of a random variable $x$ is defined as:

$$\chi = \sum x_i p_i = (x_1 p_1 + x_2 p_2 + \ldots x_n p_n)$$

*This reads that the mean value of a random variable is calculated as the sum of the individual measurements $x_i$ weighted by their probabilities of occurrence $p_i$.*

The *average squared-dispersion* $\sigma^2$ of the scatter of a random variable $x$ is defined as:

$$\sigma^2 = \sum_i (x_i^2 - \chi^2) p_i$$

This reads that the square-of-the-dispersion of the mean of a random variable is calculated as the sum of the differences between the square-of-the-means and the squares-of-the-individual-measurements weighted by their probabilities of occurrence.

## *Shape of the Distribution of a Random Variable*

Shapes of distribution curves can be learned by plotting histograms of sampled data $x_i$. The shape of a distribution curve can be (1) symmetric or (2) asymmetric. A reason for symmetry or asymmetry in the distribution curve should be sought in understanding the forces upon a phenomenon. If the shape of the distribution of the random variable $x$ is symmetric around the mean value, then the distribution is called a normal distribution.

For a continuous variable $x$, the shape of a normal distribution is graphed as the equation of $y$ as a function of $x$ as: $y = f(x)$ (Fig. 15.6)

$$y = \frac{e^{(x-\chi)^2/2\sigma^2}}{(2\pi)^{1/2}}$$

## *Bayes' Rule*

One can see that probabilities are calculated from information gained in observations (measurement) of a phenomenon. In statistics, taking measurements of the

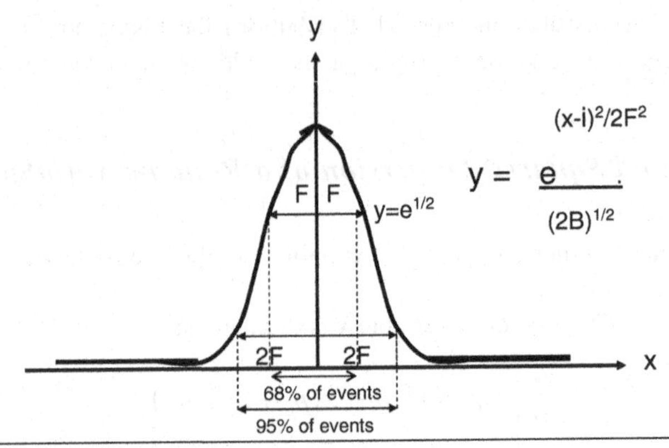


Sigma (F) = half-width of curve, where y=e1/2 Within one sigma half-width (F) of the normal distribution curve, 68% of the events occur. Within two sigma half-widths (2F) of the normal distribution curve, 95% of the events occur.

Since 95% of the events for a random variable x of a normal distribution will lie within 2F of the mean value i, then this range of 2F can be used as an estimate on the error of the measured value of the variable. $x = i \pm 2F$.

**Fig. 15.6** Normal distribution

property of an object is called sampling a population (population (elements) of the set describing the object). How does one combine samples of measurements? A proper procedure for combining estimates of probabilities from different samples of a population is called Bayes' Rule.

For example, suppose the testing of two spools of thread prepared on different machines has shown that a stand length of thread from the first spool has a probability of $P(\mathrm{A}) = 0.84$ for withstanding a standard test load. A thread from the second spool has a probability for $P(\mathrm{B}) = 0.78$ for withstanding the test load. What is the probability $P(\mathrm{A}$ and $\mathrm{B})$ that two samples taken from two spools from the different machines will both tolerate the standard test load?

$$P(\mathrm{AandB}) = P(\mathrm{A})^* P_{\mathrm{A}}(\mathrm{B}) = (0.84)^*(0.78) = 0.6552.$$

As another example, in the planting of wheat, a seed mixture is prepared mostly from stock I, with small additions from stocks II and III and IV. A seed taken from the mixture has the probabilities of being from these stocks as:

$$P(\mathrm{A}_1) = 0.96 \quad P(\mathrm{A}_2) = 0.01 \quad P(\mathrm{A}_3) = 0.02 \quad P(\mathrm{A}_4) = 0.01.$$

A desirable ear of wheat should have a tassel (fruit) bearing 50 grains or more. Let K denote an event that a tassel will bear at least 50 grains. The measured probabilities for a tassel having at least 50 grains from each stock are:

$P_{A1}(K) = 0.50$ for seeds of stock I
$P_{A2}(K) = 0.15$ for seeds of stock II

$P_{A3}(K) = 0.20$ for seeds of stock III
$P_{A4}(K) = 0.05$ for seeds of stock IV

In this notation, (K) denotes a K kind-of-event occurring and $P_{Ai}$ denotes the probability of the event occurring from the population $A_i$.

The probability of a 50-grain wheat occurring from:

Stock I is $P(A1\&K) = P(A1) \times PA1(K) = (0.96)(0.50) = 0.48$
Stock II is $P(A2\&K) = P(A2) \times PA2(K) = (0.01)(0.15) = 0.0015$
Stock III is $P(A3\&K) = P(A3) \times PA3(K) = (0.02)(0.20) = 0.004$
Stock IV is $P(A4\&K) = P(A4) \times PA4(K) = (0.01)(0.05) = 0.0005$
TOTAL 0.486

The total probability of a 50-grain wheat plant growing is thus:

$$P(K) = P(A_1 \& K) + P(A_2 \& K) + P(A_3 \& K) + P(A_4 \& K) = 0.486$$

$$P(K) = P(A_1) * P_{A1}(K) + P(A_2) * P_{A2}(K) + P(A_3) * P_{A3}(K) + P(A_4) * P_{A4}(K) = 0.486$$

Generalizing this notation, $P(K)$ denotes an K event taken from several populations $A_i$,

$P(A_i\&K) = P(A_i) * P_{Ai}(K)$ is the joint probability that event K occurs from population $A_i$.

Then the total probability of event K occurring from all the populations is:

$$P(K) = P(A_i \& K) = P(A_1) * P_{A1}(K) + P(A_2) * P_{A2}(K) + P(A_3) * P_{A3}(K) + P(A_4) * P_{A4}(K)$$

$$P(K) = P(A_i \& K) = \Sigma_i P(A_i) * P_{Ai}(K)$$

Bayes Rule rises from this. Denote the probability of event K occurring from a population $A_i$ as $P_K(A_i)$.

Then

$$P(A_i \& K) = P(A_i) * P_{Ai}(K)$$

But also

$$P(A_i \& K) = P(K) * P_K(A_i)$$

Thus

$$P(A_i) * P_{Ai}(K) = P(K) * P_K(A_i)$$

(This means that multiplication of probabilities of independent occurrences is commutative.)

So

$$P_K(A_i) = \frac{P(A_i) * P_{Ai}(K)}{\Sigma_i P(A_i) * P_{Ai}(K)} = \frac{P(A_i) * P_{Ai}(K)}{P(K)}$$

*This is called Bayes' Rule*

$$P_K(A_i) = \frac{P(A_i) * P_{Ai}(K)}{\Sigma_i P(A_i) * P_{Ai}(K)}$$

This says that the probability $P_K(A_i)$ that event K occurs from population $A_i$ is equal to the product of the probability $P(A_i)$ that the sample comes from population $A_i$ times the probability $P_{Ai}(K)$ that the event K occurs in a population $A_i$ divided by the total probability of the K event occurring:

$$(\Sigma_i P(A_i) * P_{Ai}(K)).$$

The practical use of Bayes' Rule is to revise the estimate of a probability of an event occurring from a population $A_i$ after more information (more data, more samples) have been made. All probabilities are calculated from samples. Therefore, when larger numbers of samples are measured, prior estimates of probabilities can be revised. Bayes' Rule is a systematic way to revise estimates of probabilities from more data.

For example, an artillery gun must be aimed to hit a target, whose position is estimated to be at a certain distance but less well known as to where on a horizontal line at the distance.

The estimated probabilities of the target being in segments:

$P(a) = 0.48$
$P(b) = P(b') = 0.21$
$P(c) = P(c') = 0.05$

Also the probabilities of hitting the target when aimed at the different segments are:

$P_a(K) = 0.56$
$P_{b'}(K) = 0.18$
$P_b(K) = 0.16$
$P_c(K) = 0.06$
$P_{c'}(K) = 0.02$

With this prior information, a new firing aimed at $a$ is completed, and the target is hit (event K occurs). With this new information, one can use Bayes' Rule to revise the probability estimates:

$$P_K(a) = \frac{P(a)P_a(K)}{P(a)P_a(K) + P(b)P_b(K) + P(b')P_{b'}(K) + P(c)P_c(K) + P(c')P_{c'}(K)}$$

$$P_K(a) = \frac{(0.48)(0.56)}{(0.48)(0.56) + (0.21)(0.18) + (0.21)(0.16) + (0.05)(0.06) + (0.05)(0.02)}$$

$$P_K(a) = \frac{0.2688}{0.2688 + 0.0378 + 0.0336 + 0.003 + 0.001} = \frac{0.2688}{0.3442} = 0.78$$

Thus after a new trial firing at point $a$ and hitting the target (event K), one can revise the probabilities of the location of the target $P_K(a)$ up from 0.48 to 0.78. This is how Bayes' Rule can use new trial information (new data) to revise previous estimates of probabilities of events.

## *Multiple Means in a Distribution*

When a distribution curve has more than one mean ( $\chi$ ) in the distribution of a random variable, then the curve can have several humps (maxima) in the curve (Fig. 15.7). This indicates that there are two different factors creating two distributions.

## *Correlation Between Two Random Variables X and Y*

We recall that a mathematical function between two variables $y$ and $x$ is a rule which determines $y$ given $x$ and is written as $y = f(x)$. Let both $x$ be a random variable, then $y$ will also be a random variable, associated with $x$ as $y = f(x)$. This means that $x$ is a random variable because there exist in nature secondary forces which add to the primary force that influences the phenomenon which make measurements of $x$ random. So too will that randomness in $x$ carry through to $y$, according to the function rule of $f(x)$.

## *Least-Squares Fit to a Straight-Line Function*

Sometimes the two random variables $x$ and $y$ are independently measured, and one asks if there exists a functional relationship between them. Is there an $f()$ so that $y = f(x)$? Does the random variable $y$ *correlate* with $x$?

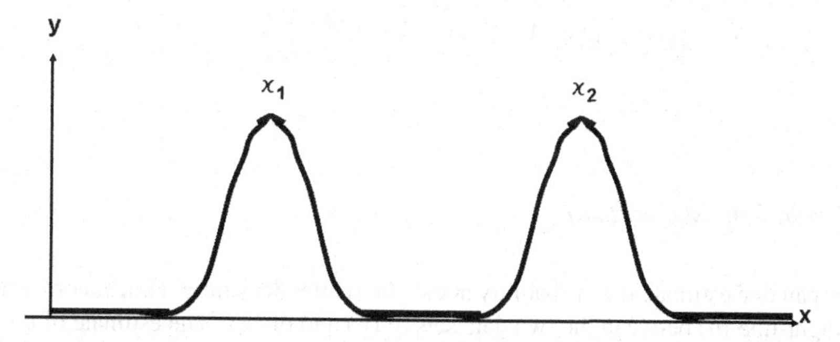


**Fig. 15.7** Multiple means in a distribution. When a distribution curve has more than one mean, ($x_1$) and ($x_2$) in the distribution of a random variable, then the curve can have several humps (maxima)

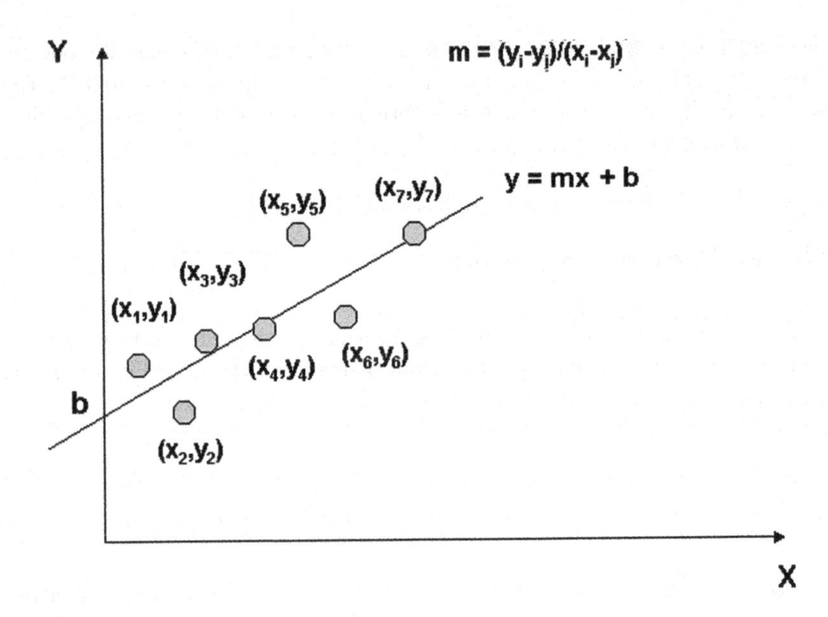


**Fig. 15.8** Correlation

The simplest case is to ask if there is a linear co-relation: $y = f(x) = mx + b$? As shown in Fig. 15.8, one tries to fit a geometric straight line to the data of $y$ and $x$ plotted on a two-dimensional geometric form.

The method of least-squares fit is to find a straight line for $m$ and for $b$ such that $\Sigma\,(y_i - y)^2$ is minimized; where $y$ is the mean of the data set $y_i$; and $x$ is the mean of the data set $x_i$.

First calculate $y = \Sigma y_i$ and then calculate $x = \Sigma x_i$.

Then the formulae for calculating a least-square fit for $m$ and $b$ are:

$$m = \sum (x_i - x)(y_i - y) / \sum (x_i - x)^2$$

$$b = \left\{\left(\Sigma x_i^2\right)\left(\Sigma y_i^2\right) - \left(\Sigma\left(x_i\right)\left(\Sigma x_i y_i\right)\right\} / \left\{n\Sigma x_i^2 - \left(\Sigma x_i\right)^2\right\}$$

## *Correlation Estimation*

How can one estimate the probability at that the scatter does infer a linear correlation (straight-line fit) between the two data sets of $(y_i)$ and of $(x_i)$? The estimate of this is called a *correlation coefficient (r)*. As shown in Fig. 15.9, the scatter (dispersion) around a straight line can be little or great.

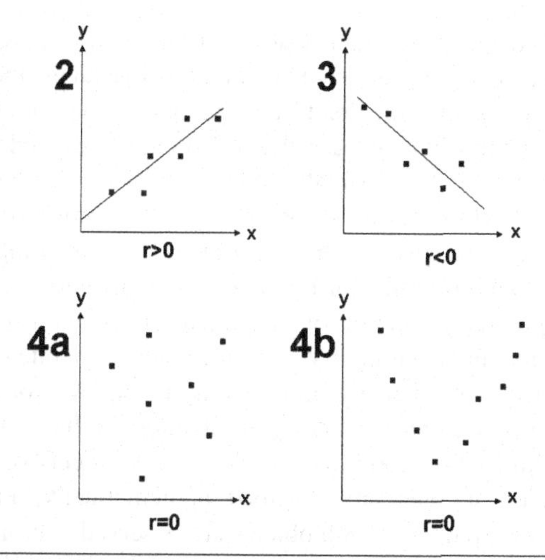


In the charts 2 and 3, there are positive and negative linear correlations between x and y.

In chart 4a, there is no correlation between x and y.
In chart 4b, there is not a linear correlation but a parabolic correlation.

**Fig. 15.9** Correlation data

The correlation coefficient $r$ between two data sets of $(y_i)$ and of $(x_i)$ can be calculated by the following formula:

$$r = \{\Sigma(x_i - x)(y_i - y)\} / \{\Sigma(x_i - x)^2 \, \Sigma(y_i - y)^2\}$$

1. The value of $r$ lies between −1 and +1: $-1 < r < 1$
2. If $r > 0$, then a positive linear relation between $y$ and $x$ is likely.
3. If $r < 0$, then a negative linear relation between $y$ and $x$ is likely.
4. If $r = 0$, then no linear relation likely exists between $y$ and $x$.

In the charts 2 and 3, there are positive and negative linear correlations between $x$ and $y$.
In chart 4a, there is no correlation between $x$ and $y$.
In chart 4b, there is not a linear correlation but a parabolic correlation.

## Summary

1. The idea of quality distinguishes one object in nature from another.
2. The *quality* of a thing in nature is defined by the *property* of an *object*.
3. The idea of quantity compares the amount of quality an object has or the amount of quality comparisons between objects.

4. Set theory is the mathematical topic that was constructed to express the idea of quality; and a set can be defined as a common property that all the members share.
5. In the concept of quantity, the first idea is that of counting, or numbers; and if one matches the cardinality of an integer in a one-to-one correspondence with the list of members of any set, one can count the numbers of the set (cardinality of any set).
6. A set of numbers operating under one operation (addition or multiplication) is called a "mathematical group"; and a set of numbers operating under two operations (addition and multiplication) is called a "mathematical algebra."
7. A quantity expressed in one bit of information is called a scalar quantity; and when information about a property must have more than one number (more than one bit of information), then a vector quantity is required for description.
8. The differential operation ($da/dt$) on an algebraic value ($a$) is defined as the mathematical limit of the ratio $(a_2 - a_1)/(t_2 - t_1)$, as $\Delta t$ gets very, very small.
9. The properties of an object are described mathematically in a quantitative perceptual space, the space in which objects are observed; as for example, a three-dimensional geometric space has three orthogonal axes ($X$, $Y$, $Z$).
10. An algebraic quantitative sentence of a subject and verb and object is constructed with the equal sign (=) as the verb between a subject $x$ and a *propositional clause* asserted about the subject (e.g., $x = a + b$*c); and an algebraic formula (quantitative sentence) can be expressed in analytical geometry as the set of points in the two-dimensional space which satisfy the functional mapping of $y = f(x)$
11. Since quantitative algebraic sentences involve a verb of equality (=), one can perform kind of deductive reasoning on the sentence; and this is called a mathematical solution to the equation.
12. When measuring properties of an object, each act measurement produces a slightly different value from other measurements; and this variation of measurements by small amounts is called a "measurement error," randomness in measurement.
13. Also a complex phenomenon which is influenced by more than one (several) phenomenal forces appears as a random variable in a measurement.
14. To quantitatively express a random variable and its histogram of measurements, one uses the mathematical topic called probability theory.
15. Probability $p_i$ is the frequency of appearance of a member of set $S_i$ in random selections from the total set $S = \Sigma S_i$.
16. The probability of events occurring as an either-or event is the sum of the individual probabilities, and the joint probability of two independent events occurring is equal to the product of the two probabilities.
17. The mathematical techniques for using probabilities in measurement constitute a mathematical topic called "statistics."

## Notes

[1]There are many, many texts on the different mathematical topics. A classical text on abstract algebra is Van der Waerden (2003). Modern texts on probability and statistics include Schervich (1996).

# Chapter 16
# Handbook of Research Methods

## Introduction

We can now summarize the questions about managing science as:

1. How does one view science as a whole?
2. What are the intellectual frameworks of science?
3. What is scientific method?
4. What role does modeling play in scientific method?
5. How does methodology differ between social and physical sciences?
6. How does one manage a research project in a university?
7. How can one identify a research issue?
8. How can one write a persuasive research proposal?
9. How should one manage a research center in a university?
10. How can science research be effectively transferred to technology research?
11. How should a research agency plan a research-funding program?
12. How should officers in a research agency manage a research-funding program?

## How Does One View Science As a Whole?

Viewed in totality in all its activities, science can be seen as a process and content of inquiry of research which asks and answers basic questions about nature. Inquiry into nature focuses upon the questions: what things exist in nature and how do these things interact? Science as inquiry can be characterized by four kinds of activities which address the following questions:

1. Of what is the universe made? This is the basic question about nature. We can call this activity about the content of nature as *scientific content (ontology)*. Scientific content expresses the current state of knowledge in science about the nature of the observable world.

F. Betz, *Managing Science*, Innovation, Technology, and Knowledge Management 9,
DOI 10.1007/978-1-4419-7488-4_16, © Springer Science+Business Media, LLC 2011

2. How do we know this? This is the basic question about method. Science now answers that question with the methods of science in experimentally-grounded theory. We can call this *scientific method (epistemology)*. Scientific method proposes the proper philosophical approach (methodology) as a set of research tasks in the process of advancing knowledge.
3. What procedures and resources are necessary to inquire into nature, to use scientific epistemology? This is the basic question about the methodological approach and organizational resources needed to conduct scientific research. We can call this *science administration (research)*. Science administration organizationally funds and/or performs research in the methodological forms of scientific inquiry – as research proposals.
4. Why is science useful to society? This is the basic question about the value of science to human civilization. Science is practical to society by providing a knowledge base for technological innovation. We can call this *scientific application (technology)*. Science administration organizationally funds and/or performs the process of advancing knowledge in terms of research tasks.

| | PROCESS OF KNOWLEDGE | STATE OF KNOWLEDGE |
|---|---|---|
| PHILOSOPHY | SCIENTIFIC METHOD (EPISTEMOLOGY) | SCIENTIFIC CONTENT (ONTOLOGY) |
| ORGANIZATION | SCIENCE ADMINISTRATION (RESEARCH) | SCIENCE APPLICATION (TECHNOLOGY) |

## What Are the Intellectual Frameworks of Science?

A "scientific paradigm" is an "intellectual-framework-behind-scientific-theories." A scientific paradigm does not describe the "details of research" at the cutting edge of disciplinary specialties. Instead, a paradigm describes the meta-theory, the framework, in which the research details (experimental formulation and theory) are constructed. A paradigm is an intellectual framework within which the scientists observe, describe, and explain nature. A paradigm is a "meta-logic" to the "logic" in a theory – a kind of "meta-theory" to the theories in a scientific discipline. The four paradigms used in science as "meta-theories" are: Mechanism, Function, System, and Logic.

| | *WORLD* | *SELF* |
|---|---|---|
| *MATTER* (SUBSTANCE) | MECHANISM | FUNCTION |
| *MIND* *(IDEA)* | SYSTEM | LOGIC |

## What Is Scientific Method?

Scientific method focuses upon the construction of theory to explain nature and grounded in the experimental observation of nature. The process of scientific inquiry is not linear, going directly either from empiricism-to-theory nor theory-to-empiricism. Instead in the history of science, scientific progress has proceeded circularly in the logic of empiricism and theory, going around and around. Yet even in a circular interaction between experiment and theory, a basic premise of scientific inquiry is that nature be observable. Or conversely, science only studies what is observable in nature. Empiricism in science is grounded in observing nature. Theory is grounded in the empirical observations of nature.

**CIRCULARITY IN HISTORICALINTERACTIONS BETWEEN RESEARCH TECHNIQUES**

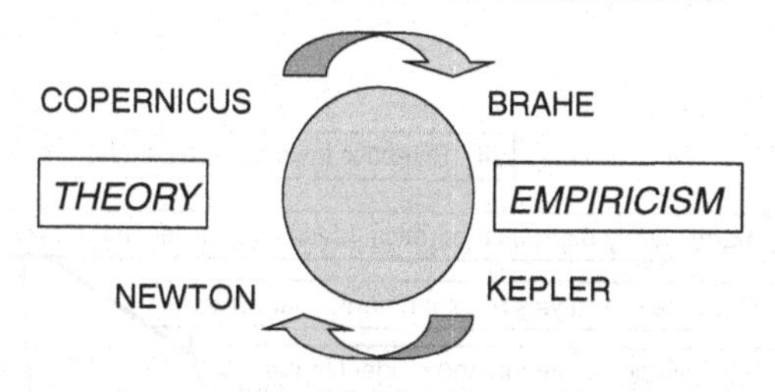


Figure 4.1 lists the different methodological techniques in the empiricism and theoretical aspects of scientific research:

*Experiment* is the controlled observation of nature, experiencing nature through the human senses aided by scientific instruments.

*Instrument* is a device which provides a means of extending the range and sensitivity of human sensing of nature.

*Measurement* is an instrumental technique in observing nature that results in quantitative data about the phenomenal thing.

*Perceptual Space* is conceptual framework within which a natural phenomenon is described.

*Analysis* is inferring a mathematical pattern in the quantitative data of a set of experiments.

*Phenomenological Law* is a generalization of relationships observed between natural objects.
*Model* is a symbolic simulation of the processes of a phenomenal object.
*Theory* is a symbolic description and explanation of a phenomenal field of objects.

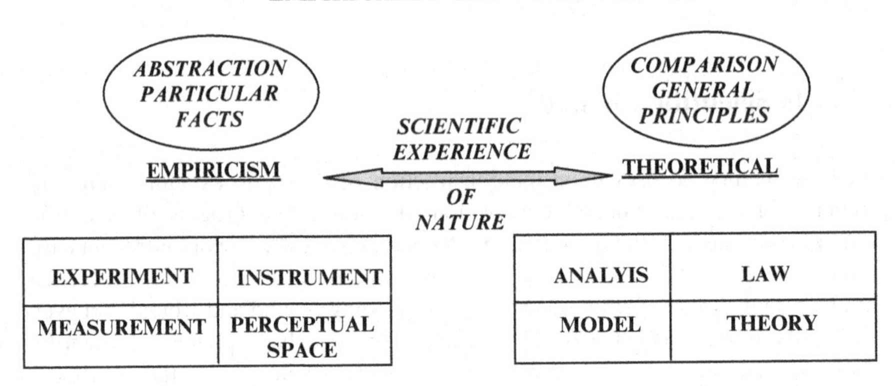


In any scientific event, one can identify the kinds or research techniques to advance science. For example, we saw in the previous case of Copernicus-to-Newton, that there were several research techniques used.

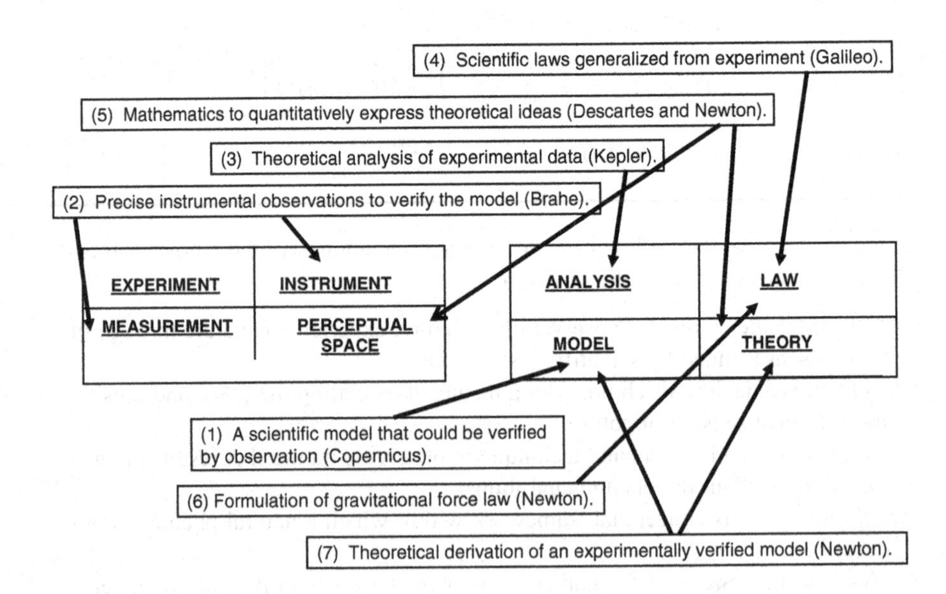

# What Role Does Modeling Play in Scientific Method?

Scientific models play an important role in theory construction. Models construct a formal abstraction and generalization (theoretical model) of the object to derive the observed properties and relationships and phenomenological laws of natural objects. The empiricism of the object interacts with the construction of its theoretical model as an intermediary in connecting empiricism to theory.

In the example of Bohr's quantized atom model and the subsequent development of quantum theory, we saw how the atomic spectrum of the hydrogen atom (Rydberg's phenomenological law) was important to Bohr in his construction and validation of his quantum model of Rutherford's atom. Next, building upon Bohr's model, Heisenberg, Born, and Jordan constructed a matrix-version of quantum theory, soon followed by Schrodinger's wave-version of quantum theory.

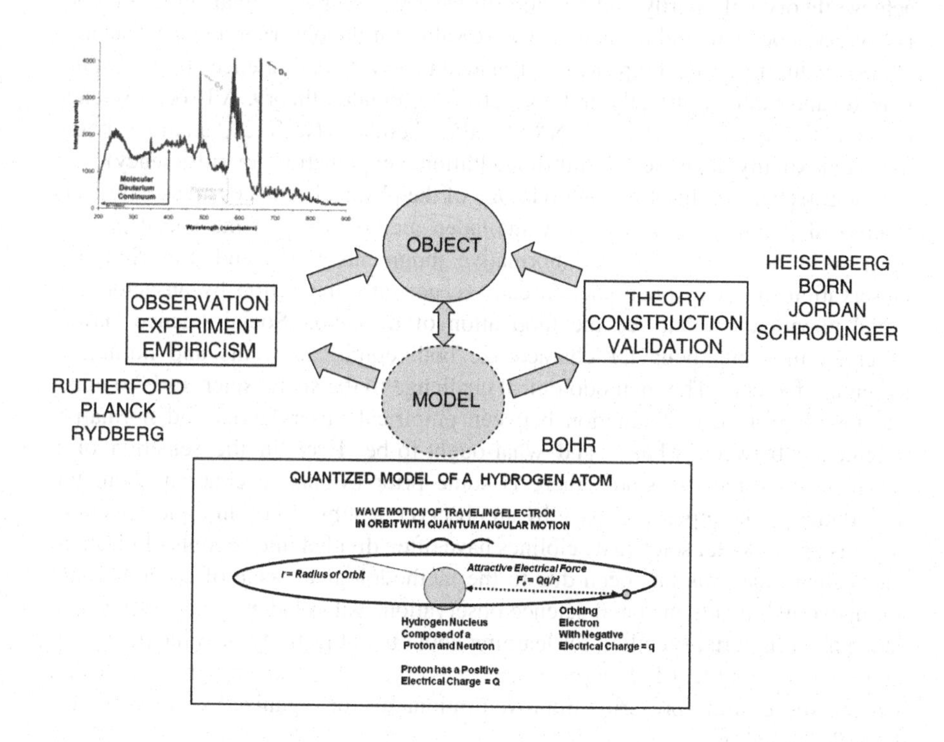


Particularly important types of social science models are: (1) topological connective, (2) topological flow, (3) dynamic systems, and (4) optimization models.

## How Does Methodology Differ Between Social and Physical Sciences?

One difference between scientific disciplines is that the scientific paradigm of mechanism is central to physics and biology but not to mathematics and social science. In contrast, the scientific paradigm of logic is central to mathematics and socials science but not to physics and biology. Moreover, the scientific paradigm of function is only shared by biology and social science, and not used in physics or mathematics. However, the scientific paradigm of Systems is used by all the disciplines. A second difference is in phenomenological laws. Only the physical and biological sciences have causal explanations, phenomenological laws of cause and effect. Mathematics and social science do not use causality in explanation, instead using prescriptive or thematic explanations. A third difference is in objectivity. The physical science disciplines can achieve theory of absolute objectivity, knowledge of a physical object independent of context and observer (the principle of invariance in the formulation of physical theory). In contrast, social science disciplines can achieve theory only partly and temporarily nearly objective. Social science theory is context dependent and influenced by the values of the observer. Physical science attains "value-free" and empirically grounded theory. Social science attains "value-loaded" and both empirically and normatively grounded theory. All social science laws and theory use prescriptive (N&S) and/or thematic (N&S) explanations, being thus contextually dependent (requiring additional explanation for sufficiency).

The causal and value-free methodology of the physical sciences gives only facts (empirical judgments), while the value-laden methodology of the social science must yield both empirical and normative judgments (facts and prescriptions). Observation of social phenomenon can see not only the action by an actor in a social phenomenon but see the motivation of the actor. Social science "laws" observed in social behavior characterize both empirical events and normative meaning of events. The methodological challenge to the social sciences is to make and maintain a clear distinction between empirical observations and normative judgments: between what-is and what-ought-to-be. Even in the selection of a social-science-object-to-study, there is some prior evaluative concern about the importance of the topic as chosen by an observer, a subjectivity in topic selection. Schools of the social science disciplines have often divided into schools of idealism and realism; and this has been due to the methodological issue of empirical and normative judgments in social science observation. Value-laden social science can be the most important method for learning about the objectivity of value (universal values) in modern life. Modern social science is essential to social progress in society, if it can successfully prescribe improved rationality for organized social activities (ethical rationality).

Social science can use "ideal-type" social theory, which is a generalization of the principles-of-order that a society thinks it should be operating in a given historical situation. An ideal type in social theory is an abstraction of the universal intentions of a society. Universal intentions in an Ideal type are expressed as principles-of-order to guide social behavior. In the social sciences, the application of social science knowledge may always potentially alter social nature through consulting practice.

In consulting practice, the observed participant values (intentions) can be temporally and partially separated from the value judgments by the social science practitioner on these intentions. And in such a partial separation, there can be a temporary distinction between participants' values and an observer's values. It is this kind of methodological partial separation of empiricism from normative judgment that makes the kinds of business and policy and practitioner operations of social science professors useful to their clients. In the empirical description of what a client is doing (action) and in thinking what the action is accomplishing (purpose) may occur as an unintended consequence for the client (unintended by the client's own intentions) and also unrecognized by the client. Empirical observations by a practitioner, on the consequences of a client's action, may show that the client's actions may not be attaining the client's intention. In such a case, a practitioner might suggest a different course-of-action to the client: a prescription for action. If such a prescription is science based, it is more likely to prove technically effective than when not science based.

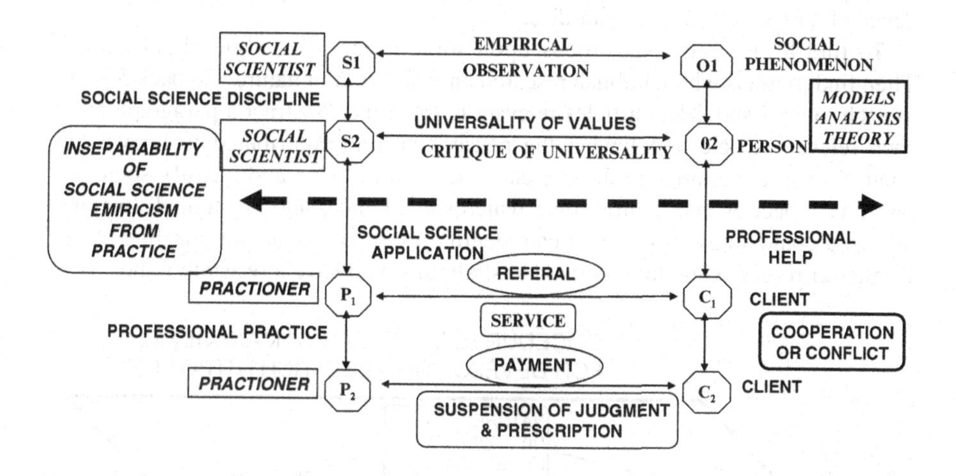


## How Does One Manage a Research Project in a University?

Since the German reform of universities in 1800, professors in research universities are paid salaries for educational services but are selected for research skills and the ability to obtain external research funding. In the university, the research manager is a professor. Research requires external funding, and most universities restrict the submission of proposals for external funding to faculty. External funding is required to support graduate students who perform the research, research labor. Managing research therefore requires a professor to: (1) methodologically conceive the experiment and write a research proposal, (2) submit the proposal to an external agency and win external research funding, (3) hire graduate students to perform the research, and (4) supervise the graduate students.

## How Can One Identify a Research Issue?

A research proposal should first define the research issue of the proposed project and should identify all the relevant recent articles in scientific journal that together describe the frontier of science (or state-of-the-art of technology) in the area of the research. The issue is an opportunity to advance that knowledge. Books on a topic do not get one to the frontiers of knowledge, only recent articles in scholarly journals depict the recent frontiers of knowledge.

Scientific communication is an essential procedural feature of science; and for this purpose scientific societies were organized. As science evolved from the 1600s in Europe, the organization of a "scientific society" also evolved. The mission of a "scientific society" is to facilitate scientific communication, through scientific meetings and journals. The formation of a society is an important step in the emergence of a new discipline or sub-discipline.

To find a research issue, focus upon some aspect of a topic about nature. Then find journals which publish research on that aspect of nature. Go back 5 years in each journal and select articles relevant to the topic. Abstract a paragraph from each relevant article about what the research therein contributed to knowledge about the topic. Summarize the present state of knowledge and identify some yet puzzling aspect of that natural topic which should and could be studied. Imagine how you could study this. If you can identify a research issue yet puzzling at the frontier of research and how you could study this, you have a research issue.

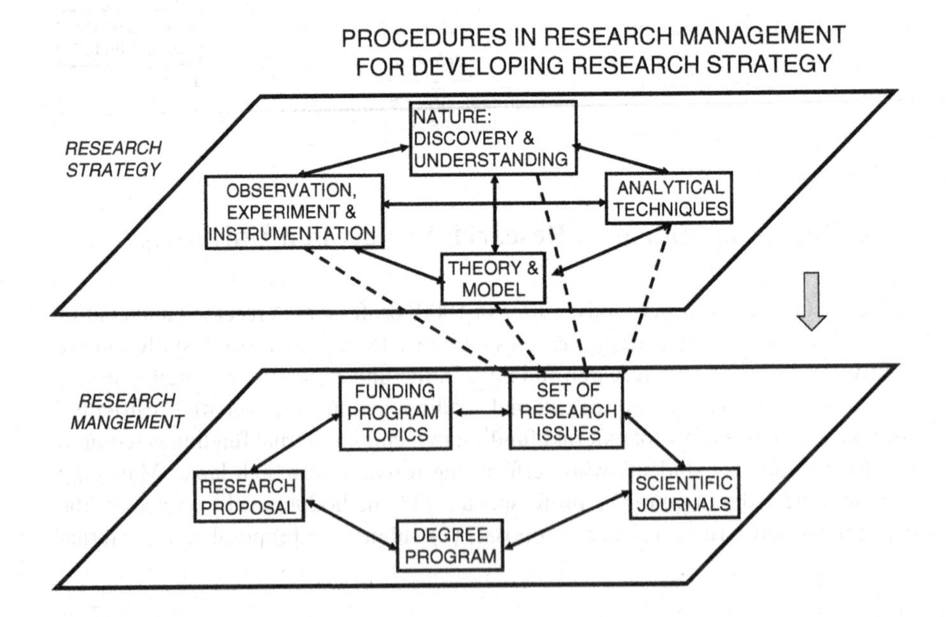

## How Can One Write a Persuasive Research Proposal?

In a proposal, the question asked about nature is the *research issue* of the proposal. The way the question will be answered is the *scientific method* of the proposal. The answer will be of what value to whom, *research impact*? A scientific research proposal for a research grant should clearly pose:

1. A scientific or technical vision of a *research issue*
2. How the research can be performed in *research methodology*
3. Potential use of the research results as *research impact*

The format of a research proposal should first address the four criteria of research vision, research methodology, research team, and research impact:

1. *Research issue – vision*
   (a) What is the frontier-of-science and/or state-of-art-in-technology in the specific research field?
   (b) What are the issues or challenges or barriers to scientific/technical progress in the field?
   (c) Which of these issues/challenges/barriers (research issues) will the research project address?
2. *Research methodology*
   (a) What research techniques will be used to address these?
3. *Research team*
   (a) What kinds of skills will be required in the methodology.
   (b) Which members of the research team will provided these.
4. *Research equipment*
   (a) What research instruments will be needed to perform the research?
   (b) How will these be accessed or purchased?
5. *Research impact*
   (a) How would successful results of the proposed project contribute to the advancements of knowledge and/or used on important problems?
   (b) How will such progress be communicated or implemented to and for whom?
6. *Research budget*
   (a) What funding amount is requested for the project and for how long?

## How Should One Manage a Research Center in a University?

A research center may be needed to add together enough "pieces" of research to attain scientific progress. There are two sizes of "research projects" now in university research: (1) doctoral-size research project and (2) a group of research projects (critical mass of projects). The first is necessary for the university integration of education and research. The second is necessary for the advancement of scientific progress in a "critical mass" of a research effort. The idea of a "critical mass" of research efforts (a group of research projects) may be necessary for significant scientific progress in a university setting, requiring a university research center.

A research center director must have a methodological capability of envisioning a research issue and the proper methodological direction of research for scientific and technological progress. A research director must transform this research vision into a concrete research plan by assembling a team of researchers who can formulate research projects along the methodological direction of the research issue, a strategic research plan for the center. Methodology in center research lies in envisioning a multidisciplinary research issue, and organization in center research lies in assembling a proper multidisciplinary research team with projects about the research issue.

Although a multidisciplinary research center has an overriding vision of its strategic research issue, this is implemented as a portfolio of research projects. The projects in a university research center are pieces of a larger research vision of the center. The management role of the faculty research-group leaders in the center is to formulate research projects and guide doctoral candidates in performing the research projects. The role of the director is to ensure that all the research projects add up to a "critical mass" of research to advance science and technology – toward the strategic research issue of the center.

In the science-technology interaction, research for technology cannot be planned until after a basic technological invention has occurred. After the basic technology invention, research can then be planned by focusing on the generic technology system, production processes, or underlying physical phenomena. Technology-focused-and-targeted basic research can be planned for:

1. Generic technology systems and subsystems for product systems
2. Generic technology systems and subsystems for production systems
3. Physical phenomena underlying the technology systems and subsystems for product systems and for production systems.

Research can be targeted to improve a technology system through improving any aspect of the system:

1. Improved components
2. Improved connections
3. Improved materials
4. Improved power and energy
5. Improved system control
6. Improved system boundary

The physical phenomena underlying a system can be focused on any of the system aspects:

1. Phenomena involved in the system boundary
2. Phenomena underlying components
3. Phenomena underlying connections
4. Phenomena underlying materials
5. Phenomena underlying power and energy
6. Phenomena underlying system control

## How Can Science Research Be Effectively Transferred to Technology Research?

Industrial researchers are very sophisticated about current technology and, in particular, about its problems and about locating and identifying the roadblocks to technical progress. However, because of the applied and developmental demands on industrial research, they have limited time and resources to explore ways to leapfrog current technical limitations. On the other hand, academic researchers have time, resources, and students to explore fundamentally new approaches and alternatives that leapfrog technologies.

Together, industry and university researchers can see how to effectively bind a technological system in order to envision a next generation of technology (NGT). This boundary is an important judgment, combining judgments (1) on technical progress and research directions, which together might produce a major advance and (2) over the domain of industrial organization that such an advance might produce a significant competitive advantage.

To effect industry and university research cooperation on next-generation-technology innovation, a bridging institution is necessary, because industries and universities live in almost completely different universes. The industrial universe is a world of technology, short-term focus, profitability, and markets. In contrast, the university universe is a world of science, long-term view, philanthropy, and students. This is the role of a university research center to bridge the two views, creating a balance between (1) technologically pulled research and scientifically pushed research, (2) short-term and long-term research focus, (3) proprietary and nonproprietary research information, and (4) vocationally relevant education and professionally skilled education. These are the issues inherent in industry and university cooperation. Properly handled, these provide creative tension:

1. Linking technology and science in real-time operation
2. Creating progress in knowledge and developing the technological competitiveness of nations

## How Should a Research Agency Plan a Research-Funding Program?

To obtain money to fund research, a government research-funding agency is budgeted through a government budget. An agency budget is then divided among agency research programs which give out research grants and/or contracts. To obtain a research budget from the government, a research-funding agency must submit annual budget plans with research strategies. A science/technology funding agency can use strategic research issues, planned as interactions between nature, problems, invention, and systems research.

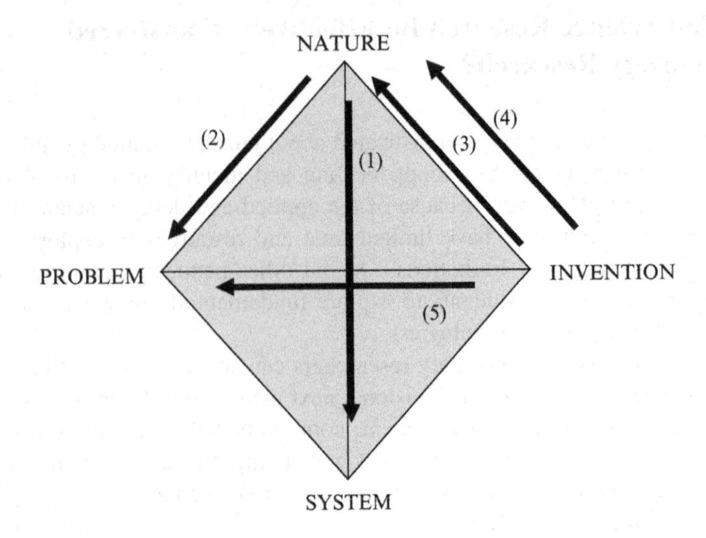


Strategic program initiatives can connect (1) progress in the science of nature for improved systems representation, (2) progress in science for improved problem analysis, (3) and (4) inventions of new technologies for improved instrumentation and techniques for science, and (5) invention of new technologies to solve problems. Science-roadmaps can facilitate the planning of research program initiatives based upon scientific opportunities to improve the understanding of nature underlying societal problems and applications. The term of "technology-roadmap" is often used by research funding agencies to describe the technical benefits from funding scientific and technology research.

## How Should Officers in a Research Agency Manage a Research-Funding Program?

A large research-funding government agency organizes into directorates and then into divisions and then into research programs. Research proposals from universities and/or companies are electronically submitted to the research programs for research grants and/or contracts. These proposals are reviewed for research quality and relevance by means of peer reviews. A program office conducts the peer review process and recommends upon an award or declination for a proposal. The professional responsibilities of the program officer are:

1. Read the proposal and identify the proper kinds of scientific expertise to review the research quality of the proposal

2. Select peer reviewers in academic, governmental, or industrial research positions who have published in the areas of expertise and are qualified to judge the research quality
3. Electronically forward a copy of the research proposals with a request to review the research proposal with a proper regard to confidentiality of information in the proposal
4. Receive the reviews and upon the advice in the review decide whether or not the proposal should be funded (graded as an excellent proposal, very good, good, fair, or poor proposal)
5. Allocate a grant budget to those of the excellent and very good proposals the program officer decides to fund and can fund within the program budget
6. In some programs, a program officer will use an external panel (selected by the program officer to offer funding advice) which will meet under the auspice of the program officer and will offer funding advice by ranking in terms of quality the reviewed research proposals, and then the program officer on this advice allocates funds to proposals (projects or centers) within the program budget constraint
7. Forward recommended proposals for grants to the division director overseeing the program
8. Upon approval by the division director, the proposals to be funded as research grants are sent to the NSF Division of Grants in the Directorate of Budget, Finance & Award Management
9. The Division of Grants then notifies the professor and university administration (for each approved proposal) that a research grant based upon the proposal will be awarded for a certain amount to begin by a certain date
10. After formal notification of the award, the university administration establishes a research budget line upon which the professor can draw to fund and begin the research project
11. As research progresses in the project, the professor sends any required progress reports to NSF and the university administration sends requests for reimbursement of research spent on the grant

For example, the review process in NSF is shown in Fig. 1.3.

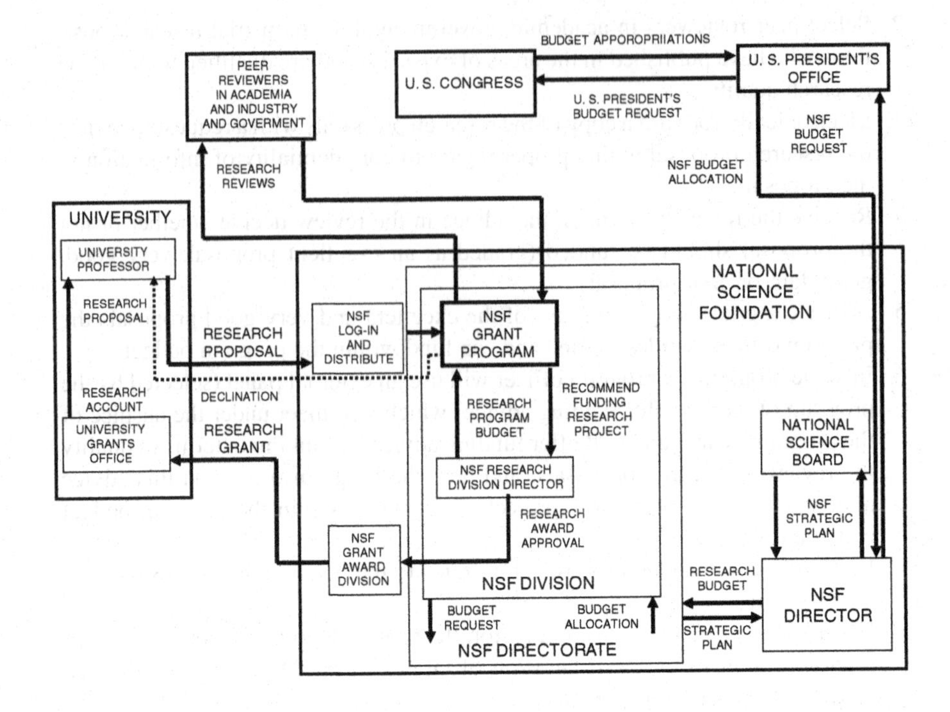
BUDGET APPOROPRIIATIONS
U. S. CONGRESS
U. S. PRESIDENT'S OFFICE
U. S. PRESIDENT'S BUDGET REQUEST
PEER REVIEWERS IN ACADEMIA AND INDUSTRY AND GOVERMENT
RESEARCH REVIEWS
NSF BUDGET REQUEST
NSF BUDGET ALLOCATION
UNIVERSITY
UNIVERSITY PROFESSOR
RESEARCH PROPOSAL
RESEARCH ACCOUNT
UNIVERSITY GRANTS OFFICE
RESEARCH PROPOSAL
DECLINATION
RESEARCH GRANT
NSF LOG-IN AND DISTRIBUTE
NSF GRANT PROGRAM
NATIONAL SCIENCE FOUNDATION
RESEARCH PROGRAM BUDGET
RECOMMEND FUNDING RESEARCH PROJECT
NSF RESEARCH DIVISION DIRECTOR
RESEARCH AWARD APPROVAL
NSF GRANT AWARD DIVISION
NSF DIVISION
NATIONAL SCIENCE BOARD
NSF STRATEGIC PLAN
RESEARCH BUDGET
NSF DIRECTOR
BUDGET REQUEST
BUDGET ALLOCATION
STRATEGIC PLAN
NSF DIRECTORATE

# References

Adler, A. 1927. Understanding Human Nature.
Austin, S. 1986. Parmenides: Being, Bounds and Logic. New Haven: Yale University Press.
Ayres, R. U. 1989. The Future of Technological Forecasting, Technology Forecasting and Social Change, 36(1–2, August), 49–60.
Banks, M. 2008. On the Way to the Web: The Secret History of the Internet and Its Founders. Berkeley: Apress.
Besanko, D., David, D., and Mark, S. 2000. Economics of Strategy, 2nd ed. New York: John Wiley & Sons.
Betz, F. 1996a. Targeted Basic Research. In Handbook of Management of Technology, Gus Gaynor (ed.). New York: McGraw-Hill.
Betz, F. 1996b. Forecasting and Planning Technology, in Handbook of Management of Technology, Gus Gaynor (ed.). New York: McGraw-Hill.
Betz, F. 2001. Executive Strategy. New York: John Wiley & Sons.
Betz, F. 2003. Managing Technological Innovation, 2nd ed. Chichester: John Wiley & Sons.
Bienkowska, B. 1973. The Scientific World of Copernicus. New York: Springer.
Boisvert, R. 1997. John Dewey: Rethinking Our Time. Albany: SUNY Press.
Brainerd, J. G., and Sharpless, T. K. 1984. "The ENIAC," Proceedings of the IEEE, Vol.72, No. 9, September, pp. 1202–1205.
Business Week. 1984. Boitech Comes of Age, January 23, pp. 84–94.
Bush, Vannevar, 1945. The Endless Frontier (http://www.nsf.gov/about/history/vbush1945.htm)
Calmes, J. and Helene, C. 2010. BP to Set Aside $20 Billion to Help Oil Spill Victims. The New York Times, June 17, p. A1.
Carnap, R. 1934. The Logical Syntax of Language (trans. Paul Kegan).
Cassidy, D. C. 1993. Uncertainty: The Life and Science of Werner Heisenberg. Berlin: Akademie Verlag.
Clark, R. W. 1971. Einstein: The Life and Times. New York: World Publishing Company.
Collins, R. 1994. Four Sociological Traditions. Oxford: Oxford University Press.
Coxon, A. H. 1986. The Fragments of Parmenides. Assen: Van Gorcum.
Curd, P. 1998. The Legacy of Parmenides. Princeton: Princeton University Press.
DeGroot, G. 2005. The Bomb: A History of Hell on Earth. London: Pimlico, ISBN 0-7126-7748-8.
Desmond, A. and James, M. 1991. Darwin. London: Michael Joseph, Penguin Group, ISBN 0-7181-3430-3.
Dewey, J. 1938. Logic: The Theory of Inquiry. New York: Henry Holt & Co.
Davies, D. G, Parsak, M. R., Pearson, J. P, Iglewski, B. H, Costerton, J. W., and Greenberg, E. P. 1999. The Use of Signal Molecules to Manipulate the Behavior of Biofilm Bacteria. Clinical Microbiology and Infection 5: 5S7–5S8.
Diesing, P. 1962. Reason in Society. Urbana: University of Illinois Press.
Dilcher, R. 1995. Studies in Heraclitus. Hildesheim: Georg Olms.
Dirac, P. A. M. 1981. The Principles of Quantum Mechanics. Oxford: Clarendon Press.

F. Betz, *Managing Science*, Innovation, Technology, and Knowledge Management 9, DOI 10.1007/978-1-4419-7488-4, 

Donovan, A. 1993. Antoine Lavoisier: Science, Administration, and Revolution. Cambridge: Cambridge University Press.
Dreyer, J. L. E. 2004. Tycho Brahe: A Picture of Scientific Life and Work in the Sixteenth Century. Whitefish: Kessinger Publishing.
Ferguson, K. 2002. He Nobleman and His Housedog: Tycho Brahe and Johannes Kepler. Cambridge: Cambridge University Press.
Ferruolo, S. 1998. The Origins of the University: The Schools of Paris and Their Critics, 1100–1215. Stanford: Stanford University Press, ISBN 0-8047-1266-2.
Flyvbjerg, B. 2001. Making Social Science Matter: Why Social Inquiry Fails and How It Can Succeed Again (trans. Steven Sampson). Cambridge: Cambridge University Press, ISBN 052177568X.
Flyvbjerg, B. 2004. A Perestroikan Straw Man Answers Back: David Laitin and Phronetic Political Science. Politics and Society, 32(3, September), 389–416.
Forrester, J. 1961. Industrial Dynamics. Cambridge: MIT Press.
Freud, S. 1900. The Interpretation of Dreams.
Giles, C. 2010. Financial Crisis Exposed Flaws in Economics. Financial Times, April 10.
Giles, C. 2010. Financial Crisis Exposed Flaws in Economics. Financial Times, April 9.
Hafner, K. 1998. Where Wizards Say Up Late: The Origins of the Internet.
Hall, S. 2006. The Contrary Map Maker. New York: Times Magazine, December 31, pp. 45–47.
Haskins, C. H. 1972. The Rise of Universities. Ithaca: Cornell University Press, ISBN 0-87968-379-1.
Heilbron, J. L. 2003. Ernest Rutherford: And the Explosion of Atoms. Oxford: Oxford University Press.
Heisenberg, W. 1983. Encounters with Einstein. Princeton: Princeton University Press.
Heppenheimer, T. A. 1990. How Von Neumann Showed the Way. Invention and Technology, 6(2, Fall), 8–17.
Holdern, C. 2002. Single Gene Dictates Ant Society. Science, 294, November 16, 1434.
Hunter, M. 2009. Boyle: Between God and Science. New Haven: Yale University Press.
Immanuel, K. 1965. Critique of Pure Reason [1781] (trans. Norman Kemp Smith). New York: St. Martins, A 51/B 75.
James, W. 1890. Principles of Psychology.
Jervis, R. 2010. At Least 11 Workers Missing after La Oil Rig Explosion. USA Today. Associated Press, April 21.
Jungk, R. 1956. Brighter Than a Thousand Suns: A Personal History of the Atomic Scientists: Harcourt: Brace.
Keeling, S. V. 1968. Descartes. Oxford: Oxford University Press.
Kell, Douglas, and Rickey Welch, G. 1991. "Perspective: No Turning Back," Times Higher Education Supplement, September 13, p. 15.
Kirk, G. S. 1954. Heraclitus: The Cosmic Fragments. Cambridge: Cambridge University Press.
Kraft, V. 1953. The Vienna Circle: The Origin of Neo-positivism. New York: Greenwood Press.
Krugman, P. 2008. The B Word. The New York Times, March 17.
Krugman, P. 2009. Its All History Now. The New York Times, March 19.
Kuhn, T. 1996. The Structure of Scientific Revolutions. Chicago: University of Chicago Press.
Laitin, D. 2003. The Perestroikan Challenge to Social Science. Politics and Society, 31(1), March 2003, 163–184.
Langford, J. 1998. Galileo, Science and the Church, 3rd ed. New York: St. Augustine's Press, ISBN 1-890318-25-6.
Lohr, S. 2001. The PC? That Old Thing? The New York Times, August 19, Section 3, pp. 1–12.
Lomask, M. 1976. A Minor Miracle: An Informal History of the National Science Foundation. US Government Printing Office. Stock No. 038-000-00288-1.
Martin, J. 2003. The Education of John Dewey. New York: Columbia University Press.
Mehar, J. and Helmut, R. 1982. The Historical Development of Quantum Theory, vols. I & II. New York: Springer-Verlag.
Miller, M. B. and Bassler, B. L. 2001. Quorum sensing in bacteria. Annual Reviews Microbiology, 55, 165–99.

Moreange, M. 1998. A History of Molecular Biology. Cambridge: Harvard University Press.
Morris, C. 1938. Foundations of the Theory of Signs. Chicago: University of Chicago Press.
Nietzsche, F. 1873. Philosophy in the Tragic Age of the Greeks. Washington: Regnery Gateway, ISBN 0-89526-944-9.
Nietzsche, F. 1984. Human, All Too Human (trans. Marion Faber and Stephen Lehmann). Lincoln: University of Nebraska Press.
Nieves, M. 1996. Francis Bacon: The History of a Character Assassination. New Haven: Yale University Press.
Nocera, J. and Tim, C. 2000. 50 Lessons. Fortune, October 30, pp. 136–137.
OECD. 2006. National Innovations Systems: Switzerland. Organization for Economic Development Co-operation and Development.
Olby, R. 1974. The Path to the Double Helix. London: Macmillan; Seattle: University of Washington Press.
Pais, A. 1982. Subtle Is the Lord. The Science and the Life of Albert Einstein. Oxford: Oxford University Press, ISBN 0-19-520438-7.
Pais, A. 1991. Niels Bohr's Times. In Physics, Philosophy and Polity. Oxford: Clarendon Press, ISBN 0-19-852049-2.
Parsons, T. 1937. The Structure of Social Action.
Parsons, T. 1939. Action, Situation and Normative Pattern. Lidz & Staubmann in German in 2004.
Parsons, T. 1951. The Social System
Pasachoff, N. 2005. Ernest Rutherford: Father of Nuclear Science.
Pickering, M. 1993. Auguste Compte: An Intellectual Biography. Cambridge: Cambridge University Press.
Poirier, J-P. 1996. Lavoisier. Philadelphia: University of Pennsylvania Press, English Edition.
Popper, K. 1934. The Logic of Scientific Discovery (English translation 1959), ISBN 04125278449.
Porter, M. 1985. Competitive Strategy. New York: The Free Press.
Portugal, F. H. and Jack S. C. 1977. A Century of DNA. Cambridge: MIT Press.
Principe, L. 1998. The Aspiring Adept: Robert Boyle and His Alchemical Quest. Princeton: Princeton University Press.
Purver, M. and Bowen, E. J. 1960. The Beginning of the Royal Society. Oxford: The Clarendon Press.
Rashdall, H. 1987. The Universities of Europe in the Middle Ages, 3 vols (rev. by F. M. Powicke and A. B. Emden). Oxford: Clarendon Press, ISBN 0-19-821431-6.
Ray, G. F. 1980. Innovation and the Long Cycle. In B. A. Vedin (ed.), Current Innovation. Stockholm: Almqvist & Wiksell.
Reeves, R. 2008. A Force of Nature: The Frontier Genius of Ernest Rutherford. New York: W.W. Norton & Company.
Rhodes, R. 1986. The Making of the Atomic Bomb. New York: Simon & Schuster, ISBN 0-671-44133-7.
Roubini, N. 2010. Interview: Solutions for a Crisis in Its Sovereign State. Financial Times, June 1, p. 11.
Sakar, S. 1996. The Emergence of Logical Empiricism: From 1900 to the Vienna Circle. New York: Garland Publishing.
Schervich, M. J. 1996. Theory of Statistics. New York: Springer.
Schumpeter, J. 2008. Capitalism, Socialism and Democracy. New York: Harper Perennial Modern Classics.
Segre, E. 1955. Fermi and Neutron Physics. Reviews of Modern Physics, 27, 257–263.
Sherman, J. 2003. The History of the Internet.
Shnayerson, R. 1996. Judgment at Nuremberg. Smithsonian Magazine, October, pp. 124–141.
Sprat, T. 2003. History of the Royal Society. London: Kessinger Publishing, ISBN 0-7661-2867-9.
Stevenson, B. 1987. Kepler's Physical Astronomy. Princeton: Princeton University Press.
Stuewer, R. H. 1985. Binging the News of Fission to America. Physics Today, October, pp. 49–56.
Sweet, P. S. 1978–1980. Wilhelm von Humboldt A Biography, vol. 2. Columbus: Ohio University Press.

Thayer, A. M. 1996. Market, Investor Attitudes Challenge Developers of Biopharmaceuticals. Chemical & Engineering News, August 1, pp. 13–21.

Tiner, J. H. 1975. Isaac Newton: Inventor, Scientist and Teacher. Milford: Mott Media.

Twentyman, J. 2010. IT Leads a Nation's Recovery. Financial Times, June 16, p. 3.

Van der Waerden, B. L. 2003. Algebra, vol. 1. New York: Springer.

von Bertalanffy, L. 1968. General System Theory: Foundations, Development, Applications. New York: George Braziller, revised edition 1976, ISBN 0-8076-0453-4.

Watson, J. 2001. The Double Helix. New York: Norton & Co.

Weber, M. 1897. Definition of Sociology. In Max Weber, Sociological Writings. Wolf Hedebrand (ed.), 1994. (Continuum. Transcribed by Andy Blunden in1998 for: http://www.sociosite.net/topics/weber, 2008).

Weber, M. 1903. Objectivity in Social Science. Archives for Social Science and Social Welfare (http://www.sociosite.net/topics/weber, 2008).

Weber, M. 1905. The Protestant Ethics and the Spirit of Capitalism (trans. Parsons in 1930).

Weber, M. 1921. The Theory of Social and Economic Organization (trans. Talcott Parsons together with Alexander Morell Henderson in 1947).

Weber, M. 1947. From Max Weber: Essays in Sociology, ISBN 0-19-500462-0.

Weber, M. 2008. The Fundamental Concepts of Sociology (http://www.sociosite.net/topics/weber, 2008).

Wundt, W. 1874. Principles of Physiological Psychology.

Zagorin, P. 1999. Francis Bacon. Princeton: Princeton University Press, ISBN 978691009667.

# Index